Problem Books in Mathematics

Series Editor:

Peter Winkler
Department of Mathematics
Dartmouth College
Hanover, NH 03755
USA

More information about this series at http://www.springer.com/series/714

Hayk Sedrakyan · Nairi Sedrakyan

Algebraic Inequalities

Hayk Sedrakyan
University Pierre and Marie Curie
Paris, France

Nairi Sedrakyan
Yerevan, Armenia

ISSN 0941-3502 ISSN 2197-8506 (electronic)
Problem Books in Mathematics
ISBN 978-3-030-08551-3 ISBN 978-3-319-77836-5 (eBook)
https://doi.org/10.1007/978-3-319-77836-5

Mathematics Subject Classification (2010): 97U40, 00A07, 26D05

Printed on acid-free paper

This Springer imprint is published by the registered company Springer International Publishing AG part of Springer Nature
The registered company address is: Gewerbestrasse 11, 6330 Cham, Switzerland

To Margarita,
a wonderful wife and a loving mother

To Ani,
a wonderful daughter and a loving sister

Preface

In mathematics one often deals with inequalities. This book is designed to teach the reader new and classical techniques for proving *algebraic inequalities*. Moreover, each chapter of the book provides a technique for proving a certain type of inequality.

The book includes techniques of using the relationship between the arithmetic, geometric, harmonic, and quadratic means, the principle of mathematical induction, the change of variable(s) method, techniques using the Cauchy–Bunyakovsky–Schwarz inequality, Jensen's inequality, and Chebyshev's properties of functions, among others. The main idea behind of the proof techniques discussed in this book is making the complicated simple, so that even a beginner can understand complicated inequalities, their proofs and applications. This approach makes it possible not only to prove a large variety of inequalities, but also to solve problems related to inequalities. To explain each technique of proof, we provide examples and problems with complete proofs or hints. At the end of each chapter there are problems for independent study. In Chapter 14 (Miscellaneous Inequalities) are included inequalities whose proofs employ various techniques not covered in the preceding chapters. In some cases, the proofs of Chapter 14 use several proof techniques from the preceding chapters simultaneously. One hundred selected inequalities and their hints are also provided in the end of Chapter 14, and interested readers are encouraged to choose and provide any methods of proofs they prefer. In each chapter we have tried to include inequalities belonging to the same topic and to present them in order of increasing difficulty, using principles similar to those in [11]. This allows the reader to try to prove these inequalities step by step and to refer to the provided proofs only when difficulties arise. We recommend to use the proofs provided in the book, paying more attention to the choice of the mathematical proof technique.

Most of the inequalities in this book were created by the authors. Nevertheless, some of the inequalities were proposed in different mathematical olympiads in different countries or have been published elsewhere (including author-created inequalities). However, the provided solutions are different from the original ones. Most such inequalities are included in the books [2, 5, 6, 7, 8, 9], and since the name of the author of individual inequalities is unknown to us, we cite these books as the main references. However, for well-known inequalities we have tried to provide the name of the authors. This book was published in Seoul in Korean [13, 14] and is based on [16], which was later published in Moscow in Russian [15]. The historical origins provided at the beginning of some chapters are mostly based on [10] or our personal knowledge.

It was considered appropriate to give the proofs of each chapter at the end of the same chapter.

Paris, France
Yerevan, Armenia

Hayk Sedrakyan
Nairi Sedrakyan

Contents

Contents

About the Authors

Hayk Sedrakyan is an IMO medal winner, a professor of mathematics in Paris, France, and a professional math olympiad coach in the greater Boston area, Massachusetts, USA. He received his doctorate in mathematics at the Université Pierre et Marie Curie, Paris, France. Hayk is a Doctor of Mathematical Sciences in USA, France, Armenia. He has been awarded master's degrees in mathematics from Germany, Austria, and Armenia and completed part of his doctoral studies in Italy. Hayk has authored several books on the topic of problem-solving and olympiad-style mathematics published in USA and South Korea.

Nairi Sedrakyan has long been involved in national and international mathematical olympiads, having served as an *International Mathematical Olympiad* (IMO) problem selection committee member and the president of Armenian Mathematics Olympiads. He is the author of one of the hardest problems ever proposed in the history of International Mathematical Olympiads (the fifth problem of the 37th IMO). He has been the leader of the Armenian IMO Team, jury member of the IMO, jury member and problem selection committee member of the Zhautykov International Mathematical Olympiad (ZIMO), jury member and problem selection committee member of the International Olympiad of Metropolises, and the president of the International Mathematical Olympiad Tournament of the Towns in Armenia. He is also the author of a large number of problems proposed in these olympiads and has authored several books on the topic of problem-solving and olympiad-style mathematics published in United States, Russia, Armenia, and South Korea. The students of Nairi Sedrakyan have obtained 20 medals (1 gold medal, 4 silver medals, 15 bronze medals) in IMO. For his outstanding teaching, Nairi Sedrakyan received the title of *best teacher* of the Republic of Armenia and was awarded special recognition by the prime minister.

About the Authors

Overview

This book is designed to teach the reader new and classical mathematical proof techniques for proving inequalities, in particular, to prove *algebraic inequalities.*

These proof techniques and methods are applied to prove inequalities of various types. The main idea behind this book and the proof techniques discussed is making the complicated simple, so that even a beginner can understand complicated inequalities, their proofs and applications. The book *Algebraic Inequalities* is also devoted to the topic of *inequalities* and can be considered a continuation of the book *Geometric Inequalities*: *Methods of proving* [12].

It can serve teachers, high-school students, and mathematical competitors.

Chapter 1
Basic Inequalities and Their Applications

Historical origins. According to [3], mathematical inequalities first were expressed verbally. Later, the following symbols were introduced:

Less than **and** ***greater than***: The mathematical symbols $<$ and $>$ first appear in the book *Artis Analyticae Praxis ad Aequationes Algebraicas Resolvendas* (*The Analytical Arts Applied to Solving Algebraic Equations*) posthumously published in 1631 and written by the English astronomer and mathematician *Thomas Harriot* (c. 1560–1621), who was born in Oxford, England, and died in London, England. Harriot initially used triangular symbols, but the editor modified them slightly, so that they resemble the modern *less than* and *greater than* symbols. In his book is stated the following: "The mark of the majority (*signum majoritatis*) as $a > b$, signifies a greater than b and the mark of the minority (*signum minoritatis*) to $a < b$ signifies a lesser than b."

Less than or equal to **and** ***greater than or equal to***: The double-bar style of the less than or equal to $\leqq$ and greater than or equal to $\geqq$ symbols was first employed in 1734 by the French mathematician, geophysicist, and astronomer *Pierre Bouger* (1698–1758) born in Le Croisic, France, and died in Paris, France.

In 1670, a similar reduced single-bar notation was employed by the English mathematician *John Wallis*. Wallis used a single horizontal bar above instead of below the inequality symbols, leading to $\overline{<}$ and $\overline{>}$. Wallis also introduced the symbol for infinity ∞. His academic advisor was the English mathematician *William Oughtred* (1574–1660), who introduced the multiplication symbol $\times$ and the abbreviations sin and cos for the sine and cosine. Oughtred was born in Eton, England, and died in Albury, England.

Not equal to, not greater than, not less than: These symbols were employed by the Swiss mathematician, physicist, and astronomer *Leonhard Euler* (1707–1783), who was born in Basel, Switzerland, and died in Saint Petersburg, Russian Empire (now Russia). Euler introduced many modern mathematical notations, for example the notation for a mathematical function. He is considered one of the greatest mathematicians of all time. His doctoral advisor was the prominent Swiss mathematician *Johann Bernoulli* (1667–1748), who was born and died in Basel, Switzerland. Euler

H. Sedrakyan and N. Sedrakyan, *Algebraic Inequalities*, Problem Books in Mathematics, https://doi.org/10.1007/978-3-319-77836-5_1

was the doctoral advisor of six students, including the well-known Italian mathematician *Joseph Louis Lagrange* (1736–1813). He was born Giuseppe Lodovico Lagrangia in Turin, Kingdom of Sardinia (now Turin, Italy) and died in Paris, France, and his son *Johann Euler* (1734–1800) born and died in Saint Petersburg, Russia.

In this chapter, basic but very useful inequalities are presented. We recommend that the reader pay special attention to these inequalities, which will lay a foundation for more challenging inequalities in the forthcoming sections. The main idea behind the proof techniques and methods presented here is *making the complex simple*, so that even a beginner can understand complex inequalities, their proofs and applications. A background in high-school-level algebra is sufficient for solving the inequalities presented below.

Problems

Prove the following inequalities (1.1–1.26).

1.1. $a^2+b^2 \geq 2ab$.

1.2. $\frac{a+b}{2} \geq \sqrt{ab}$, where $a \geq 0,\ b \geq 0$.

1.3. $\sqrt{ab} \geq \frac{2}{\frac{1}{a}+\frac{1}{b}}$, where $a > 0,\ b > 0$.

1.4. $\sqrt{\frac{a^2+b^2}{2}} \geq \frac{a+b}{2}$.

1.5. $\frac{a+b}{2} \geq \frac{2}{\frac{1}{a}+\frac{1}{b}}$, where $a > 0,\ b > 0$.

1.6. $\frac{a^2+b^2}{2} \geq \left(\frac{a+b}{2}\right)^2$.

1.7. $a+b > 1+ab$, where $b < 1 < a$.

1.8. $a^2+b^2 > c^2+(a+b-c)^2$, where $b < c < a$.

1.9. (a) $2 \leq \frac{a}{b}+\frac{b}{a}$, where $ab > 0$,
(b) $\frac{a}{b}+\frac{b}{a} \leq -2$, where $ab < 0$.

1.10. $x_1 \leq \frac{x_1+\cdots+x_n}{n} \leq x_n$, where $x_1 \leq \cdots \leq x_n$.

1.11. $\frac{x_1}{y_1} \leq \frac{x_1+\cdots+x_n}{y_1+\cdots+y_n} \leq \frac{x_n}{y_n}$, where $\frac{x_1}{y_1} \leq \cdots \leq \frac{x_n}{y_n}$ and $y_i > 0,\ i=1,\ldots,n$.

1.12. $x_1 \leq (x_1\cdots x_n)^{\frac{1}{n}} \leq x_n$, where $n \geq 2,\quad 0 \leq x_1 \leq \cdots \leq x_n$.

1.13. $|a_1|+\cdots+|a_n| \geq |a_1+a_2+\cdots+a_n|$.

1.14. $\frac{a_1+\cdots+a_n}{n} \geq \frac{n}{\frac{1}{a_1}+\cdots+\frac{1}{a_n}}$, where $n \geq 2,\quad a_i > 0,\ i=1,\ldots,n$.

1.15. $(a+b)\sqrt{\frac{a+b}{2}} \geq a\sqrt{b}+b\sqrt{a}$, where $a > 0,\quad b > 0$.

1.16. $\frac{1}{2}(a+b)+\frac{1}{4} \geq \sqrt{\frac{a+b}{2}}$, where $a > 0,\quad b > 0$.

1.17. $a(x+y-a) \geq xy$, where $x \leq a \leq y$.

1.18. $\frac{1}{x-1}+\frac{1}{x+1} > \frac{2}{x}$, where $x > 1$.

1.19. $\frac{1}{3k+1}+\frac{1}{3k+2}+\frac{1}{3k+3} > \frac{1}{2k+1}+\frac{1}{2k+2}$, where $k \in \mathbb{N}$.

1.20. $\frac{ab}{(a+b)^2} \leq \frac{(1-a)(1-b)}{((1-a)+(1-b))^2}$, where $0 < a \leq \frac{1}{2},\ 0 < b \leq \frac{1}{2}$.

1.21. $\frac{1}{\sqrt{3k+1}} \cdot \frac{2k+1}{2k+2} < \frac{1}{\sqrt{3k+4}}$, where $k \in \mathbb{N}$.
1.22. $2^{n-1} \geq n$, where $n \in \mathbb{N}$.
1.23. $\frac{1}{3} + \frac{2}{3} \cdot \frac{1}{5} + \frac{2}{3} \cdot \frac{4}{5} \cdot \frac{1}{7} + \cdots + \frac{2}{3} \cdot \frac{4}{5} \cdot \frac{6}{7} \cdots \frac{100}{101} \cdot \frac{1}{103} < 1$.
1.24. (a) $\frac{1-a}{1-b} + \frac{1-b}{1-a} \leq \frac{a}{b} + \frac{b}{a}$, where $0 < a, b \leq \frac{1}{2}$,

(b) $\sum_{i=1}^{n} \frac{1}{1-a_i} \sum_{i=1}^{n} (1 - a_i) \leq \sum_{i=1}^{n} \frac{1}{a_i} \sum_{i=1}^{n} a_i$, where $0 < a_1, \ldots, a_n \leq \frac{1}{2}$.
1.25. $1 + \frac{1}{2^3} + \cdots + \frac{1}{n^3} < \frac{5}{4}$, where $n \in \mathbb{N}$.
1.26. $\frac{1}{1+a+b} \leq 1 - \frac{a+b}{2} + \frac{ab}{3}$, where $0 \leq a \leq 1, 0 \leq b \leq 1$.

Proofs

1.1. We have $a^2 + b^2 - 2ab = (a - b)^2 \geq 0$. Note that equality holds if and only if $a = b$.
1.2. The proof is similar to the proof of Problem 1.1. We have $a + b - 2\sqrt{ab} = \left(\sqrt{a} - \sqrt{b}\right)^2 \geq 0$. Note that equality holds if and only if $a = b$.
1.3. Multiplying both sides of the inequality of Problem 1.2 by $\frac{2\sqrt{ab}}{a+b}$, we obtain the required inequality. Note that equality holds if and only if $a = b$.
1.4. We have $2ab \leq a^2 + b^2$ (see Problem 1.1). We obtain the equivalent inequality $a^2 + b^2 + 2ab \leq 2a^2 + 2b^2$. Thus, it follows that $\frac{(a+b)^2}{4} \leq \frac{a^2+b^2}{2}$. The last inequality can be rewritten as $\left|\frac{a+b}{2}\right| \leq \sqrt{\frac{a^2+b^2}{2}}$. Therefore, $\sqrt{\frac{a^2+b^2}{2}} \geq \frac{a+b}{2}$. Note that equality holds if and only if $a = b$.
1.5. This inequality follows from the inequalities of Problems 1.2 and 1.3. Note that equality holds if and only if $a = b$.
1.6. See the proof of Problem 1.4. Note that equality holds if and only if $a = b$.
1.7. Using that $a + b - 1 - ab = (a - 1)(1 - b)$ and the condition $b < 1 < a$, it follows that $(a - 1)(1 - b) > 0$.
1.8. Let us evaluate the difference between the left-hand side and the right-hand side of the given inequality:

$$a^2 + b^2 - c^2 - (a + b - c)^2 = \left(a^2 - c^2\right) - \left((a + b - c)^2 - b^2\right) =$$
$$= (a - c)(a + c) - (a - c)(a + 2b - c) = 2(a - c)(c - b) > 0,$$

since according to the assumption, we have $b < c < a$.
1.9. (a) Multiplying both sides of the inequality by ab, where $ab > 0$, we deduce that $a^2 + b^2 \geq 2ab$.

(b) Dividing both sides of the inequality $a^2+b^2 \geq -2ab$ by ab, where $ab < 0$, we obtain the required inequality.
1.10. We have $x_1 \leq x_1, \ldots, x_n \leq x_n$, whence $nx_1 \leq x_1 + \cdots + x_n \leq nx_n$. Thus, it follows that $x_1 \leq \frac{x_1 + \cdots + x_n}{n} \leq x_n$.

1.11. From the assumptions of the problem, it follows that $y_i \frac{x_1}{y_1} \leq x_i \leq y_i \frac{x_n}{y_n}$, $i = 1, \ldots, n$. Summing up these inequalities, we deduce that

$$\frac{x_1}{y_1}(y_1 + \cdots + y_n) \leq x_1 + \cdots + x_n \leq \frac{x_n}{y_n}(y_1 + \cdots + y_n).$$

Hence, it follows that $\frac{x_1}{y_1} \leq \frac{x_1 + \cdots + x_n}{y_1 + \cdots + y_n} \leq \frac{x_n}{y_n}$.

1.12. We have that $x_1 \leq x_i \leq x_n$, $i = 1, \ldots, n$. Multiplying these inequalities, we obtain $x_1^n \leq x_1 \cdots x_n \leq x_n^n$. Thus, it follows that $x_1 \leq (x_1 \cdots x_n)^{\frac{1}{n}} \leq x_n$.

1.13. If $a_1 + \cdots + a_n \geq 0$, then $|a_1 + \cdots + a_n| = a_1 + \cdots + a_n$.
Using $a \leq |a|$, we deduce that

$$|a_1 + \cdots + a_n| = a_1 + \cdots + a_n \leq |a_1| + \cdots + |a_n|.$$

If $a_1 + \cdots + a_n < 0$, then $|a_1 + \cdots + a_n| = -a_1 - \cdots - a_n$. Using the inequality $-a \leq |a|$, it follows that $|a_1 + \cdots + a_n| = -a_1 - \cdots - a_n \leq |a_1| + \cdots + |a_n|$.

1.14. We have that

$$(a_1 + \cdots + a_n)\left(\frac{1}{a_1} + \cdots + \frac{1}{a_n}\right) = \left(\underbrace{\left(\frac{a_1}{a_2} + \frac{a_2}{a_1}\right) + \cdots + \left(\frac{a_{n-1}}{a_n} + \frac{a_n}{a_{n-1}}\right)}_{n(n-1)/2}\right) + n,$$

and using Problem 1.9 (a), it follows that

$$(a_1 + \cdots + a_n)\left(\frac{1}{a_1} + \cdots + \frac{1}{a_n}\right) \geq n + 2 \cdot \frac{n(n-1)}{2} = n^2.$$

1.15. The given inequality is equivalent to the inequality $(a+b)\sqrt{\frac{a+b}{2}} \geq 2\sqrt{ab} \cdot \frac{\sqrt{a}+\sqrt{b}}{2}$, which can be obtained by multiplying the inequalities $a + b \geq 2\sqrt{ab}$ (Problem 1.2) and $\sqrt{\frac{a+b}{2}} \geq \frac{\sqrt{a}+\sqrt{b}}{2}$ (see Problem 1.6).

1.16. We have

$$\frac{1}{2}(a+b) + \frac{1}{4} - \sqrt{\frac{a+b}{2}} = \left(\sqrt{\frac{a+b}{2}} - \frac{1}{2}\right)^2 \geq 0.$$

Therefore, $\frac{1}{2}(a+b) + \frac{1}{4} \geq \sqrt{\frac{a+b}{2}}$.

1.17. Since $a(x + y - a) - xy = ax - xy + a(y - a) = (y - a)(a - x)$ and $y \geq a \geq x$, it follows that $(y - a)(a - x) \geq 0$. Therefore, $a(x + y - a) \geq xy$.

1.18. Using the inequality of Problem 1.5, it follows that

$$\frac{\frac{1}{x-1}+\frac{1}{x+1}}{2} > \frac{2}{(x-1)+(x+1)}, \text{ or } \frac{1}{x-1}+\frac{1}{x+1} > \frac{2}{x}.$$

1.19. According to the inequality of Problem 1.18, we have that $\frac{1}{3k+1}+\frac{1}{3k+3}=\frac{1}{(3k+2)-1}+\frac{1}{(3k+2)+1} > \frac{2}{3k+2}$. Therefore, $\frac{1}{3k+1}+\frac{1}{3k+2}+\frac{1}{3k+3} > \frac{3}{3k+2}$.
Now let us prove that $\frac{3}{3k+2} > \frac{1}{2k+1}+\frac{1}{2k+2}$.
Indeed, $\frac{1}{2k+1}+\frac{1}{2k+2}-\frac{3}{3k+2}=\frac{-k}{(2k+1)(2k+2)(3k+2)} < 0$.

1.20. The given inequality is equivalent to the following inequality:

$$\left(\frac{2}{a+b}-1\right)^2 \leq \left(\frac{1}{a}-1\right)\left(\frac{1}{b}-1\right).$$

We have that

$$\left(\frac{1}{a}-1\right)\left(\frac{1}{b}-1\right)-\left(\frac{2}{a+b}-1\right)^2=\frac{1}{ab}-\frac{1}{a}-\frac{1}{b}-\frac{4}{(a+b)^2}+\frac{4}{a+b}=$$
$$=\frac{1}{ab}-\frac{4}{(a+b)^2}+\frac{4}{a+b}-\frac{a+b}{ab}=\frac{(a-b)^2}{ab(a+b)^2}-\frac{(a-b)^2}{ab(a+b)}=\frac{(a-b)^2(1-(a+b))}{ab(a+b)^2}$$

and $0 < a \leq \frac{1}{2}$, $0 < b \leq \frac{1}{2}$. Then $\frac{(a-b)^2(1-(a+b))}{ab(a+b)^2} \geq 0$, and therefore, $\left(\frac{2}{a+b}-1\right)^2 \leq \left(\frac{1}{a}-1\right)\left(\frac{1}{b}-1\right)$.

1.21. The given inequality is equivalent to the inequality $(2k+1)\sqrt{3k+4} < (2k+2)\sqrt{3k+1}$ or to the following inequality: $(2k+1)^2(3k+4) < (2k+2)^2(3k+1)$.
The last inequality holds because

$$(2k+2)^2(3k+1)-(2k+1)^2(3k+4)=k>0.$$

1.22. Since $1 < 2 < 2^2 < \cdots < 2^{n-1}$ and the number of positive integers $1, 2, 2^2, \ldots, 2^{n-1}$ is equal to n, it follows that $2^{n-1} \geq n$.

1.23. Consider a unit line segment and suppose on the first day, we paint $\frac{1}{3}$ of thegiven segment, the second day $\frac{1}{5}$ of the rest of the segment, on the 51st day, $\frac{1}{103}$ of the rest of the segment. Since every day there remains a part of the given segment, the sum of the painted parts must be less than 1.
The first day we have painted $\frac{1}{3}$ of the given segment, on the second day $\frac{2}{3}\cdot\frac{1}{5}$, on the 51st day $\frac{2}{3}\cdot\frac{4}{5}\cdots\frac{100}{101}\cdot\frac{1}{103}$. Hence, we deduce that

$$\frac{1}{3}+\frac{2}{3}\cdot\frac{1}{5}+\frac{2}{3}\cdot\frac{4}{5}\cdot\frac{1}{7}+\cdots+\frac{2}{3}\cdot\frac{4}{5}\cdots\frac{98}{99}\cdot\frac{100}{101}\cdot\frac{1}{103} < 1.$$

1.24 (a) According to the inequality of Problem 1.20, we have that

$$\frac{((1-a)+(1-b))^2}{(1-a)(1-b)} \leq \frac{(a+b)^2}{ab} \text{ or } \frac{1-a}{1-b}+\frac{1-b}{1-a} \leq \frac{a}{b}+\frac{b}{a}.$$

An alternative proof. Note that $1-a \geq a$ and $1-b \geq b$, and therefore,

$$\frac{1-a}{1-b}+\frac{1-b}{1-a} = \frac{(1-a)^2+(1-b)^2}{(1-a)(1-b)} = \frac{((1-a)-(1-b))^2+2(1-a)(1-b)}{(1-a)(1-b)} =$$

$$= \frac{(a-b)^2}{(1-a)(1-b)}+2 \leq \frac{(a-b)^2}{ab}+2 = \frac{a}{b}+\frac{b}{a}.$$

Hence, we deduce that $\frac{1-a}{1-b}+\frac{1-b}{1-a} \leq \frac{a}{b}+\frac{b}{a}$.

(b) Since $\sum\limits_{i=1}^{n} \frac{1}{1-a_i} \sum\limits_{i=1}^{n}(1-a_i) = \underbrace{\left(\frac{1-a_1}{1-a_2}+\frac{1-a_2}{1-a_1}\right)+\cdots+\left(\frac{1-a_{n-1}}{1-a_n}+\frac{1-a_n}{1-a_{n-1}}\right)}_{n(n-1)/2} + n$, using the inequality of Problem 1.24 (a), we obtain that

$$\sum_{i=1}^{n}\frac{1}{1-a_i}\sum_{i=1}^{n}(1-a_i) \leq \underbrace{\left(\frac{a_1}{a_2}+\frac{a_2}{a_1}\right)+\cdots+\left(\frac{a_{n-1}}{a_n}+\frac{a_n}{a_{n-1}}\right)}_{n(n-1)/2}+n = \sum_{i=1}^{n}\frac{1}{a_i}\sum_{i=1}^{n}a_i.$$

1.25. Note that if $n \geq 4$, then

$$1+\frac{1}{2^3}+\cdots+\frac{1}{n^3} = 1+\frac{2-1}{2^3}+\cdots+\frac{n-(n-1)}{n^3}$$

$$= \frac{5}{4}-\left(\frac{1}{2^3}-\frac{1}{3^2}\right)-\left(\frac{2}{3^3}-\frac{1}{4^2}\right)-\cdots-\frac{n-1}{n^3} < \frac{5}{4},$$

since $\frac{k}{(k+1)^3} > \frac{1}{(k+2)^2}$, where $k \in \mathbb{N}$.

1.26. We have that $(1-a)(1-b) \geq 0$, and therefore $a+b-1 \leq ab$. Then

$$\frac{1}{1+a+b}-\left(1-\frac{a+b}{2}\right) = \frac{a+b}{2(1+a+b)}(a+b-1) \leq \frac{1}{3}ab.$$

Problems for Independent Study

Prove the following inequalities (1–32).

1. $|x-y| < |1-xy|$, where $|x|<1,\ |y|<1$.
2. $\frac{\sin x-1}{\sin x-2}+\frac{1}{2} \geq \frac{2-\sin x}{3-\sin x}$.
3. $\frac{a}{bc}+\frac{b}{ca}+\frac{c}{ab} \geq \frac{2}{a}+\frac{2}{b}-\frac{2}{c}$, where $a>0,\ b>0,\ c>0$.
4. $\frac{1}{a}+\frac{1}{b}-\frac{1}{c} < \frac{1}{abc}$, where $a^2+b^2+c^2=\frac{5}{3}$ and $a>0,\ b>0,\ c>0$.
5. $3\left(1+a^2+a^4\right) \geq \left(1+a+a^2\right)^2$.

6. $(ac+bd)^2+(ad-bc)^2 \geq 144$, where $a+b=4$, $c+d=6$.
7. $x_1^2+x_2^2+\cdots+x_{2n}^2+na^2 \geq a\sqrt{2}(x_1+x_2+\cdots+x_{2n})$.
8. $\frac{1}{a+b}+\frac{1}{b+c}+\frac{1}{a+c} \leq \frac{\sqrt{a}+\sqrt{b}+\sqrt{c}}{2\sqrt{abc}}$, where $a>0$, $b>0$, $c>0$.
9. $a^3(b^2-c^2)+b^3(c^2-a^2)+c^3(a^2-b^2)<0$, where $0<a<b<c$.
10. $a^3b+b^3c+c^3a \geq a^2b^2+b^2c^2+a^2c^2$, where $a \geq b \geq c > 0$.
11. $\frac{y}{x}+\frac{y}{z}+\frac{x+z}{y} \leq \frac{(x+z)^2}{xz}$, where $0<x \leq y \leq z$.
12. $\sqrt{1+\sqrt{a}}+\sqrt{1+\sqrt{a+\sqrt{a^2}}}+\cdots+\sqrt{1+\sqrt{a+\cdots+\sqrt{a^n}}}<na$,
    where $n \geq 2$, $a \geq 2$, $n \in \mathbb{N}$.
13. (a) $\tan^2(\alpha-\beta) \leq \frac{(n-1)^2}{4n}$, where $\tan\alpha = n\tan\beta$, $n>0$,
    (b) $1+\cos(\alpha-\beta) \geq \cos\alpha+\cos\beta$, where $0 \leq \alpha \leq \frac{\pi}{2}$, $0 \leq \beta \leq \frac{\pi}{2}$.
14. $[5x] \geq [x]+\frac{[2x]}{2}+\frac{[3x]}{3}+\frac{[4x]}{4}+\frac{[5x]}{5}$,
    where $[a]$ is the integer part of the real number a.
15. $(n!)^2 \geq n^n$, where $n \in \mathbb{N}$.
16. $x^6+x^5+4x^4-12x^3+4x^2+x+1 \geq 0$.
17. $\log^2\alpha \geq \log\beta\,\log\gamma$, where $\alpha>1$, $\beta>1$, $\gamma>1$, $\alpha^2 \geq \beta\gamma$.
18. $\log_4 5+\log_5 6+\log_6 7+\log_7 8>4,4$.
19. $\frac{1}{3}+\cdots+\frac{n}{3\cdot 5\cdots(2n+1)}<\frac{1}{2}$, where $n \in \mathbb{N}$.
20. $\frac{2^3+1}{2^3-1}\cdots\frac{n^3+1}{n^3-1}<\frac{3}{2}$, where $n \geq 2$, $n \in \mathbb{N}$.
21. $1\cdot 1!+\cdots+n\cdot n!<(n+1)!$, where $n \in \mathbb{N}$.
22. $(1+\frac{1}{2^2})(1+\frac{1}{3^2})\ldots(1+\frac{1}{n^2})<2$, where $n \geq 2$, $n \in \mathbb{N}$.
23. $\left(1-\frac{1}{p_1^2}\right)\left(1-\frac{1}{p_2^2}\right)\ldots\left(1-\frac{1}{p_n^2}\right)>\frac{1}{2}$, where $1<p_1<p_2<\cdots<p_n$, $p_i \in \mathbb{N}$, $i=1,\ldots,n$.
24. $\frac{1}{2}-\frac{1}{3}+\frac{1}{4}-\frac{1}{5}+\cdots-\frac{1}{999}+\frac{1}{1000}<\frac{2}{5}$.
25. $(\sin x+2\cos 2x)(2\sin 2x-\cos x)<4,5$.
26. (a) $\frac{a+b}{1+a+b} \leq \frac{a}{1+a}+\frac{b}{1+b}$, where $a \geq 0$, $b \geq 0$.
    (b) $\frac{a+b}{2+a+b} \geq \frac{1}{2}\left(\frac{a}{1+a}+\frac{b}{1+b}\right)$, where $a \geq 0$, $b \geq 0$.
    (c) $n \leq \frac{a_1+b_1}{a_1+b_1+2}+\ldots+\frac{a_n+b_n}{a_n+b_n+2}+\frac{\frac{1}{a_1}+\frac{1}{b_{i_1}}}{\frac{1}{a_1}+\frac{1}{b_{i_1}}+2}+\ldots+\frac{\frac{1}{a_n}+\frac{1}{b_{i_n}}}{\frac{1}{a_n}+\frac{1}{b_{i_n}}+2}<2n$, where $i_1,\ldots,i_n$ is some permutation of the numbers $1,\ldots,n$, $a_i, b_i>0$, $i=1,\ldots,n$.
27. $\sum_{i=1}^{n}\frac{a_1+2a_2+\cdots+ia_i}{i^2} \leq 2\sum_{i=1}^{n}a_i$, where $a_i \geq 0$, $i=1,\ldots,n$.
28. $\frac{1}{a}+\frac{1}{b}+\frac{1}{c} \leq \frac{41}{42}$, where $\frac{1}{a}+\frac{1}{b}+\frac{1}{c}<1$, $a,b,c \in \mathbb{N}$.
29. $\frac{4x}{y+z}+\frac{y}{x+z}+\frac{z}{x+y}>2$, where $x,y,z>0$.
30. $1<\frac{a}{a+b+d}+\frac{b}{a+b+c}+\frac{c}{b+c+d}+\frac{d}{a+c+d}<2$, where $a>0, b>0, c>0, d>0$.
31. $a+b>c+d$, where $a,b,c,d \geq \frac{1}{2}$ and $a^2+b>c^2+d$, $a+b^2>c+d^2$.
    Hint. If $a+b \leq c+d$, then $a \leq c$ or $b \leq d$. If $a \leq c$, then $b-d>(c-a)(c+a) \geq c-a$.
32. $(b-a)(9-a^2)+(c-a)(9-b^2)+(c-b)(9-c^2) \leq 24\sqrt{2}$, where $0 \leq a \leq b \leq c \leq 3$.

Hint. We have that

$$(b-a)(9-a^2)+(c-a)(9-b^2)+(c-b)(9-c^2) \le 9b+c(9-b^2)+(c-b)(9-c^2) =$$
$$= 18c - c^3 + bc(c-b) \le 18c - c^3 + \frac{1}{4}c^3 = 18c - \frac{3}{4}c^3.$$

33. If $0 < a, b, c < 1$, then one of the numbers $(1-a)b, \quad (1-b)c, \quad (1-c)a$ is not greater than $\frac{1}{4}$.
34. Let $a > 0, \ b > 0, \ c > 0$, and $a+b+c=1$. Prove that

    (a) $\sqrt{a+\frac{1}{4}(b-c)^2}+\sqrt{b+\frac{1}{4}(c-a)^2}+\sqrt{c+\frac{1}{4}(b-a)^2} \le 2$,

    (b) $\sqrt{a+\frac{1}{4}(b-c)^2}+\sqrt{b}+\sqrt{c} \le \sqrt{3}$.

    Hint. (a) $\sqrt{x+\frac{1}{4}(y-z)^2} \le x + \frac{y+z}{2}$ if $x, y, z > 0$ and $x+y+z=1$.
35. Find the smallest possible value of the following expression: $\frac{a^4}{b^4}+\frac{b^4}{a^4}-\frac{a^2}{b^2}-\frac{b^2}{a^2}+\frac{a}{b}+\frac{b}{a}$, where $a > 0, \quad b > 0$.

Chapter 2
Sturm's Method

Historical origins. *Sturm's method* was proposed by the Prussian mathematician *Friedrich Otto Rudolf Sturm*, born 6 January 1841 in Breslau, Prussia (now Wrocław, Poland), died 12 April 1919 in Breslau, Germany. Sturm obtained his doctorate in mathematics from the University of Breslau (now University of Wrocław) in 1863 under the supervision of the well-known Prussian mathematician *Heinrich Eduard Schröter*. Sturm was the advisor of 19 doctoral students, including the well-known Prussian mathematician *Otto Toeplitz*, born 1 August 1881 in Breslau, Prussia, died 15 February 1940 in Jerusalem, Mandatory Palestine (now Israel). Toeplitz studied mathematics at the University of Breslau, where under Sturm's supervision he obtained his doctorate in 1905.

Besides its various applications, Sturm's method provides an opportunity to prove a large number of different inequalities under certain conditions.

Example 2.1 Prove that if the product of positive numbers $x_1, \ldots, x_n \quad (n \geq 2)$ is equal to 1, then $x_1 + \cdots + x_n \geq n$.

Proof If $x_1 = \cdots = x_n = 1$, then $x_1 + \cdots + x_n = n$.

Suppose that among the considered numbers there are at least two different numbers. Then among the numbers $x_1, \ldots, x_n$ there are two numbers such that one of them is greater than 1 and the other one is less than 1 (Problem 1.12). Without loss of generality one can assume that those numbers are x_1 and x_2, and that $x_1 < 1 < x_2$. Note that $x_1 + x_2 > 1 + x_1x_2$ (Problem 1.7).

If one substitutes the given numbers by numbers $1, x_1x_2, x_3, \ldots, x_n$, then their product is again equal to 1, and the sum satisfies $1+x_1x_2+x_3+\cdots+x_n < x_1+\cdots+x_n$.

Doing the same with the numbers $1, x_1x_2, x_3, \ldots, x_n$, in a similar way we obtain a new sequence such that two numbers in it are equal to 1. Doing the same at most $n - 1$ times, we obtain a sequence such that $n - 1$ numbers in it are equal to 1, and the nth number is equal to $x_1 \cdots x_n$.

On the other hand, $x_1 \cdots x_n = 1$.

H. Sedrakyan and N. Sedrakyan, *Algebraic Inequalities*, Problem Books in Mathematics, https://doi.org/10.1007/978-3-319-77836-5_2

Hence, we obtain that $n < x_1 + x_2 + \cdots + x_n$.
From the proof, it follows that equality holds if and only if $x_1 = \cdots = x_n = 1$.

Example 2.2 Prove that if the sum of the numbers $x_1, \ldots, x_n \quad (n \geq 2)$ is equal to 1, then $x_1^2 + \cdots + x_n^2 \geq \frac{1}{n}$.

Proof If $x_1 = \cdots = x_n = \frac{1}{n}$, then $x_1^2 + \cdots + x_n^2 = \frac{1}{n}$.

Suppose that among the considered numbers there are at least two different numbers.

Then among the numbers $x_1, \ldots, x_n$ there are two numbers such that one of them is greater than $\frac{1}{n}$ and the other one is less than $\frac{1}{n}$ (Problem 1.10). Without loss of generality, one can assume that those numbers are x_1 and x_2, and that $x_1 < \frac{1}{n}$ and $x_2 > \frac{1}{n}$. Therefore, substituting x_1 by $\frac{1}{n}$, and x_2 by $x_1 + x_2 - \frac{1}{n}$, we obtain a new sequence of numbers $\frac{1}{n}, x_1 + x_2 - \frac{1}{n}, x_3, \ldots, x_n$ such that their sum is again equal to 1.

On the other hand, we have that
$x_1^2 + x_2^2 > \left(\frac{1}{n}\right)^2 + \left(x_1 + x_2 - \frac{1}{n}\right)^2$ (see Problem 1.8), whence

$$x_1^2 + \cdots + x_n^2 > \left(\frac{1}{n}\right)^2 + \left(x_1 + x_2 - \frac{1}{n}\right)^2 + x_3^2 + \cdots + x_n^2.$$

Repeating these steps a finite number of times, we obtain a sequence such that its all terms are equal to $\frac{1}{n}$, and the sum of their squares is less than the sum of the squares of the numbers $x_1, \ldots, x_n$, that is, $x_1^2 + \cdots + x_n^2 > \left(\frac{1}{n}\right)^2 + \cdots + \left(\frac{1}{n}\right)^2 = \frac{1}{n}$.

From the proof, it follows that equality holds if and only if $x_1 = \cdots = x_n = \frac{1}{n}$.

Problems

Prove the following inequalities (2.1–2.6).

2.1 *AM-GM* (*arithmetic mean–geometric mean inequality*):
$\frac{x_1 + \cdots + x_n}{n} \geq \sqrt[n]{x_1 \cdots x_n}$, where $n \geq 2, \quad x_i > 0, \quad i = 1, \ldots, n$.

2.2 *QM-AM* (*quadratic mean–geometric mean inequality* or *RMS/root mean square inequality*): $\sqrt{\frac{x_1^2 + \cdots + x_n^2}{n}} \geq \frac{x_1 + \cdots + x_n}{n}$.

2.3 $\frac{(1 - x_1) \cdots (1 - x_n)}{x_1 \cdots x_n} \geq (n - 1)^n$, where $n \geq 2, \quad x_i > 0, \quad i = 1, \ldots, n$, and $x_1 + \cdots + x_n = 1$.

2.4 $\frac{1}{1+x_1} + \cdots + \frac{1}{1+x_n} \geq \frac{n}{1+\sqrt[n]{x_1 \ldots x_n}}$, where $n \geq 2, \quad x_1 \geq 1, \ldots, x_n \geq 1$.

2.5 $abc + bcd + cda + dab \leq \frac{1}{27} + \frac{176}{27}abcd$, where $a \geq 0, \ b \geq 0, \ c \geq 0, \ d \geq 0$, and $a + b + c + d = 1$.

2.6 $0 \leq xy + yz + zx - 2xyz \leq \frac{7}{27}$, where $x \geq 0, \ y \geq 0, \ z \geq 0$ and $x + y + z = 1$.

2.7 Among all triangles with no angle greater than 75° inscribed in a given circle, find the triangle such that its perimeter is

(a) the greatest,
(b) the smallest.

2.8 (a) *Schur's inequality:* Prove that if for some numbers α and β one has the inequality $[\alpha f(a)+\beta f(b)\leq f(\alpha a+\beta b)]\alpha f(a)+\beta f(b)\geq f(\alpha a+\beta b)$, where $\alpha\geq 0,\ \beta\geq 0,\ \alpha+\beta=1,\ a,\ b$ are any numbers belonging to $D(f)=I$,[1] and $x_1,\ldots,x_n,y_1,\ldots,y_n\in I$ such that $y_1\geq\ldots\geq y_n, y_1\leq x_1, y_1+y_2\leq x_1+x_2,\ldots,y_1+\cdots+y_{n-1}\leq x_1+\cdots+x_{n-1}, y_1+\cdots+y_n=x_1+\cdots+x_n$, then $f(y_1)+f(y_2)+\cdots+f(y_n)\leq f(x_1)+f(x_2)+\cdots+f(x_n)$

$$[f(y_1)+f(y_2)+\cdots+f(y_n)\geq f(x_1)+f(x_2)+\cdots+f(x_n)].$$

(b) *Popoviciu's inequality*: Prove that if for numbers α and β one has the inequality $\alpha f(a)+\beta f(b)\geq f(\alpha a+\beta b)$, where $\alpha,\beta\geq 0,\alpha+\beta=1$, and a,b are any numbers belonging to the interval I, then for all numbers x,y,z from the interval I, one has the inequality

$$f(x)+f(y)+f(z)+3f\left(\frac{x+y+z}{3}\right)\geq 2f\left(\frac{x+y}{2}\right)+2f\left(\frac{y+z}{2}\right)+2f\left(\frac{z+x}{2}\right).$$

2.9 Suppose that for numbers $x_1,\ldots,x_{1997}$, the following conditions hold:

(a) $-\frac{1}{\sqrt{3}}\leq x_i\leq\sqrt{3},\quad i=1,\ldots,1997,$
(b) $x_1+\cdots+x_{1997}=-318\sqrt{3}.$

Find the greatest possible value of the expression $x_1^{12}+\cdots+x_{1997}^{12}$.

2.10 Prove that $\cos\alpha_1\cos\alpha_2\cdots\cos\alpha_n(\tan\alpha_1+\cdots+\tan\alpha_n)\leq\frac{(n-1)^{\frac{n-1}{2}}}{n^{\frac{n-2}{2}}}$, where $n\geq 2$ and $0\leq\alpha_i<\frac{\pi}{2}, i=1,2,\ldots,n.$

2.11. Prove that $\sum\limits_{i=1}^{n}x_i^k(1-x_i)\leq a_k$, where $k\geq 2,\quad k\in\mathbb{N}$, and $a_k=\max\limits_{[0;1]}(x^k(1-x)+(1-x)^kx), x_i\geq 0,\quad i=1,\ldots,n,\quad x_1+\ldots+x_n=1,\quad n\geq 2.$

2.12. Prove the following inequalities:

(a) $2(n-1)(x_1x_2+x_1x_3+\cdots+x_1x_n+x_2x_3+\cdots+x_2x_n+\cdots+x_{n-1}x_n)-n^{n-1}x_1x_2\cdots x_n\leq n-2$, where $n\geq 2, x_1\geq 0,\ldots,x_n\geq 0$ and $x_1+x_2+\cdots+x_n=1$,

(b) $\frac{x_1+x_2+\cdots+x_n}{n}-\sqrt[n]{x_1x_2\cdots x_n}\leq$
$\leq\frac{(\sqrt{x_1}-\sqrt{x_2})^2+(\sqrt{x_1}-\sqrt{x_3})^2+\cdots+(\sqrt{x_1}-\sqrt{x_n})^2+\cdots+(\sqrt{x_{n-1}}-\sqrt{x_n})^2}{n}$, where $n\geq 2$, $x_1\geq 0,\ldots,x_n\geq 0.$

[1] f is any real-valued function, D is the domain of function f, $D(f)$ is denoted by I.

(c) For n=4, the following inequality is called *Turkevici's inequality*: $(n-1)(x_1^2+x_2^2+\cdots+x_n^2)+n\sqrt[n]{x_1^2x_2^2\cdots x_n^2} \geq (x_1+x_2+\cdots+x_n)^2$, where $n \geq 2, x_1 \geq 0, \ldots, x_n \geq 0$.

Proofs

2.1 Consider the numbers $\frac{x_1}{\sqrt[n]{x_1\cdots x_n}}, \ldots, \frac{x_n}{\sqrt[n]{x_1\cdots x_n}}$, and note that their product is equal to 1. Then according to Example 2.1, we have that $\frac{x_1}{\sqrt[n]{x_1\cdots x_n}}+\cdots+\frac{x_n}{\sqrt[n]{x_1\cdots x_n}} \geq n$, or

$$\frac{x_1+\cdots+x_n}{n} \geq \sqrt[n]{x_1\cdots x_n}. \tag{2.1}$$

Note that in (2.1), the equality holds if $x_1=\cdots=x_n$.
Inequality (2.1) is known as Cauchy's inequality.

2.2 Consider the numbers $\frac{x_1}{x_1+\cdots+x_n}, \ldots, \frac{x_n}{x_1+\cdots+x_n}$, and note that their sum is equal to 1. According to Example 2.2, we have that $\left(\frac{x_1}{x_1+\cdots+x_n}\right)^2+\cdots+\left(\frac{x_n}{x_1+\cdots+x_n}\right)^2 \geq \frac{1}{n}$, or

$$\frac{x_1^2+x_2^2+\cdots+x_n^2}{n} \geq \left(\frac{x_1+x_2+\cdots+x_n}{n}\right)^2. \tag{2.2}$$

This ends the proof.
Note that in (2.2), equality holds if and only if $x_1=\cdots=x_n$.

2.3 Since $x_1+\cdots+x_n=1$, it follows by Problem 1.10 that there are two numbers such that one of them is not greater than $\frac{1}{n}$, and the other one is not less than $\frac{1}{n}$. Without loss of generality, one can assume that $x_1 \leq \frac{1}{n} \leq x_2$.
Let us substitute x_1 by $\frac{1}{n}$, x_2 by $x_1+x_2-\frac{1}{n}$. Then we obtain numbers $\frac{1}{n}, x_1+x_2-\frac{1}{n}, x_3, \ldots, x_n$ such that $\frac{(1-x_1)\cdots(1-x_n)}{x_1\cdots x_n} \geq \frac{(1-\frac{1}{n})(1-x_1-x_2+\frac{1}{n})\cdots(1-x_n)}{\frac{1}{n}(x_1+x_2-\frac{1}{n})\cdots x_n}$, since $\frac{(1-x_1)(1-x_2)}{x_1x_2} \geq \frac{(1-\frac{1}{n})(1-x_1-x_2+\frac{1}{n})}{\frac{1}{n}(x_1+x_2-\frac{1}{n})}$ (Problem 1.17) and $\frac{(1-x_1)(1-x_2)}{x_1x_2}=1+\frac{1-(x_1+x_2)}{x_1x_2}$.
Repeating these steps a finite number of times, we obtain n numbers equal to $\frac{1}{n}$. For these numbers, the left-hand side of the inequality is equal to $(n-1)^n$, and it is not greater than $\frac{(1-x_1)\cdots(1-x_n)}{x_1\cdots x_n}$.

2.4 Let us set $\sqrt[n]{x_1\cdots x_n}=m$. According to Problem 1.12, without loss of generality one can assume that $x_1 \leq m \leq x_2$, and therefore, for numbers $m, \frac{x_1x_2}{m}, x_3, \ldots, x_n$, one has

$$\frac{1}{1+x_1}+\cdots+\frac{1}{1+x_n} \geq \frac{1}{1+m}+\frac{1}{1+\frac{x_1x_2}{m}}+\cdots+\frac{1}{1+x_n}, \text{ since}$$

$\frac{1}{1+x_1} + \frac{1}{1+x_2} \geq \frac{1}{1+m} + \frac{1}{1+\frac{x_1x_2}{m}}$ (Problem 1.17) and $\frac{1}{1+x_1} + \frac{1}{1+x_2} = 1 + \frac{1-x_1x_2}{1+x_1+x_2+x_1x_2}$.

After a finite number of steps, we deduce that $\frac{1}{1+x_1} + \cdots + \frac{1}{1+x_n} \geq \underbrace{\frac{1}{1+m} + \cdots + \frac{1}{1+m}}_{n} = \frac{n}{1+\sqrt[n]{x_1 \cdots x_n}}$.

2.5 If $a = b = c = d = \frac{1}{4}$, then one has equality.
Let $a < \frac{1}{4} < b$. Consider the following two cases.

(a) If $c + d - \frac{176}{27}cd < 0$, then
$A = ab\big(c + d - \frac{176}{27}cd\big) + cd(a + b) \leq cd(a + b) \leq \left(\frac{c+d+(a+b)}{3}\right)^3 = \frac{1}{27}$, using Problem 2.1.

(b) If $c + d - \frac{176}{27}cd \geq 0$, then note that

$$A \leq \frac{1}{4}\left(a + b - \frac{1}{4}\right)\left(c + d - \frac{176}{27}cd\right) + cd\left(\frac{1}{4} + \left(a + b - \frac{1}{4}\right)\right).$$

Hence, we must prove the inequality for numbers $a_1 = \frac{1}{4},\ b_1 = a + b - \frac{1}{4}$,

$$c_1 = c,\ d_1 = d.$$

In a similar way, either one can prove the inequality for numbers a_1, b_1, c_1, d_1, or it will be sufficient to prove the inequality for the case that among a_1, b_1, c_1, d_1, two numbers are equal to $\frac{1}{4}$. Continuing in this way, we obtain that it is sufficient to prove the inequality for numbers $\frac{1}{4}, \frac{1}{4}, \frac{1}{4}, \frac{1}{4}$. Note that in this case, the inequality obviously holds.

2.6 Let $x \geq y \geq z$. Then $y \leq \frac{1}{2}$, and therefore,

$$0 \leq y(x + z) + xz(1 - 2y) \leq y\left(\frac{1}{3} + \left(x + z - \frac{1}{3}\right)\right) + \frac{1}{3}\left(x + z - \frac{1}{3}\right)(1 - 2y).$$

Therefore, if one substitutes the numbers x, y, z by the numbers $\frac{1}{3}, y, x + z - \frac{1}{3}$ in the expression $xy + yz + xz - 2xyz$, then its value does not decrease. Continuing in a similar way, one can substitute the numbers $\frac{1}{3}, y, x + z - \frac{1}{3}$ by the numbers $\frac{1}{3}, \frac{1}{3}, \frac{1}{3}$. Hence, we deduce that $xy + yz + xz - 2xyz \leq \frac{1}{9} + \frac{1}{9} + \frac{1}{9} - \frac{2}{27} = \frac{7}{27}$.

2.7 According to the law of sines, we have that $p = a + b + c = 2R(\sin\alpha + \sin\beta + \sin\gamma)$.

(a) Let us prove that the greatest perimeter is that of an equilateral triangle.
If a triangle is not equilateral, then without loss of generality one can assume that $\alpha > \frac{\pi}{3} > \beta$.
Now let us prove that

$$\sin\alpha + \sin\beta + \sin(\alpha+\beta) < \sin\frac{\pi}{3} + \sin\left(\alpha+\beta-\frac{\pi}{3}\right) + \sin(\alpha+\beta).$$

We have that

$$\sin\alpha + \sin\beta - \sin\frac{\pi}{3} - \sin\left(\alpha+\beta-\frac{\pi}{3}\right) = 2\sin\frac{\alpha-\pi/3}{2}\cos\frac{\alpha+\pi/3}{2} -$$
$$-2\sin\frac{\alpha-\pi/3}{2}\cdot\cos\frac{\alpha+2\beta-\pi/3}{2} =$$
$$= 2\sin\frac{\alpha-\pi/3}{2}\left(-2\sin\frac{2\pi/3-2\beta}{4}\sin\frac{\alpha+\beta}{2}\right) =$$
$$= 4\sin\frac{\alpha-\pi/3}{2}\sin\frac{\beta-\pi/3}{2}\sin\frac{\alpha+\beta}{2} < 0.$$

Therefore, the perimeter of a triangle with angles α, β, γ is less than the perimeter of a triangle with angles $\frac{\pi}{3}$, $\alpha+\beta-\frac{\pi}{3}$, γ. If $\gamma \neq \frac{\pi}{3}$, then we obtain that the perimeter of a triangle with angles $\frac{\pi}{3}, \alpha+\beta-\frac{\pi}{3}, \gamma$ is less than the perimeter of a triangle with angles $\frac{\pi}{3}, \frac{\pi}{3}, \frac{\pi}{3}$.

(b) Let us prove that the smallest perimeter is that of a triangle with angles $75^\circ, 75^\circ, 30^\circ$. It is sufficient to prove that the perimeter of a triangle with angles α, β, γ is not less than the perimeter of a triangle with angles $75^\circ, \alpha+\beta-75^\circ, \gamma$.
Indeed, $\sin 75^\circ + \sin(\alpha+\beta-75^\circ) \leq \sin\alpha + \sin\beta$, since

$$\sin\alpha + \sin\beta - \sin 75^\circ - \sin\left(\alpha+\beta-75^\circ\right) = \left(\sin\alpha - \sin 75^\circ\right) + \left(\sin\beta - \sin\left(\alpha+\beta-75^\circ\right)\right) =$$
$$= 2\sin\frac{\alpha-75^\circ}{2}\cos\frac{\alpha+75^\circ}{2} -$$
$$-2\sin\frac{\alpha-75^\circ}{2}\cos\frac{\alpha+2\beta-75^\circ}{2} =$$
$$= 4\sin\frac{\alpha-75^\circ}{2}\sin\frac{\beta-75^\circ}{2}\sin\frac{\alpha+\beta}{2} \geq 0, \text{ and}$$
$$0^\circ < \alpha \leq 75^\circ,\ 0 < \beta \leq 75^\circ.$$

2.8 (a) If $x_i = y_i$, $i = 1, 2, \ldots, n$, then $f(x_1) + \cdots + f(x_n) = f(y_1) + \cdots + f(y_n)$. Suppose that for some i, one has $x_i \neq y_i$. From the assumptions, it follows that if m is the smallest number such that $x_m \neq y_m$, then $y_m < x_m$. Let j be the greatest number such that $y_j < x_j$.
Let k be the smallest number greater than j such that $x_k < y_k$. Note that there exists such a number k, for otherwise, we would have the following inequalities:

$$x_1 + \cdots + x_{j-1} + x_j > y_1 + \cdots + y_j,$$

$$x_{j+1} \geq y_{j+1},$$

$$\cdots\cdots\cdots$$

$$x_n \geq y_n.$$

Summing these inequalities leads us to the contradiction $x_1 + \cdots + x_n > y_1 + \cdots + y_n$.
Hence, it follows that $x_j > y_j \geq y_k > x_k$.
Let $\delta = \min(x_j - y_j, y_k - x_k)$ and $\lambda = 1 - \frac{\delta}{x_j - x_k}$.
Consider the numbers $x_1, \ldots, x_{j-1}, \lambda x_j + (1-\lambda)x_k = x_j^*, x_{j+1}, \ldots, x_{k-1}, \lambda x_k + (1-\lambda)x_j = x_k^*, x_{k+1}, \ldots, x_n$, and let us verify that for these numbers, the assumptions of the problem hold.
We have $x_1 + \cdots + x_i \geq y_1 + \cdots + y_i, \quad i = 1, \ldots, j-1$.
The inequality $y_1 + \cdots + y_j \leq x_1 + \cdots + x_{j-1} + \lambda x_j + (1-\lambda)x_k$ holds, since $\lambda x_j + (1-\lambda)x_k = x_j - \delta \geq x_j - (x_j - y_j) = y_j$.
Since $x_{j+1} \geq y_{j+1}, \ldots, x_{k-1} \geq y_{k-1}$, it follows that $x_1 + \ldots + x_i \geq y_1 + \ldots + y_i$, $i = j+1, \ldots, k-1$.
For $i \geq k$, we have $x_1 + \cdots + x_i = x_1 + \cdots + x_{j-1} + (\lambda x_j + (1-\lambda)x_k) + x_{j+1} + \cdots + x_{k-1} + (\lambda x_k + (1-\lambda)x_j) + \cdots + x_i$,
and therefore, for the numbers $x_1, \ldots, x_{j-1}, x_j^*, x_{j+1}, \ldots, x_{k-1}, x_k^*, x_{k+1}, \ldots, x_n$, the assumptions of the problem again hold.
On the other hand,

$$f(x_1) + \cdots + f(x_{j-1}) + f(x_j) + \cdots + f(x_k) + \cdots + f(x_n) \geq$$
$$\geq f(x_1) + \cdots + f(x_{j-1}) + f\left(x_j^*\right) + f(x_{j+1}) + \cdots + f(x_{k-1}) + f(x_k^*) + \cdots + f(x_n),$$

since $f\left(x_j^*\right) = f(\lambda x_j + (1-\lambda)x_k) \leq \lambda f(x_j) + (1-\lambda) f(x_k)$ and $f(x_k^*) = f(\lambda x_k + (1-\lambda)x_j) \leq \lambda f(x_k) + (1-\lambda) f(x_j)$. Therefore, $f\left(x_j^*\right) + f(x_k^*) \leq f(x_j) + f(x_k)$.
On the other hand, $\delta = x_j - y_j$ or $\delta = y_k - x_k$, whence $x_j^* = y_j$ or $x_k^* = y_k$. Therefore, we have substituted the numbers $x_1, x_2, \ldots, x_j, \ldots, x_n$ by the numbers $x_1, \ldots, x_{j-1}, x_j^*, x_{j+1}, \ldots, x_{k-1}, x_k^*, x_{k+1}, \ldots, x_n$. Hence, the assumptions of the problem again hold. On the other hand, the sum $f(x_1) + \cdots + f(x_n)$ did not increase, and the number of indices i satisfying the condition $x_i = y_i$ increased by 1. Therefore, after at most $n-1$ steps, we obtain that $f(x_1) + \cdots + f(x_n) \geq f(y_1) + \cdots + f(y_n)$.
(b) Let $x \geq y \geq z$. Since for the numbers $y_1 = y_2 = \frac{x+y}{2}, y_3 = y_4 = \frac{z+x}{2}, y_5 = y_6 = \frac{y+z}{2}$ and $x_1 = x, x_2 = x_3 = x_4 = \frac{x+y+z}{3}, x_5 = y, x_6 = z$ or $x_1 = x, x_2 = y, x_3 = x_4 = x_5 = \frac{x+y+z}{3}, x_6 = z$, the assumptions of Problem 2.8 (a) hold true, then it follows that

$$f(x) + f(y) + f(z) + 3f(\tfrac{x+y+z}{3}) = f(x_1) + f(x_2) + f(x_3) + f(x_4) + f(x_5) + f(x_6) \geq$$
$$\geq f(y_1) + f(y_2) + f(y_3) + f(y_4) + f(y_5) + f(y_6) = 2f(\tfrac{x+y}{2}) + 2f(\tfrac{y+z}{2}) + 2f(\tfrac{z+x}{2}).$$

2.9 Let us first prove the following lemma.

Lemma *If $a \leq b$ and $x > 0$, then $(a-x)^{12} + (b+x)^{12} > a^{12} + b^{12}$.*

Note that

$$(a-x)^{12}+(b+x)^{12}-a^{12}-b^{12}=C_{12}^{1}x\left(b^{11}-a^{11}\right)+C_{12}^{2}x^{2}\left(b^{10}+a^{10}\right)+\cdots+2x^{12}>0.$$

Set $y_i=\sqrt{3}x_i \quad i=1,\ldots,1997$. We have

$$-1\leq y_i\leq 3, \tag{1}$$

$$y_1+\cdots+y_{1997}=-954, \tag{2}$$

and $x_1^{12}+\cdots+x_{1997}^{12}=\frac{y_1^{12}+\cdots+y_{1997}^{12}}{3^6}$.

If any two numbers among the numbers $y_1,\ldots,y_{1997}$ belong to $(-1,3)$, then according to the lemma, one can substitute these two numbers by numbers such that one of them is equal to either -1 or 3. Then conditions (1), (2) hold, and $y_1^{12}+\cdots+y_{1997}^{12}$ increases.

Therefore, the sum $y_1^{12}+\cdots+y_{1997}^{12}$ is the greatest possible if one substitutes the numbers $y_1,\ldots,y_{1997}$ by either $-1,\ldots,-1,3,\ldots,3$, or $-1,\ldots,-1,3,\ldots,3,a$, where $a\in(-1,3)$.

Taking into consideration (2), we obtain that only the second case is possible, so that $k=\frac{a+2}{4}+1735$, where k is the number of -1's. Since $\frac{a+2}{4}\in\mathbb{Z}$ and $a\in(-1,3)$, we must have $a=2$.

Therefore, the greatest possible value of $x_1^{12}+\ldots+x_{1997}^{12}$ is equal to $\frac{1736+260\cdot 3^{12}+2^{12}}{3^6}=189548$.

2.10 Let $\frac{\alpha_1+\cdots+\alpha_n}{n}=\varphi$ (1). If $\alpha_1=\alpha_2=\cdots=\alpha_n=\varphi$, then

$$\cos\alpha_1\cos\alpha_2\cdots\cos\alpha_n(\tan\alpha_1+\cdots+\tan\alpha_n)=\cos^n\varphi\cdot n\cdot\tan\varphi=n\cdot\sin\varphi\cdot cos^{n-1}\varphi=$$

$$=n\sqrt{\sin^2\varphi\cdot(cos^2\varphi)^{n-1}}=n\sqrt{(n-1)^{n-1}\sin^2\varphi\cdot\frac{cos^2\varphi}{n-1}\cdots\frac{cos^2\varphi}{n-1}}\leq$$

$$\leq n\sqrt{(n-1)^{n-1}\left(\frac{\sin^2\varphi+\frac{cos^2\varphi}{n-1}+\cdots+\frac{cos^2\varphi}{n-1}}{n}\right)^n}=n\sqrt{(n-1)^{n-1}\cdot\frac{1}{n^n}}=\frac{(n-1)^{\frac{n-1}{2}}}{n^{\frac{n-2}{2}}}$$

If $\alpha_i\neq\alpha_j$ for some i and j $(i\neq j)$, then among those numbers there are two numbers such that one of them is greater than φ and the other one is less than φ (see Problem 1.10).

Let it be the numbers α_1 and α_2 such that $\alpha_1<\varphi<\alpha_2$. Thus, substituting α_1 by φ and α_2 by $\alpha_1+\alpha_2-\varphi$, we obtain a new sequence of numbers $\varphi,\alpha_1+\alpha_2-\varphi,\alpha_3,\ldots,\alpha_n$ such that (1) holds. On the other hand, since

$$\cos\alpha_1 \cos\alpha_2 = \frac{1}{2}(\cos(\alpha_1+\alpha_2)+\cos(\alpha_1-\alpha_2)) <$$
$$< \frac{1}{2}(\cos(\alpha_1+\alpha_2)+\cos(2\varphi-(\alpha_1+\alpha_2))) =$$
$$= \cos\varphi\cos(\alpha_1+\alpha_2-\varphi), \text{ we have}$$
$$\cos\alpha_1\cos\alpha_2\cdots\cos\alpha_n(\tan\alpha_1+\cdots+\tan\alpha_n) =$$
$$= \sin(\alpha_1+\alpha_2)\cos\alpha_3\cdots\cos\alpha_n + \cos\alpha_1\cos\alpha_2\cos\alpha_3\cdots\cos\alpha_n(\tan\alpha_3+\cdots+\tan\alpha_n) <$$
$$< \sin(\alpha_1+\alpha_2)\cos\alpha_3\cdots\cos\alpha_n +$$
$$+\cos\varphi\cos(\alpha_1+\alpha_2-\varphi)\cos\alpha_3\cdots\cos\alpha_n(\tan\alpha_3+\cdots+\tan\alpha_n) =$$
$$= \cos\varphi\cos(\alpha_1+\alpha_2-\varphi)\cos\alpha_3\ldots\cos\alpha_n(\tan\varphi+\tan(\alpha_1+\alpha_2-\varphi)+$$
$$+\tan\alpha_3+\cdots+\tan\alpha_n).$$

Continuing in the same way for the numbers $\varphi, \alpha_1+\alpha_2-\varphi, \alpha_3, \ldots, \alpha_n$, we obtain a new sequence two of whose terms are equal to φ. Let us repeat these steps at most $n-1$ times. Then we obtain a sequence $n-1$ of whose terms are equal to φ, and the nth term is equal to $n\varphi-(n-1)\varphi=\varphi$.
Hence, we have obtained

$$\cos\alpha_1\cos\alpha_2\cdots\cos\alpha_n(\tan\alpha_1+\cdots+\tan\alpha_n) < \cos^n\varphi\cdot n\cdot\tan\varphi \le \frac{(n-1)^{\frac{n-1}{2}}}{n^{\frac{n-2}{2}}}.$$

From the proof, it follows that equality holds if and only if $\alpha_1=\alpha_2=\cdots=\alpha_n=\varphi$, where $\varphi=\arctan\frac{1}{\sqrt{n-1}}$.

2.11 Prove that if $x\ge 0,\ y\ge 0,\ x+y\le\frac{2}{3}$, and $k\ge 2,\ \ k\in\mathbb{N}$, then

$$x^k(1-x)+y^k(1-y)\le(x+y)^k(1-x-y). \tag{2.6}$$

If $x+y=0$, then (2.6) holds, while if $x+y\ne 0$, then

$$\frac{x^k}{(x+y)^k}\cdot(1-x)+\frac{y^k}{(x+y)^k}\cdot(1-y)\le\left(\frac{x}{x+y}\right)^2\cdot(1-x)+\left(\frac{y}{x+y}\right)^2\cdot(1-y)=$$
$$=\frac{(x+y)^2(1-x-y)+xy(3(x+y)-2)}{(x+y)^2}\le 1-x-y.$$

Let $x_{i+1}\ge x_i\ge 0,\ \ i=1,\ldots,n-1,\ \ x_1+\cdots+x_n=1$ and $n\ge 3$. Then $(n-2)x_1+(n-2)x_2\le(x_3+\cdots+x_n)+(x_3+\cdots+x_n)=2-2x_1-2x_2$, and therefore,

$$x_1+x_2\le\frac{2}{n}\le\frac{2}{3}.$$

Therefore, if we substitute the numbers $x_1,\ldots,x_n$ by $0, x_1+x_2, x_3,\ldots,x_n$, then their sum will be equal to 1. Note that

$$\sum_{i-1}^{n}x_i^k(1-x_i)\le(x_1+x_2)^k(1-x_1-x_2)+x_3^k(1-x_3)+\cdots+x_n^k(1-x_n).$$

Repeating these steps a finite number of times, we end up with the case of $n = 2$, that is, $\sum_{i=1}^{n} x_i^k(1-x_i) \leq x^k(1-x) + (1-x)^k x$, and therefore, $\sum_{i=1}^{n} x_i^k(1-x_i) \leq a_k$.

Remark Note that $a_1 = \max_{[0;1]}(2x(1-x)) = \frac{1}{2}$, $a_2 = \max_{[0;1]}(x(1-x)) = \frac{1}{2}$,

$$a_3 = \max_{[0;1]}(x(1-x)(1-2x(1-x))) = \frac{1}{8}, \quad a_4 = \max_{[0;1]}(x(1-x)(1-3x(1-x))) = \frac{1}{12},$$

$$a_5 = \max_{[0;1]}(x(1-x) - 4(x(1-x))^2 + 2(x(1-x))^3) = \frac{5\sqrt{10}-14}{27}$$

(see Problem 12.12).

2.12 (a) Let $x_1 \leq x_2 \leq \cdots \leq x_n$, $n \geq 3$. Consider the following two cases. If $x_2 \cdots x_{n-1} \geq \frac{2(n-1)}{n^{n-1}}$, then we have that

$$A = 2(n-1)(x_1x_2 + x_1x_3 + \cdots + x_1x_n + x_2x_3 + \cdots + x_2x_n + \cdots + x_{n-1}x_n) - n^{n-1}x_1x_2\cdots x_n =$$
$$= 2(n-1)((x_1+x_n)(x_2+\cdots+x_{n-1}) + x_2x_3 + \cdots + x_2x_{n-1} + \cdots + x_{n-2}x_{n-1}) +$$
$$+ x_1x_n(2(n-1) - n^{n-1}x_2\cdots x_{n-1}) \leq 2(n-1)(x(1-x) + x_2x_3 + \cdots + x_2x_{n-1} + \cdots + x_{n-2}x_{n-1}),$$

where $x = x_2 + \cdots + x_{n-1}$.
According to Problem 2.2, it follows that $\frac{x_2^2+\cdots+x_{n-1}^2}{n-2} \geq \left(\frac{x}{n-2}\right)^2$, and therefore, $x_2x_3 + \cdots + x_2x_{n-1} + \cdots + x_{n-2}x_{n-1} \leq \frac{n-3}{2(n-2)}x^2$, whence

$$A \leq 2(n-1)(x(1-x) + \frac{n-3}{2(n-2)}x^2) = 4(n-2)\cdot\frac{n-1}{2(n-2)}x\cdot(1-\frac{n-1}{2(n-2)}x) \leq n-2.$$

If $x_2 \cdots x_{n-1} < \frac{2(n-1)}{n^{n-1}}$, then for $x_1 = x_2 = \cdots = x_n = \frac{1}{n}$, we have $A = n-2$. Otherwise, if $x_i \neq \frac{1}{n}$ for some value of i, then $x_1 < \frac{1}{n} < x_n$.
Substituting x_1 by $\frac{1}{n}$ and x_n by $x_1 + x_n - \frac{1}{n}$, we see that the value of the expression

$$A = 2(n-1)((x_1+x_n)(x_2+\cdots+x_{n-1}) + x_2x_3 + \cdots + x_2x_{n-1} + \cdots + x_{n-2}x_{n-1}) +$$
$$+ x_1x_n(2(n-1) - n^{n-1}x_2\cdots x_{n-1})$$

increases (see Problem 1.17).
Continuing in a similar way, either one can end the proof of the inequality or it will be sufficient to prove the inequality for $x_1 = x_2 = \cdots = x_n = \frac{1}{n}$.
(b) If $y_1, y_2, \ldots, y_n \geq 0$ and $y_1 + y_2 + \cdots + y_n > 0$, then according to Problem 2.12 (a), for $x_i = \frac{y_i}{y_1+y_2+\cdots+y_n}$, $i = 1, 2, \ldots, n$, we obtain that $2(n-1)q_n p_n^{n-2} - n^{n-1}y_1\cdot y_2\cdots y_n \leq (n-2)p_n^n$, where $p_n = y_1+y_2+\cdots+y_n$, $q_n = y_1y_2 + y_1y_3 + \cdots + y_1y_n + \cdots + y_{n-1}y_n$. Therefore, for $y_1\cdot y_2\cdots y_n = 1$ we have

$$q_n \leq \frac{(n-2)p_n^n + n^{n-1}}{2(n-1)p_n^{n-2}}. \tag{1}$$

If $x_1 = 0$, then we obtain the following inequality:

$$\frac{x_2 + \cdots + x_n}{n} \leq \frac{\left(\sqrt{x_2} - \sqrt{x_3}\right)^2 + \cdots + \left(\sqrt{x_2} - \sqrt{x_n}\right)^2 + \cdots + \left(\sqrt{x_{n-1}} - \sqrt{x_n}\right)^2 + x_2 + \cdots + x_n}{n},$$

or $2\sqrt{x_2x_3} + 2\sqrt{x_2x_4} + \cdots + 2\sqrt{x_2x_n} + \cdots + 2\sqrt{x_{n-1}x_n} \leq (n-2)(x_2 + \cdots + x_n)$.
The last inequality can be proved using that $2\sqrt{ab} \leq a + b (a, b \geq 0)$.
If $x_i > 0$, let $y_i = \frac{\sqrt{x_i}}{\sqrt[2n]{x_1 \cdot x_2 \cdots x_n}}, i = 1, 2, \ldots, n$. Then $y_1 \cdot y_2 \cdots y_n = 1$, and one needs to prove that $\frac{p_n^2 - 2q_n}{n} - 1 \leq \frac{(n-1)(p_n^2 - 2q_n) - 2q_n}{n}$, or $q_n \leq \frac{(n-2)p_n^2 + n}{2(n-1)}$.
The last inequality holds because by (1), it follows that (see Problem 2.1) $q_n \leq \frac{(n-2)p_n^n + n^{n-1}}{2(n-1)p_n^{n-2}} \leq \frac{(n-2)p_n^2 + n}{2(n-1)}$.
(c) For numbers $y_1 = x_1^2, \ldots, y_n = x_n^2$, using the inequality of Problem 2.12 (b), we deduce that

$$\frac{x_1^2 + \cdots + x_n^2}{n} - \sqrt[n]{x_1^2 \cdot \ldots \cdot x_n^2} \leq \frac{(|x_1| - |x_2|)^2 + \cdots + (|x_1| - |x_n|)^2 + \cdots + \left(|x_{n-1}| - |x_n|\right)^2}{n},$$

or $(n-1)(x_1^2 + \cdots + x_n^2) + n\sqrt[n]{x_1^2 \cdots x_n^2} \geq (|x_1| + \cdots + |x_n|)^2 \geq (x_1 + \cdots + x_n)^2$ (see Problem 1.13).

Problems for Independent Study

Prove the following inequalities (1–4).

1. $\frac{1}{1+x_1} + \cdots + \frac{1}{1+x_n} \leq \frac{n}{1+\sqrt[n]{x_1 \cdots x_n}}$, where $n \geq 2, \quad 0 < x_i \leq 1, \quad i = 1, \ldots, n$.
2. $\sqrt{\frac{a^2+b^2+c^2+d^2}{4}} \geq \sqrt[3]{\frac{abc+abd+acd+bcd}{4}}$, where $a > 0, \ b > 0, \ c > 0, \ d > 0$.
3. $9 \leq xy + yz + zx \leq \frac{9+x^2y^2z^2}{4}$, where $x + y + z = xyz$ and $x > 0, \ y > 0, \ z > 0$.
4. (a) $\frac{1+x_1}{x_1} \cdots \frac{1+x_n}{x_n} \geq (n+1)^n$, where $n \geq 2, \quad x_i > 0, \quad i = 1, \ldots, n$, and $x_1 + \cdots + x_n = 1$,
   (b) $\frac{1+x_1}{1-x_1} \cdots \frac{1+x_n}{1-x_n} \geq \left(\frac{n+1}{n-1}\right)^n$, where $n \geq 2, \quad x_i > 0, \quad i = 1, \ldots, n$, and $x_1 + \cdots + x_n = 1$,
   (c) $(s - (n-1)x_1) \cdots (s - (n-1)x_n) \leq x_1 \cdots x_n$, where $0 \leq x_i \leq \frac{s}{n-1}$, for $i = 1, \ldots, n$, and

$$x_1 + \cdots + x_n = s.$$

5. Prove that among all convex n-gons inscribed in a given circle, the greatest area it that of a regular n-gon.
6. Prove that among all convex n-gons inscribed in a given circle, the greatest perimeter is that of a regular n-gon.
7. Prove that among all convex polygons inscribed in a given circle, the greatest sum of the squares of the sides is that of an equilateral triangle.

8. Let $n \geq 4$ and $x_i > 0$, $i = 1, \ldots, n$, be real numbers such that $x_1 + \cdots + x_n = 1$. Prove that for all λ $(0 < \lambda < n)$, one has the inequality

$$x_1 \cdots x_n \left(\frac{1}{x_1} + \cdots + \frac{1}{x_n} \right) - \lambda x_1 \cdots x_n \leq \frac{n^2 - \lambda}{n^n}.$$

9. Prove that for two triangles with angles α, β, γ and $\alpha_1, \beta_1, \gamma_1$, one always has

$$\frac{\cos \alpha_1}{\sin \alpha} + \frac{\cos \beta_1}{\sin \beta} + \frac{\cos \gamma_1}{\sin \gamma} \leq \cot \alpha + \cot \beta + \cot \gamma.$$

10. Prove that $\sum\limits_{(i_1,\ldots,i_n)} y_{i_1}^{a_1} \cdots y_{i_n}^{a_n} \leq \sum\limits_{(i_1,\ldots,i_n)} y_{i_1}^{b_1} \cdots y_{i_n}^{b_n}$, (the summation is over all permutations of the numbers $(1, \ldots, n)$), where $y_1 > 0, \ldots, y_n > 0$, and $a_1 \geq \ldots \geq a_n$, $a_1 \leq b_1$, $a_1 + a_2 \leq b_1 + b_2, \ldots, a_1 + \cdots + a_{n-1} \leq b_1 + \cdots + b_{n-1}$, $a_1 + \cdots + a_n = b_1 + \cdots + b_n$.
11. Prove that
$n(a_1 b_1 + \cdots + a_n b_n) \geq (a_1 + \cdots + a_n)(b_1 + \cdots + b_n)$ if given that from the condition $a_i < a < a_j$, it follows that $b_i \leq b_j$, where $a = \frac{a_1 + a_2 + \cdots + a_n}{n}$.
12. Let f be an odd function defined and decreasing on $(-\infty, +\infty)$.
Prove that $f(a)f(b) + f(b)f(c) + f(c)f(a) \leq 0$, where $a + b + c = 0$.
13. Suppose that for the numbers $a_1, \ldots, a_n$ $(n \geq 2)$, the following conditions hold:

    (a) $a_1 \leq \cdots \leq a_n$,
    (b) $a_1 + \cdots + a_n = 0$,
    (c) $|a_1| + \cdots + |a_n| = S$.

    Prove that $a_n - a_1 \geq \frac{2S}{n}$.
14. Let n be a given integer such that $n \geq 2$.

    (a) Find the smallest constant C such that the inequality

    $$\sum_{1 \leq i < j \leq n} x_i x_j (x_i^2 + x_j^2) \leq C \left(\sum_{1 \leq i \leq n} x_i \right)^4$$

    holds for all nonnegative numbers $x_1, \ldots, x_n$.
    (b) Find when the equality holds for the obtained constant C.

15. Prove that

    (a) $\frac{1}{1+s-x_1} + \cdots + \frac{1}{1+s-x_n} \leq 1$, where $s = x_1 + \cdots + x_n$, $x_1 \cdots x_n = 1$ and $x_1, \ldots, x_n > 0$.
    (b) $\frac{1}{a_1^n + \cdots + a_{n-1}^n + na_1 \cdots a_n} + \frac{1}{a_1^n + \cdots + a_{n-2}^n + na_1 \cdots a_n} + \cdots + \frac{1}{a_2^n + \cdots + a_n^n + na_1 \cdots a_n} \leq \frac{1}{a_1 \cdots a_n}$,

    where $a_1, \ldots, a_n > 0$.
16. Prove that $5(a^2 + b^2 + c^2) \leq 6(a^3 + b^3 + c^3) + 1$, where $a, b, c > 0$ and $a + b + c = 1$.

Chapter 3
The HM-GM-AM-QM Inequalities

For solving some problems and in order to prove a large number of inequalities, one often needs to use various *means* and the relationships among them.

For positive real numbers a and b, the expression $\frac{2}{\frac{1}{a}+\frac{1}{b}}$ is called their *harmonic mean* (HM), $\sqrt{ab}$ is called their *geometric mean* (GM), $\frac{a+b}{2}$ is called their *arithmetic mean* (AM), and $\sqrt{\frac{a^2+b^2}{2}}$ is called their *quadratic mean* (QM) or *root mean square.* One has the following relationship among the means. In this chapter we consider the so called the *HM-GM-AM-QM* inequalities, first for two positive numbers and afterward more generally for n arbitrary positive numbers.

Lemma *(HM-GM-AM-QM inequalities for two positive numbers) Let a and b be positive real numbers. Then*

$$\frac{2}{\frac{1}{a}+\frac{1}{b}} \leq \sqrt{ab} \leq \frac{a+b}{2} \leq \sqrt{\frac{a^2+b^2}{2}}. \tag{3.1}$$

Moreover, in (3.1) equality holds if and only if $a = b$.

Proof Proofs of the inequalities (3.1) are given in Chapter 1 (Problems 1.2–1.4).

Let us consider the following examples in order to see how these inequalities can be applied.

Example 3.1 Prove that $\frac{a}{b}+\frac{b}{c}+\frac{c}{a} \geq 3$, where $a > 0,\ b > 0,\ c > 0$.

Proof Without loss of generality one can assume that $c \geq a, \quad c \geq b$.

Using that

$$\frac{a+b}{2} \geq \sqrt{ab}, \tag{3.2}$$

we deduce that

H. Sedrakyan and N. Sedrakyan, *Algebraic Inequalities*, Problem Books in Mathematics, https://doi.org/10.1007/978-3-319-77836-5_3

$$\frac{a}{b} + \frac{b}{a} \geq 2. \tag{3.3}$$

Note that $\frac{b}{c} + \frac{c}{a} - \frac{b}{a} - 1 = \frac{(c-a)(c-b)}{ac} \geq 0$, whence

$$\frac{b}{c} + \frac{c}{a} - \frac{b}{a} - 1 \geq 0. \tag{3.4}$$

Summing inequalities (3.3) and (3.4), we obtain $\frac{a}{b} + \frac{b}{c} + \frac{c}{a} \geq 3$.
This ends the proof.

Example 3.2 Prove that $\sqrt{\frac{a}{b+c}} + \sqrt{\frac{b}{c+a}} + \sqrt{\frac{c}{a+b}} > 2$, where $a > 0,\ b > 0,\ c > 0$.

Proof Without loss of generality, one can assume that $a \geq b \geq c$. Thus

$$\sqrt{\frac{a}{b+c}} + \sqrt{\frac{b}{c+a}} + \sqrt{\frac{c}{a+b}} = \sqrt{\frac{a}{b+c}} + \sqrt{\frac{b+c}{a}} + \sqrt{\frac{b}{c+a}} + \sqrt{\frac{c}{a+b}} - \sqrt{\frac{b+c}{a}} \geq$$
$$\geq 2 + \sqrt{\frac{b}{c+a}} + \sqrt{\frac{c}{a+b}} - \sqrt{\frac{b+c}{a}} \geq 2 + \sqrt{\frac{b}{c+a}} + \sqrt{\frac{c}{2a}} - \sqrt{\frac{b+c}{a}} =$$
$$= 2 + \frac{1}{\sqrt{a}}\left(\sqrt{\frac{b}{1+\frac{c}{a}}} + \sqrt{\frac{c}{2}} - \sqrt{b+c}\right) \geq 2 + \frac{1}{\sqrt{a}}\left(\sqrt{\frac{b}{1+\frac{c}{b}}} + \sqrt{\frac{c}{2}} - \sqrt{b+c}\right) =$$
$$= 2 + \frac{\sqrt{c}}{\sqrt{2a(b+c)}}\left(\sqrt{b+c} - \sqrt{2c}\right) \geq 2.$$

Equality holds if $a = b+c,\quad b = a = c$, which leads to a contradiction. Therefore, $\sqrt{\frac{a}{b+c}} + \sqrt{\frac{b}{c+a}} + \sqrt{\frac{c}{a+b}} > 2$.
This ends the proof.

Alternative proof Let us use the inequality $\sqrt{\frac{a}{b+c} \cdot 1} \geq \frac{2}{1+\frac{b+c}{a}} = \frac{2a}{a+b+c}$, $\sqrt{\frac{b}{a+c} \cdot 1} \geq \frac{2}{1+\frac{a+c}{b}} = \frac{2b}{a+b+c}$, $\sqrt{\frac{c}{a+b} \cdot 1} \geq \frac{2}{1+\frac{a+b}{c}} = \frac{2c}{a+b+c}$. Summing these inequalities, we obtain $\sqrt{\frac{a}{b+c}} + \sqrt{\frac{b}{c+a}} + \sqrt{\frac{c}{a+b}} \geq 2$. Equality holds if $1 = \frac{a}{b+c} = \frac{b}{a+c} = \frac{c}{a+b}$, which leads to a contradiction. Thus, it follows that $\sqrt{\frac{a}{b+c}} + \sqrt{\frac{b}{c+a}} + \sqrt{\frac{c}{a+b}} > 2$.
This ends the proof.

Generalization of inequalities (3.1).
Note that (3.1) can be generalized for positive numbers $a_1, \ldots, a_n$.

The expressions $\frac{n}{\frac{1}{a_1} + \cdots + \frac{1}{a_n}}$, $\sqrt[n]{a_1 \cdots a_n}$, $\frac{a_1 + \cdots + a_n}{n}$, $\sqrt{\frac{a_1^2 + \cdots + a_n^2}{n}}$ are called the *harmonic mean* (HM), *geometric mean* (GM), *arithmetic mean* (AM), and *quadratic mean* (QM) or *root mean square* of the numbers $a_1, \ldots, a_n$

Lemma *(HM-GM-AM-QM inequalities for n arbitrary positive numbers) Let $a_1, \ldots, a_n$ be positive real numbers. Then*

$$\frac{n}{\frac{1}{a_1} + \cdots + \frac{1}{a_n}} \le \sqrt[n]{a_1 \cdots a_n} \le \frac{a_1 + \cdots + a_n}{n} \le \sqrt{\frac{a_1^2 + \cdots + a_n^2}{n}}. \quad (3.5)$$

Moreover, in (3.5), inequality holds if and only if $a_1 = \cdots = a_n$.

Proof See Example 3.3, Problems 2.1 and 2.2.

Example 3.3 (HM-GM) Prove that $\frac{n}{\frac{1}{a_1}+\cdots+\frac{1}{a_n}} \le \sqrt[n]{a_1 \cdots a_n}$, where $n \ge 2$, $a_1 > 0, \ldots, a_n > 0$.

Proof Using the inequality $\frac{x_1+\cdots+x_n}{n} \ge \sqrt[n]{x_1 \cdots x_n}$ for positive numbers $x_i = \frac{1}{a_i}$, $i = 1, \ldots, n$ (see Problem 2.1), we obtain the following inequality: $\frac{\frac{1}{a_1}+\cdots+\frac{1}{a_n}}{n} \ge \frac{1}{\sqrt[n]{a_1 \cdots a_n}}$.

This ends the proof.

Remark $\frac{a_1+\cdots+a_n}{n} \ge \sqrt[n]{a_1 \cdots a_n} \ge \frac{n}{\frac{1}{a_1}+\cdots+\frac{1}{a_n}}$, whence

$$(a_1 + \cdots + a_n)\left(\frac{1}{a_1} + \cdots + \frac{1}{a_n}\right) \ge n^2. \quad (3.6)$$

Let us consider the following examples in order to see how these inequalities can be applied.

Example 3.4 Solve the following system:

$$\begin{cases} \sqrt{1 + x_1} + \cdots + \sqrt{1 + x_{100}} = 100\sqrt{1 + \frac{1}{100}}, \\ \sqrt{1 - x_1} + \cdots + \sqrt{1 - x_{100}} = 100\sqrt{1 - \frac{1}{100}}. \end{cases}$$

Solution. Using the inequality

$$\frac{a_1 + \cdots + a_n}{n} \le \sqrt{\frac{a_1^2 + \cdots + a_n^2}{n}} \quad (3.7)$$

(see Problem 2.2) for the numbers $\sqrt{1+x_1}, \ldots, \sqrt{1+x_{100}}$ and $\sqrt{1-x_1}, \ldots, \sqrt{1-x_{100}}$, we deduce that

$$\frac{\sqrt{1 + x_1} + \cdots + \sqrt{1 + x_{100}}}{100} \le \sqrt{\frac{100 + (x_1 + \cdots + x_{100})}{100}}$$

$$\text{and} \quad \frac{\sqrt{1 - x_1} + \cdots + \sqrt{1 - x_{100}}}{100} \le \sqrt{\frac{100 - (x_1 + \cdots + x_{100})}{100}}. \quad (3.8)$$

From the obtained inequalities and the given system, we deduce that

$$\sqrt{1+\frac{1}{100}} \leq \sqrt{1+\frac{x_1+\cdots+x_{100}}{100}}, \quad \text{and} \quad \sqrt{1-\frac{1}{100}} \leq \sqrt{1-\frac{x_1+\cdots+x_{100}}{100}}.$$

Therefore, $x_1 + \cdots + x_{100} = 1$, and thus it follows that inequality (3.8) is an equality.

Hence, we obtain that $\sqrt{1+x_1} = \cdots = \sqrt{1+x_{100}}$, or $x_1 = \cdots = x_{100} = 0.01$. It is not difficult to verify that these numbers satisfy the given system.

The relationship between the arithmetic mean and the geometric mean helps in solving problems the involve finding maximum and minimum values.

Remark 1 Consider all n-tuples $(a_1, \ldots, a_n)$ of positive numbers having the same geometric mean. Prove that among those n-tuples, the minimum arithmetic mean has the n-tuple with $a_1 = \cdots = a_n$.

Remark 2 Consider all n-tuples $(a_1, \ldots, a_n)$ of positive numbers having the same arithmetic mean. Prove that among those n-tuples, the minimum geometric mean has the n-tuple with $a_1 = \cdots = a_n$.

Example 3.5 Among all triangles with a given area, find the triangle that has the minimum perimeter.

Proof Let us denote the side lengths of that triangle by a, b, c, and the perimeter by $2p$.

By Heron's formula and the inequality of Problem 2.1, we obtain $S = \sqrt{p(p-a)(p-b)(p-c)} \leq \sqrt{p\left(\frac{(p-a)+(p-b)+(p-c)}{3}\right)^3} = \frac{\sqrt{3}}{9}p^2$, and therefore, $2p \geq 2\sqrt{3\sqrt{3}\cdot S}$.

Equality holds if $a = b = c$. Therefore, among all triangles with a given area, the one with the minimum perimeter is an equilateral triangle.

Example 3.6 Among the triangles with perimeter equal to $2p$, find the triangle with the maximum area.

Proof Denote the area of that triangle by S, and the side lengths by a, b, c.

By Heron's formula, $S = \sqrt{p}\cdot\sqrt{(p-a)(p-b)(p-c)}$. Let us find the maximum value of the expression $(p-a)(p-b)(p-c)$. Since $(p-a)+(p-b)+(p-c) = p$ is a constant, it follows from Remark 2 that the product $(p-a)(p-b)(p-c)$ attains its maximum value when $p-a = p-b = p-c$. Therefore, $a = b = c$, and so among all triangles with perimeter equal to $2p$, the one with maximum area is an equilateral triangle.

This ends the proof.

Problems

Prove the following inequalities (3.1–3.28).

3.1. $(a+b)(b+c)(c+a) \geq 8\,abc$, where $a > 0,\ b > 0,\ c > 0$.

3.2. $(a+b+c-d)(b+c+d-a)(c+d+a-b)(d+a+b-c) \leq (a+b)(b+c)(c+d)(d+a)$,
where $a > 0, b > 0,\ c > 0,\ d > 0$.

3.3. (a) *Schur's inequality*: $a^3+b^3+c^3+3abc \geq a^2b+ab^2+b^2c+c^2b+a^2c+ac^2$, where $a > 0,\ b > 0,\ c > 0$.
(b) $\left(1+\frac{4a}{b+c}\right)\left(1+\frac{4b}{c+a}\right)\left(1+\frac{4c}{a+b}\right) > 25$, where $a > 0,\ b > 0,\ c > 0$.

3.4. $\frac{\log(a-1)}{\log a} < \frac{\log a}{\log(a+1)}$, where $a > 1$.

3.5. *Schur's inequality (alternative form)*: $abc \geq (a+b-c)(a+c-b)(b+c-a)$, where $a > 0,\ b > 0,\ c > 0$.

3.6. $x^8+y^8 \geq \frac{1}{128}$, if $x+y=1$.

3.7. $\left(a+\frac{1}{a}\right)^2+\left(b+\frac{1}{b}\right)^2 \geq 12.5$, if $a > 0,\ b > 0$ and $a+b=1$.

3.8. $\left(x_1+\frac{1}{x_1}\right)^2+\cdots+\left(x_n+\frac{1}{x_n}\right)^2 \geq \frac{(n^2+1)^2}{n}$, if $n \geq 2,\ x_1 > 0,\ldots,x_n > 0$ and $x_1+\cdots+x_n=1$.

3.9. $a^4+b^4+c^4 \geq abc\,(a+b+c)$.

3.10. $x^2+y^2 \geq 2\sqrt{2}(x-y)$, if $xy=1$.

3.11. $\sqrt{6a_1+1}+\sqrt{6a_2+1}+\sqrt{6a_3+1}+\sqrt{6a_4+1}+\sqrt{6a_5+1} \leq \sqrt{55}$, if $a_1 > 0,\ \ldots,\ a_5 > 0$ and $a_1+\cdots+a_5=1$.

3.12. $6a+4b+5c \geq 5\sqrt{ab}+3\sqrt{bc}+7\sqrt{ca}$, where $a \geq 0,\ b \geq 0,\ c \geq 0$.

3.13. $2\left(a^4+b^4\right)+17 > 16\,ab$.

3.14. (a) $\frac{1}{b+c-a}+\frac{1}{a-b+c}+\frac{1}{a+b-c} \geq \frac{1}{a}+\frac{1}{b}+\frac{1}{c}$, where $a,\ b,\ c$ are the side lengths of some triangle.
(b) $\frac{a}{\sqrt{2b^2+2c^2-a^2}}+\frac{b}{\sqrt{2c^2+2a^2-b^2}}+\frac{c}{\sqrt{2a^2+2b^2-c^2}} \geq \sqrt{3}$, where $a,\ b,\ c$ are the side lengths of some triangle.

3.15. $\left(\frac{1+nb}{n+1}\right)^{n+1} \geq b^n$, where $n \in \mathbb{N},\ b > 0$.

3.16. (a) $\left(1+\frac{1}{n}\right)^n < \left(1+\frac{1}{n+1}\right)^{n+1}$, where $n \in \mathbb{N}$,
(b) $\left(1+\frac{1}{n}\right)^{n+1} > \left(1+\frac{1}{n+1}\right)^{n+2}$, where $n \in \mathbb{N}$,
(c) $\left(1+\frac{m}{n-1}\right)^{\frac{n-1}{m}} < \left(1+\frac{m}{n}\right)^{\frac{n}{m}} < \left(1+\frac{m-1}{n}\right)^{\frac{n}{m-1}}$, where $m > 1,\ n > 1$, and $m \in \mathbb{N},\ n \in \mathbb{N}$.

3.17. $n! < \left(\frac{n+1}{2}\right)^n$, where $n = 2, 3, 4, \ldots$.

3.18. (a) $n(n+1)^{\frac{1}{n}} < n+S_n$, where $S_n = \frac{1}{1}+\cdots+\frac{1}{n},\ n = 2, 3, 4, \ldots,$
(b) $n-S_n > (n-1)n^{\frac{1}{1-n}}$, where $S_n = \frac{1}{1}+\cdots+\frac{1}{n},\ n = 3, 4, \ldots$.

3.19. $(q^n-1)\left(q^{n+1}+1\right) \geq 2nq^n(q-1)$, where $q > 1,\ n \in \mathbb{N}$.

3.20. (a) $a^2+b^2+c^2+d^2+ab+ac+ad+bc+bd+cd \geq 10$, where $a>0,\ b>0,\ c>0,\ d>0$, and $abcd=1$.
(b) $\left(a-1+\frac{1}{b}\right)\left(b-1+\frac{1}{c}\right)\left(c-1+\frac{1}{a}\right) \leq \left(\frac{1+abc}{2\sqrt{abc}}\right)^3$, where $a, b, c > 0$.
(c) $\left(a+\frac{1}{b}-t\right)\left(b+\frac{1}{c}-t\right)\left(c+\frac{1}{a}-t\right) \leq (a+b+c)\left(\frac{1}{a}+\frac{1}{b}+\frac{1}{c}\right)(1-t)^2 + 4-3t$, where $a, b, c, t > 0$ and $abc=1$.

3.21. $n\sqrt[n]{a_1\cdots a_n}-(n-1)\sqrt[n-1]{a_1\cdots a_{n-1}} \leq a_n$, where $a_i>0,\ i=1,\ldots,n,\ n=3,4,\ldots$.

3.22.
$$\sqrt[n]{a_1\cdots a_n}+\sqrt[n]{b_1\cdots b_n}+\ldots+\sqrt[n]{k_1\cdots k_n} \leq \sqrt[n]{(a_1+b_1+\cdots+k_1)(a_2+b_2+\cdots+k_2)\cdots(a_n+b_n+\cdots+k_n)},$$
where $a_1>0,\ldots,a_n>0,\ b_1>0,\ldots,b_n>0,\ldots,k_1>0,\ldots,k_n>0$.

3.23. $a_1+\sqrt{a_1a_2}+\cdots+\sqrt[n]{a_1\cdots a_n} \leq e(a_1+\cdots+a_n)$, where $n\geq 2,\ a_1\geq 0,\ldots,a_n\geq 0$.

3.24. (a) $na^k-ka^n \leq n-k$, where $n>k,\ \ n,k\in\mathbb{N},\ \ a>0$.
(b) $\frac{x_1^2}{x_2}+\frac{x_2^3}{x_3^2}+\cdots+\frac{x_n^{n+1}}{x_1^n} \geq x_1+\cdots+x_n$, where $n\geq 2,\ n\in\mathbb{N},\ x_1=\min(x_1,\ldots,x_n)>0$.

3.25. $\frac{a^{x_1-x_2}}{x_1+x_2}+\frac{a^{x_2-x_3}}{x_2+x_3}+\cdots+\frac{a^{x_n-x_1}}{x_n+x_1} \geq \frac{n^2}{2\sum\limits_{i=1}^{n}x_i}$, where $a>0,\ \ x_i>0,\ i=1,\ldots,n$.

3.26. $\sqrt[p]{x_1+1}+\cdots+\sqrt[p]{x_n+1} \leq n+1$, where $n\geq 2, x_1>0,\ldots,x_n>0, x_1+\cdots+x_n=p, p\in\mathbb{N}, p\geq 2$.

3.27. (a) $(1+\alpha)^n \geq 1+n\alpha$, where $\alpha\geq -1,\ \ n\in\mathbb{N}$ (*Bernoulli's inequality*).
(b) $(1+\alpha)^n > 1+n\alpha$, where $\alpha\neq 0,\ \ \alpha\geq -1,\ \ n\in\mathbb{N},\ \ n>1$.

3.28. (a) $\cos^3 t\,\sin t \leq \frac{3\sqrt{3}}{16}$,
(b) $x^k(1-x^m) \leq \frac{k^{\frac{k}{m}}\cdot m}{(k+m)^{1+\frac{k}{m}}}$, where $0\leq x\leq 1,\ \ k,m\in\mathbb{N}$,
(c) $\frac{x}{1-x^2}+\frac{y}{1-y^2}+\frac{z}{1-z^2} \geq \frac{3\sqrt{3}}{2}$, where $x,y,z>0$ and $x^2+y^2+z^2=1$,
(d) $\frac{1}{1-x}+\frac{1}{1-y}+\frac{1}{1-z} \geq \frac{9+3\sqrt{3}}{2}$, where $x,y,z>0$ and $x^2+y^2+z^2=1$.

3.29. Find the minimum value of the function $f(x)=\frac{1}{\sqrt[n]{1+x}}+\frac{1}{\sqrt[n]{1-x}}$ in $[0,1)$, where $n\in\mathbb{N},\ \ n>1$.

3.30. Find the minimum value of the function $f(x)=ax^m+\frac{b}{x^n}$ in $(0,\infty)$, where $a>0,\ b>0,\ \ m,n\in\mathbb{N}$.

3.31. Find in $[a,b]$ $(0<a<b)$ a point x_0 such that the function $f(x)=(x-a)^2(b^2-x^2)$ attains its maximum value in $\lfloor a,b\rfloor$ at x_0.

3.32. Find the greatest possible value of the product xyz given $x>0,\ y>0,\ z>0$, and $2x+\sqrt{3}y+\pi z=1$.

3.33. Find the maximum and minimum values of the function $y=\frac{x}{ax^2+b}$, where $a>0,\ b>0$.

3.34. (a) Find the maximum value of the function $y = \frac{5\sqrt{x^2+6x+8}+12}{x+3}$.

(b) Find the maximum value of the function $y = \frac{\sqrt[3]{(x^2+1)^2(x^2+3)}}{3x^2+4}$.

3.35. Suppose the sum of the six edges of a triangular pyramid (tetrahedron) $PABC$ is equal to S, and $\angle APB = \angle BPC = \angle CPA = 90^0$. Find among such pyramids one with the greatest volume.

3.36. Solve the system of equations

$$\begin{cases} x+y=2, \\ xy-z^2=1. \end{cases}$$

3.37. Solve the system of equations

$$\begin{cases} x+y+z=3, \\ x^2+y^2+z^2=3. \end{cases}$$

3.38. Given numbers a, b, c, d, e such that

$$\begin{cases} a+b+c+d+e=8, \\ a^2+b^2+c^2+d^2+e^2=16, \end{cases}$$

find the greatest possible value of e.

3.39. Find the minimum value of the expression $\frac{x_1}{x_2}+\frac{x_3}{x_4}+\frac{x_5}{x_6}$ if $1 \leq x_1 \leq x_2 \leq x_3 \leq x_4 \leq x_5 \leq x_6 \leq 1000$.

3.40. Solve the equation $x^4+y^4+2=4xy$.

3.41. Find all integer solutions of the equation $\frac{xy}{z}+\frac{xz}{y}+\frac{yz}{x}=3$.

3.42. Prove that

(a) $x_1^\alpha+\cdots+x_n^\alpha \geq x_1^\beta+\cdots+x_n^\beta$, where $n \geq 2$, $x_1>0, \ldots, x_n>0$, $\alpha > \beta \geq 0$, and $x_1 \cdots x_n = 1$,

(b) $x_1^\alpha+\cdots+x_n^\alpha \geq x_1^\beta+\cdots+x_n^\beta$, where $n \geq 2$, $x_1>0, \ldots, x_n>0$, $\alpha \geq (n-1)|\beta|$, and $x_1 \cdots x_n = 1$,

3.43. Prove that $x^2y+y^2z+z^2x \leq \frac{4}{27}$, where $x \geq 0$, $y \geq 0$, $z \geq 0$, and $x+y+z=1$.

3.44. Prove that

(a) $\frac{1+a}{1+ab}+\frac{1+b}{1+bc}+\frac{1+c}{1+cd}+\frac{1+d}{1+da} \geq 4$, where $a>0$, $b>0$, $c>0$, $d>0$, and $abcd=1$,

(b) $\frac{1+ab}{1+a}+\frac{1+bc}{1+b}+\frac{1+cd}{1+c}+\frac{1+da}{1+d} \geq 4$, where $a>0$, $b>0$, $c>0$, $d>0$, and $abcd=1$.

3.45. Prove that $2ST > \sqrt{3(S+T)(S(bd+df+fb)+T(ac+ce+ea))}$, where $0 < a < b < c < d < e < f$ and $a+c+e=S, \quad b+d+f=T$.

3.46. Prove that $\frac{a+\sqrt{ab}+\sqrt[3]{abc}+\sqrt[4]{abcd}}{4} \le \sqrt[4]{a\cdot\frac{a+b}{2}\cdot\frac{a+b+c}{3}\cdot\frac{a+b+c+d}{4}}$, where $a > 0,\ b > 0,\ c > 0,\ d > 0$.

3.47. Prove that $a^{12}+(ab)^6+(abc)^4+(abcd)^3 \le 1,43(a^{12}+b^{12}+c^{12}+d^{12})$.

Proofs

3.1. Let us use the inequality (3.2) for positive numbers a and b, b and c, c and a. Then we obtain $\frac{a+b}{2} \ge \sqrt{ab}$, $\frac{b+c}{2} \ge \sqrt{bc}$, $\frac{c+a}{2} \ge \sqrt{ca}$, and on multiplying these inequalities, we deduce that $(a+b)(b+c)(c+a) \ge 8abc$.

3.2. Note that three factors on the left-hand side of the given inequality are positive. If only one factor on the left-hand side is not positive, then the proof is obvious. Consider the case in which all four factors are positive. In this case, according to inequality (3.2) for the numbers $a+b+c-d$ and $b+c+d-a$, $a+b+c-d$ and $d+a+b-c$, $b+c+d-a$ and $c+d+a-b$, $c+d+a-b$ and $d+a+b-c$, we obtain the following inequalities:

$$\sqrt{(a+b+c-d)(b+c+d-a)} \le \frac{(a+b+c-d)+(b+c+d-a)}{2} = b+c,$$
$$\sqrt{(a+b+c-d)(d+a+b-c)} \le a+b, \quad \sqrt{(b+c+d-a)(c+d+a-b)} \le c+d,$$
$$\sqrt{(c+d+a-b)(d+a+b-c)} \le a+d,$$

and on multiplying these inequalities, we obtain the required inequality.

3.3. (a) The given inequality is equivalent to the following inequality: $abc \ge (a+b-c)(a+c-b)(b+c-a)$. See Problem 3.5.

(b) The given inequality is equivalent to the following inequality: $7abc+a^3+b^3+c^3 > a^2b+b^2a+c^2a+a^2c+b^2c+c^2b$, and the proof follows from the inequality of Problem 3.3(a).

3.4. The given inequality is equivalent to the following inequality: $\log(a-1)\log(a+1) < \log^2 a$. If $\log(a-1) \le 0$, then $\log(a-1)\log(a+1) \le 0 < \log^2 a$.
Otherwise, if $\log(a-1) > 0$, then using inequality (3.2), we deduce that $2\sqrt{\log(a-1)\log(a+1)} \le \log(a-1)+\log(a+1) = \log\big(a^2-1\big) < \log a^2$. This inequality is equivalent to the inequality $\log(a-1)\log(a+1) < \log^2 a$.

Alternative proof. The given inequality is equivalent to the following inequality:
$1-\frac{\log(a-1)}{\log a} > 1-\frac{\log a}{\log(a+1)}$, or $\frac{\log\frac{a}{a-1}}{\log a} > \frac{\log\frac{a+1}{a}}{\log(a+1)}$. This proof follows from the inequalities $\frac{a}{a-1} > \frac{a+1}{a} > 1$ and $a+1 > a > 1$.

3.5. Since $a > 0,\ b > 0,\ c > 0$, at least two factors on the right-hand side of the inequality are positive. If only one factor is not positive, then the proof is obvious. Consider the case in which all three factors on the right-hand side are positive.
Since $\frac{x+y}{2} \geq \sqrt{xy}$ for nonnegative numbers, it follows that

$$\sqrt{(a+b-c)(a+c-b)} \leq \frac{(a+b-c)+(a+c-b)}{2} = a, \quad \sqrt{(a+b-c)(b+c-a)} \leq b,$$

$$\sqrt{(a+c-b)(b+c-a)} \leq c.$$

On multiplying these inequalities, we obtain the required inequality.

3.6. We have

$$\frac{x^8+y^8}{2} \geq \left(\frac{x^4+y^4}{2}\right)^2 \geq \left(\frac{x^2+y^2}{2}\right)^4 \geq \left(\frac{x+y}{2}\right)^8 = \frac{1}{128}.$$

3.7. Since $a+b=1$, we deduce from (3.2) that $\frac{1}{ab} \geq 4$. Hence, according to inequality (3.7), we obtain

$$\frac{\left(a+\frac{1}{a}\right)^2+\left(b+\frac{1}{b}\right)^2}{2} \geq \left(\frac{a+\frac{1}{a}+b+\frac{1}{b}}{2}\right)^2 = \left(\frac{1+\frac{1}{ab}}{2}\right)^2 \geq \left(\frac{1+4}{2}\right)^2 = \frac{25}{4}.$$

3.8. Since $x_1+\cdots+x_n=1$, we deduce from (3.6) that $\frac{1}{x_1}+\cdots+\frac{1}{x_n} \geq n^2$. Using inequality (3.7), we obtain

$$\left(x_1+\frac{1}{x_1}\right)^2+\cdots+\left(x_n+\frac{1}{x_n}\right)^2 \geq n\left(\frac{x_1+\frac{1}{x_1}+\cdots+x_n+\frac{1}{x_n}}{n}\right)^2 =$$

$$= n\left(\frac{1+\frac{1}{x_1}+\cdots+\frac{1}{x_n}}{n}\right)^2 \geq n\left(\frac{1+n^2}{n}\right)^2 = \frac{\left(1+n^2\right)^2}{n}.$$

3.9. According to inequality (3.2), we have $a^4+b^4 \geq 2a^2b^2,\quad b^4+c^4 \geq 2b^2c^2,\quad c^4+a^4 \geq 2c^2a^2$.
Summing these inequalities, we obtain

$$a^4+b^4+c^4 \geq a^2b^2+b^2c^2+c^2a^2. \tag{3.9}$$

According to inequality (3.2), we have

$$a^2b^2+b^2c^2 \geq 2ab^2c, \quad b^2c^2+c^2a^2 \geq 2abc^2, \quad c^2a^2+a^2b^2 \geq 2a^2bc.$$

Summing these inequalities, we deduce that

$$a^2b^2+b^2c^2+c^2a^2 \geq abc(a+b+c). \tag{3.10}$$

From inequalities (3.9) and (3.10) it follows that $a^4+b^4+c^4 \geq abc(a+b+c)$.

3.10. Using that $x^2+y^2=(x-y)^2+2xy$ and $xy=1$, we deduce that $x^2+y^2=(x-y)^2+\left(\sqrt{2}\right)^2 \geq 2\sqrt{2}(x-y)$.

3.11. According to inequality (3.2), for every value of λ, one has the following inequalities:

$$(6a_i+1)+\lambda^2 \geq 2\lambda\sqrt{6a_i+1},\ i=1,\ldots,5.$$

Summing these inequalities and using that $a_1+\cdots+a_5=1$, for $\lambda>0$ we obtain $\frac{11+5\lambda^2}{2\lambda} \geq \sum\limits_{i=1}^{n} \sqrt{6a_i+1}$.

Taking $\lambda=\sqrt{\frac{11}{5}}$, it follows that $\sqrt{55} \geq \sum\limits_{i=1}^{5} \sqrt{6a_i+1}$.

3.12. According to inequality (3.2), we have $5\sqrt{ab}+7\sqrt{ac}+3\sqrt{bc} \leq \frac{5(a+b)}{2}+\frac{7(a+c)}{2}+\frac{3(b+c)}{2}=6a+4b+5c$.

3.13. Since $a^4+b^4 \geq 2a^2b^2$, it follows that $2\left(a^4+b^4\right)+17 \geq 4a^2b^2+17 > 4\left(a^2b^2+4\right) \geq 16ab$, and therefore $2\left(a^4+b^4\right)+17>16ab$.

3.14. (a) Using inequality (3.6) for the numbers $\frac{1}{b+c-a}$ and $\frac{1}{a-b+c}$, $\frac{1}{b+c-a}$ and $\frac{1}{a+b-c}$, $\frac{1}{a-b+c}$ and $\frac{1}{a+b-c}$, we obtain that $\frac{\frac{1}{b+c-a}+\frac{1}{a-b+c}}{2} \geq \frac{1}{c}$, $\frac{\frac{1}{b+c-a}+\frac{1}{a+b-c}}{2} \geq \frac{1}{b}$, $\frac{\frac{1}{a-b+c}+\frac{1}{a+b-c}}{2} \geq \frac{1}{a}$. Summing these inequalities, we obtain the required inequality.

(b) According to the inequality of Problem 1.2, we have

$$\frac{a}{\sqrt{2b^2+2c^2-a^2}}+\frac{b}{\sqrt{2c^2+2a^2-b^2}}+\frac{c}{\sqrt{2a^2+2b^2-c^2}}=$$

$$=\sqrt{3}\left(\frac{a^2}{\sqrt{3a^2(2b^2+2c^2-a^2)}}+\frac{b^2}{\sqrt{3b^2(2c^2+2a^2-b^2)}}+\frac{c^2}{\sqrt{3c^2(2a^2+2b^2-c^2)}}\right) \geq$$

$$\geq \sqrt{3}\left(\frac{a^2}{\frac{3a^2+(2b^2+2c^2-a^2)}{2}}+\frac{b^2}{\frac{3b^2+(2c^2+2a^2-b^2)}{2}}+\frac{c^2}{\frac{3c^2+(2a^2+2b^2-c^2)}{2}}\right)=\sqrt{3},$$

and therefore, $\frac{a}{\sqrt{2b^2+2c^2-a^2}}+\frac{b}{\sqrt{2c^2+2a^2-b^2}}+\frac{c}{\sqrt{2a^2+2b^2-c^2}} \geq \sqrt{3}$.

3.15. One has $1+\underbrace{b+\cdots+b}_{n} \geq (n+1)\sqrt[n+1]{b^n}$, and thus it follows that $\left(\frac{1+nb}{n+1}\right)^{n+1} \geq b^n$.

3.16. (a) Using the inequality of Problem 2.1 for the numbers $\underbrace{(1+1/n),\ldots,(1+1/n)}_{n}$, 1, we obtain $\frac{(1+\frac{1}{n})+\ldots+(1+\frac{1}{n})+1}{n+1} > \sqrt[n+1]{\left(1+\frac{1}{n}\right)^n}$, or $\left(1+\frac{1}{n+1}\right)^{n+1} > \left(1+\frac{1}{n}\right)^n$.

(b) Using the inequality of Problem 2.1 for the numbers $\underbrace{1-\frac{1}{n+1},\dots,1-\frac{1}{n+1}}_{n+1}$, 1, we obtain $\left(1-\frac{1}{n+2}\right)^{n+2} > \left(1-\frac{1}{n+1}\right)^{n+1}$ or $\left(1+\frac{1}{n}\right)^{n+1} > \left(1+\frac{1}{n+1}\right)^{n+2}$.

(c) Using the inequality of Problem 2.1 for the numbers $\underbrace{1+\frac{m}{n-1},\dots,1+\frac{m}{n-1}}_{n-1}$, 1, we obtain $\left(1+\frac{m}{n-1}\right)^{\frac{n-1}{m}} < \left(1+\frac{m}{n}\right)^{\frac{n}{m}}$.

Using the inequality of Problem 2.1 for the numbers $\underbrace{1+\frac{m}{n},\dots,1+\frac{m}{n}}_{m-1}$, 1, we obtain $\left(1+\frac{m}{n}\right)^{\frac{n}{m}} < \left(1+\frac{m-1}{n}\right)^{\frac{n}{m-1}}$.

3.17. Using the inequality of Problem 2.1, we obtain $\sqrt[n]{n!} = \sqrt[n]{1\cdots n} < \frac{1+\cdots+n}{n} = \frac{n+1}{2}$.

3.18. (a) $S_n+n = n+\frac{1}{1}+\cdots+\frac{1}{n} = 2+\frac{3}{2}+\cdots+\frac{n+1}{n} > n\cdot\sqrt[n]{2\cdot\frac{3}{2}\cdots\frac{n+1}{n}} = n\cdot\sqrt[n]{n+1}$.

(b) $n - S_n = \frac{1}{2}+\cdots+\frac{n-1}{n} > (n-1)\cdot\sqrt[n-1]{\frac{1}{2}\cdots\frac{n-1}{n}} = (n-1)n^{\frac{1}{1-n}}$.

3.19. We have $q^n - 1 = (q-1)\left(q^{n-1}+\cdots+1\right)$ and $q > 1$. Therefore we obtain an inequality equivalent to the given inequality:

$$\left(q^{n-1}+\cdots+1\right)\left(q^{n+1}+1\right) \geq 2nq^n. \tag{3.11}$$

According to Problem 2.1 for the case $n > 1$, we have $q^{n-1}+\cdots+1 \geq n\cdot\sqrt[n]{q^{n-1}\cdots 1} = nq^{\frac{n-1}{2}}$ and $q^{n+1}+1 \geq 2q^{\frac{n+1}{2}}$. Multiplying these inequalities, we obtain (3.11).

3.20. (a) $a^2+b^2+c^2+d^2+ab+ac+ad+bc+bd+cd \geq 10\cdot\sqrt[10]{(abcd)^5} = 10$.

(b) Since $a, b, c > 0$, at least two factors on the left-hand side are positive (see Problem 1.9(a)). If only one factor on the left-hand side is nonpositive, then the proof is obvious. Consider the case in which all three factors on the left-hand side are positive. In this case, note that

$$3+3abc = b\left(a-1+\frac{1}{b}\right)+c\left(b-1+\frac{1}{c}\right)+a\left(c-1+\frac{1}{a}\right)+bc\left(a-1+\frac{1}{b}\right)+$$
$$+\,ac\left(b-1+\frac{1}{c}\right)+ab\left(c-1+\frac{1}{a}\right).$$

According to Problem 2.1, we deduce that

$$3+3abc \geq 6\sqrt[3]{a^3b^3c^3\left(a-1+\frac{1}{b}\right)^2\left(b-1+\frac{1}{c}\right)^2\left(c-1+\frac{1}{a}\right)^2}.$$

This is equivalent to the given inequality.

(c) Define $a+b+c=A$ and $\frac{1}{a}+\frac{1}{b}+\frac{1}{c}=B$. Then one needs to prove that

$$t^3+(AB-A-B)t^2+\left(\frac{a}{c}+\frac{b}{a}+\frac{c}{b}+A+B-2AB\right)t+AB+2-A-B\geq 0. \quad (1)$$

For $t=1$, inequality (1) holds, since for $t=1$ it is equivalent to Problem 3.2(b) for $abc=1$.
Therefore, for $t=1$, from (1) it follows that $\frac{a}{c}+\frac{b}{a}+\frac{c}{b}\geq A+B-3$.
Since $t>0, A=a+b+c\geq 3\sqrt[3]{abc}=3$, $B=\frac{1}{a}+\frac{1}{b}+\frac{1}{c}\geq 3\sqrt[3]{\frac{1}{abc}}=3$, and $AB+2-A-B=(A-1)(B-1)+1>0$, it follows that

$$t^3+(AB-A-B)t^2+(\frac{a}{c}+\frac{b}{a}+\frac{c}{b}+A+B-2AB)t+AB+2-A-B\geq$$
$$\geq t^3+(AB-A-B)t^2+(2A+2B-2AB-3)t+AB+2-A-B=$$
$$=(t-1)^2(t+AB+2-A-B)\geq 0,$$

3.21. $a_n+\underbrace{\sqrt[n-1]{a_1\cdots a_{n-1}}+\cdots+\sqrt[n-1]{a_1\cdots a_{n-1}}}_{n-1}\geq$ $n\cdot\sqrt[n]{a_n\cdot\underbrace{\sqrt[n-1]{a_1\cdots a_{n-1}}\cdots\sqrt[n-1]{a_1\cdots a_{n-1}}}_{n-1}}$, and therefore $n\sqrt[n]{a_1\cdots a_n}-(n-1)\sqrt[n-1]{a_1\cdots a_{n-1}}\leq a_n$.

3.22.

$$\sqrt[n]{\frac{a_1}{a_1+b_1+\cdots+k_1}\cdot\frac{a_2}{a_2+b_2+\cdots+k_2}\cdots\frac{a_n}{a_n+b_n+\cdots+k_n}}+$$
$$+\sqrt[n]{\frac{b_1}{a_1+b_1+\cdots+k_1}\cdot\frac{b_2}{a_2+b_2+\cdots+k_2}\cdots\frac{b_n}{a_n+b_n+\ldots+k_n}}+\cdots+$$
$$+\sqrt[n]{\frac{k_1}{a_1+b_1+\cdots+k_1}\cdot\frac{k_2}{a_2+b_2+\cdots+k_2}\cdots\frac{k_n}{a_n+b_n+\cdots+k_n}}\leq$$
$$\leq\frac{1}{n}\left(\frac{a_1}{a_1+b_1+\cdots+k_1}+\frac{a_2}{a_2+b_2+\cdots+k_2}+\cdots+\frac{a_n}{a_n+b_n+\cdots+k_n}\right)+$$
$$+\frac{1}{n}\left(\frac{b_1}{a_1+b_1+\cdots+k_1}+\frac{b_2}{a_2+b_2+\cdots+k_2}+\cdots+\frac{b_n}{a_n+b_n+\cdots+k_n}\right)+\cdots+$$
$$+\frac{1}{n}\left(\frac{k_1}{a_1+b_1+\cdots+k_1}+\frac{k_2}{a_2+b_2+\cdots+k_2}+\ldots+\frac{k_n}{a_n+b_n+\cdots+k_n}\right)=$$
$$=\frac{1}{n}\left(\frac{a_1}{a_1+b_1+\cdots+k_1}+\frac{b_1}{a_1+b_1+\cdots+k_1}+\cdots+\frac{k_1}{a_1+b_1+\cdots+k_1}\right)+$$
$$+\frac{1}{n}\left(\frac{a_2}{a_2+b_2+\cdots+k_2}+\frac{b_2}{a_2+b_2+\cdots+k_2}+\cdots+\frac{k_2}{a_2+b_2+\cdots+k_2}\right)+\cdots+$$
$$+\frac{1}{n}\left(\frac{a_n}{a_n+b_n+\cdots+k_n}+\frac{b_n}{a_n+b_n+\cdots+k_n}+\ldots+\frac{k_n}{a_n+b_n+\cdots+k_n}\right)=1.$$

3.23. Note that

$$a_1 \geq \frac{1}{1\cdot 2}a_1 + \frac{1}{2\cdot 3}a_1 + \cdots + \frac{1}{n(n+1)}a_1,$$
$$a_2 \geq \frac{2}{2\cdot 3}a_2 + \frac{2}{3\cdot 4}a_2 + \cdots + \frac{2}{n(n+1)}a_2,$$
$$\ldots$$

$a_n \geq \frac{n}{n(n+1)}a_n$, and therefore,
$a_1 + a_2 + \cdots + a_n \geq \frac{a_1}{2} + \frac{a_1+2a_2}{2\cdot 3} + \cdots + \frac{a_1+2a_2+\cdots+ka_k}{k(k+1)} + \cdots + \frac{a_1+2a_2+\ldots+na_n}{n(n+1)}$.
Let us prove that $e\frac{a_1+2a_2+\cdots+ka_k}{k(k+1)} \geq (a_1\cdots a_k)^{\frac{1}{k}}, \quad k = 1, \ldots, n.$
According to Problem 2.1, we have

$$\frac{a_1+2a_2+\cdots+ka_k}{k(k+1)} \geq \frac{\sqrt[k]{a_1\cdot 2a_2\cdots ka_k}}{k+1} = \frac{\sqrt[k]{k!}}{k+1}(a_1\cdots a_k)^{\frac{1}{k}}.$$

In order to complete the proof, one needs to prove that
$\frac{\sqrt[k]{k!}}{k+1} > \frac{1}{e}$ or $k! > \left(\frac{k+1}{e}\right)^k$ (see Problem 7.18).
Hence, we deduce that

$$e(a_1 + \cdots + a_n) \geq \sum_{k=1}^{n} e\frac{a_1 + 2a_2 + \cdots + ka_k}{k(k+1)} \geq \sum_{k=1}^{n} (a_1 \cdots a_k)^{\frac{1}{k}}.$$

Now let us prove that it is not possible to substitute e in the given inequality by a number smaller than e. Indeed, let $c(a_1 + \cdots + a_n) \geq \sum_{k=1}^{n} (a_1 \cdots a_k)^{\frac{1}{k}}$. Then taking $a_k = \frac{1}{k}, \quad k = 1, \ldots, n$, it follows that

$$c\sum_{k=1}^{n}\frac{1}{k} \geq \sum_{k=1}^{n}\frac{1}{(k!)^{\frac{1}{k}}} > \sum_{k=8}^{n}\frac{e}{k\cdot\sqrt[k]{k}} \tag{3.12}$$

(here we have used the inequality from Problem 14.18).
On the other hand, according to Problem 7.17, we have $\left(1+\sqrt{\frac{2}{k}}\right)^k > 1 + k\sqrt{\frac{2}{k}} + \frac{k(k-1)}{2}\left(\sqrt{\frac{2}{k}}\right)^2 > k$, whence $\sqrt[k]{k} < 1 + \sqrt{\frac{2}{k}}$, and therefore,

$$\sum_{k=8}^{n}\frac{e}{k\cdot\sqrt[k]{k}} > \sum_{k=8}^{n}\frac{e}{k\left(1+\sqrt{\frac{2}{k}}\right)}. \tag{3.13}$$

From (3.12) and (3.13), it follows that

$e\sum\limits_{k=8}^{n}\frac{1}{k\left(1+\sqrt{\frac{2}{k}}\right)} < c\sum\limits_{k=1}^{n}\frac{1}{k}$, or

$e\cdot\sum\limits_{k=8}^{n}\frac{1}{k} - e\cdot\sum\limits_{k=8}^{n}\frac{\sqrt{\frac{2}{k}}}{k\left(1+\sqrt{\frac{2}{k}}\right)} < c\sum\limits_{k=1}^{n}\frac{1}{k}$, and therefore,

$e\cdot\sum\limits_{k=8}^{n}\frac{1}{k} - e\cdot\sum\limits_{k=8}^{n}\frac{\sqrt{2}}{k\sqrt{k}} < c\sum\limits_{k=1}^{n}\frac{1}{k}$, or

$\alpha_n\cdot e - \sqrt{2}e\cdot\frac{\sum\limits_{k=1}^{n}\frac{1}{k\sqrt{k}}}{\sum\limits_{k=1}^{n}\frac{1}{k}} < c$, where $\alpha_n = \frac{\sum\limits_{k=8}^{n}\frac{1}{k}}{\sum\limits_{k=1}^{n}\frac{1}{k}} = 1 - \frac{\sum\limits_{k=1}^{7}\frac{1}{k}}{\sum\limits_{k=1}^{n}\frac{1}{k}}$.

Passing to the limit, we obtain $e \leq c$ (see Problems 7.15 and 7.16).

3.24. (a) Note that $n - k + ka^n = \underbrace{1+\cdots+1}_{n-k}+\underbrace{a^n+\cdots+a^n}_{k} \geq n\cdot\sqrt[n]{1\cdots 1\cdot\underbrace{a^n\cdots a^n}_{k}} = na^k$,

whence $n - k \geq na^k - ka^n$.

(b) Note that $\frac{x^k}{y^{k-1}} + (k-1)y \geq k\sqrt[k]{\frac{x^k}{y^{k-1}}\cdot y^{k-1}} = kx$, where $k \in \mathbb{N}, k \geq 2$, whence

$$\frac{x^k}{y^{k-1}} \geq kx - (k-1)y. \quad (1)$$

Using inequality (1), we obtain

$$\frac{x_1^2}{x_2} + \frac{x_2^3}{x_3^2} + \cdots + \frac{x_n^{n+1}}{x_1^n}$$
$$\geq (2x_1 - x_2) + (3x_2 - 2x_3) + \cdots + (nx_{n-1} - (n-1)x_n) +$$
$$+ ((n+1)x_n - nx_1) = 2(x_1 + \cdots + x_n) - nx_1 \geq x_1 + \cdots + x_n,$$

whence $\frac{x_1^2}{x_2} + \frac{x_2^3}{x_3^2} + \cdots + \frac{x_n^{n+1}}{x_1^n} \geq x_1 + \cdots + x_n$.

3.25. Note that $\frac{a^{x_1-x_2}}{x_1+x_2} + \frac{a^{x_2-x_3}}{x_2+x_3} + \cdots + \frac{a^{x_n-x_1}}{x_n+x_1} \geq n\cdot\sqrt[n]{\frac{a^{x_1-x_2}}{x_1+x_2}\cdot\frac{a^{x_2-x_3}}{x_2+x_3}\cdots\frac{a^{x_n-x_1}}{x_n+x_1}} = \frac{n}{\sqrt[n]{(x_1+x_2)\cdot(x_2+x_3)\cdots(x_n+x_1)}} \geq \frac{n}{\frac{(x_1+x_2)+\cdots+(x_n+x_1)}{n}} = \frac{n^2}{2\sum\limits_{i=1}^{n}x_i}$.

3.26. Using the inequality of Problem 2.1 for the numbers $x_i + 1$ and $\underbrace{1, 1, \ldots, 1}_{p-1}$, where $i = 1, \ldots, n$, we obtain $\sqrt[p]{x_i+1} = \sqrt[p]{(x_i+1)\cdot\underbrace{1\cdots 1}_{p-1}} \leq \frac{(x_i+1)+(p-1)\cdot 1}{p} = 1 + \frac{x_i}{p}$, whence $\sqrt[p]{x_1+1} + \cdots + \sqrt[p]{x_n+1} \leq \left(1+\frac{x_1}{p}\right) + \cdots + \left(1+\frac{x_n}{p}\right) = n + \frac{x_1+\cdots+x_n}{p} = n+1$.

3.27. (a) Since $1+\alpha \geq 0$, it is enough to prove the inequality for the case $1+n\alpha \geq 0$. According to Problem 2.1, for $n > 1$, we obtain $\sqrt[n]{(1+n\alpha)\cdot\underbrace{1\cdots 1}_{n-1}} \leq \frac{1+n\alpha+1+\cdots+1}{n} = 1+\alpha$. Therefore, $1+n\alpha \leq (1+\alpha)^n$.

(b) See the proof of Problem 3.27(a).

3.28. (a) $\cos^6 t \sin^2 t = 27 \cdot \frac{\cos^2 t}{3} \cdot \frac{\cos^2 t}{3} \cdot \frac{\cos^2 t}{3}(1-\cos^2 t) \leq 27\left(\frac{\frac{\cos^2 t}{3}+\frac{\cos^2 t}{3}+\frac{\cos^2 t}{3}+1-\cos^2 t}{4}\right)^4 = \frac{27}{4^4}$, whence $\left|\cos^3 t \cdot \sin t\right| \leq \frac{3\sqrt{3}}{16}$, and therefore $\cos^3 t \sin t \leq \frac{3\sqrt{3}}{16}$.

(b) We have

$$x^k(1-x^m) = \left(x^{km}(1-x^m)^m\right)^{\frac{1}{m}} = \left(mx^k \cdots mx^k(k-kx^m)\cdots(k-kx^m)\right)^{\frac{1}{m}} \cdot \frac{1}{m^{\frac{m}{k}} \cdot k} \leq$$

$$\leq \frac{1}{m^{\frac{m}{k}} \cdot k} \cdot \left(\frac{mx^k+\cdots+mx^k+(k-kx^m)+\cdots+(k-kx^m)}{m+k}\right)^{\frac{m+k}{m}} = \frac{k^{\frac{k}{m}} \cdot m}{(m+k)^{\frac{k}{m}+1}}.$$

(c) According to Problem 3.28(b), it follows that $x(1-x^2) \leq \frac{2\sqrt{3}}{9}$, whence

$$\frac{x}{1-x^2}+\frac{y}{1-y^2}+\frac{z}{1-z^2} = \frac{x^2}{x(1-x^2)}+\frac{y^2}{y(1-y^2)}+\frac{z^2}{z(1-z^2)}$$

$$\geq \frac{9}{2\sqrt{3}}\left(x^2+y^2+z^2\right) = \frac{3\sqrt{3}}{2}.$$

Therefore, $\frac{x}{1-x^2}+\frac{y}{1-y^2}+\frac{z}{1-z^2} \geq \frac{3\sqrt{3}}{2}$.

(d) We have

$$\frac{1}{1-x}+\frac{1}{1-y}+\frac{1}{1-z} = \frac{1+x}{1-x^2}+\frac{1+y}{1-y^2}+\frac{1+z}{1-z^2} = \frac{1}{1-x^2}+\frac{1}{1-y^2}+\frac{1}{1-z^2}+\frac{x}{1-x^2}+$$

$$+\frac{y}{1-y^2}+\frac{z}{1-z^2} \geq \frac{9}{1-x^2+1-y^2+1-z^2}+\frac{x}{1-x^2}+\frac{y}{1-y^2}+\frac{z}{1-z^2} \geq 4.5+1.5\sqrt{3},$$

(see Problem 3.28b). Therefore, $\frac{1}{1-x}+\frac{1}{1-y}+\frac{1}{1-z} \geq 4.5+1.5\sqrt{3}$.

3.29. From inequality (3.2), it follows that
$f(x) = (1+x)^{-\frac{1}{n}}+(1-x)^{-\frac{1}{n}} \geq 2\sqrt{(1+x)^{-\frac{1}{n}}(1-x)^{-\frac{1}{n}}} = \frac{2}{\sqrt[2n]{1-x^2}} \geq 2$,
Since $x \in [0; 1)$.
On the other hand, $f(0) = 2$. Thus the minimum value of the function $f(x)$ is equal to 2.

3.30. Let us represent the function $f(x)$ as

$$f(x) = (m+n) \cdot \frac{n \cdot \frac{ax^m}{n} + m \cdot \frac{b}{mx^n}}{m+n}.$$

By Problem 2.1, we obtain
$f(x) \geq (m+n)\sqrt[m+n]{\left(\frac{ax^m}{n}\right)^n \cdot \left(\frac{b}{mx^n}\right)^m} = (m+n)\sqrt[m+n]{\frac{a^n b^m}{n^n m^m}}$, where equality holds if $\frac{ax^m}{n} = \frac{b}{mx^n}$.
Therefore, $x = \sqrt[m+n]{\frac{bn}{am}}$.

Answer. $\min\limits_{(0,+\infty)} f(x) = f(x_0) = (m+n)\sqrt[m+n]{\frac{a^n b^m}{n^n m^m}}$, where $x_0 = \sqrt[m+n]{\frac{bn}{am}}$.

3.31. Let us represent the function $f(x)$ as
$f(x) = \frac{(x-a)\cdot(x-a)\cdot\alpha(b-x)\cdot\beta(b+x)}{\alpha\beta}$, where $\alpha > 0, \quad \beta > 0$.
Using Problem 2.1, we deduce that

$$4\sqrt[4]{(x-a)\cdot(x-a)\cdot\alpha(b-x)\cdot\beta(b+x)} \leq (x-a)+(x-a)+\alpha(b-x)+\beta(b+x) =$$
$$= (2-\alpha+\beta)x + (\alpha+\beta)b - 2a.$$

Note that the right-hand side does not depend on x if $\alpha - \beta = 2$, and equality holds if $x - a = \alpha(b-x) = \beta(b+x)$. Thus, it follows that $\alpha = \frac{x-a}{b-x}$, $\beta = \frac{x-a}{b+x}$.
Hence, from the equation $\alpha - \beta = 2$, we deduce that $2x^2 - ax - b^2 = 0$. This equation has only one positive root, $x_0 = \frac{a+\sqrt{a^2+8b^2}}{4}$. One can easily prove that $x_0 \in [a,b]$. Therefore,$f(x)$ attains its maximum value in $[a,b]$ at the point x_0.
Answer. $x_0 = \frac{a+\sqrt{a^2+8b^2}}{4}$.

3.32. Let us represent the product xyz as
$xyz = \frac{1}{2\sqrt{3}\pi}\cdot 2x\cdot\sqrt{3}y\cdot\pi z$, and according to Problem 2.1, for the numbers $2x, \sqrt{3}y, \pi z$, we have $xyz = \frac{1}{2\sqrt{3}\pi}\cdot 2x\cdot\sqrt{3}y\cdot\pi z \leq \frac{1}{2\sqrt{3}\pi}\cdot\left(\frac{2x+\sqrt{3}y+\pi z}{3}\right)^3 = \frac{1}{54\sqrt{3}\pi}$, where inequality holds if $2x = \sqrt{3}y = \pi z$. From $2x+\sqrt{3}y+\pi z = 1$, it follows that $x = \frac{1}{6}, \quad y = \frac{\sqrt{3}}{9}, z = \frac{1}{3\pi}$.
Answer. $\frac{1}{54\sqrt{3}\pi}$.

3.33. Note that $f(0) = 0$, and if $x < 0$, then $f(x) < 0$. On the other hand, if $x > 0$, then $f(x) > 0$.
Therefore, the function $f(x)$ attains its maximum value in $(0,+\infty)$, and its minimum value in $(-\infty, 0)$.
First proof. From inequality (3.2) for the numbers ax^2 and b, we have

$$\frac{ax^2+b}{2} \geq |x|\cdot\sqrt{ab}, \tag{3.14}$$

where equality holds if $ax^2 = b$.
From inequality (3.14), it follows that $\frac{x}{ax^2+b} \leq \frac{1}{2\sqrt{ab}}$.

Therefore, the maximum value of the function $f(x)$ is equal to $\frac{1}{2\sqrt{ab}}$, and it attains this value at the point $x = \sqrt{\frac{b}{a}}$. Since $f(x)$ is an odd function, its minimum value is equal to $-\frac{1}{2\sqrt{ab}}$, which the function $f(x)$ attains at the point $x = -\sqrt{\frac{b}{a}}$.

Alternative proof. In $(0, +\infty)$, the function $f(x)$ coincides with the function $g(x) = \frac{1}{ax+\frac{b}{x}}$. This function attains its maximum value at the point where the function $h(x) = ax + \frac{b}{x}$ attains its minimum value.

Since the product $ax \cdot \frac{b}{x} = ab$ is constant, the sum $ax + \frac{b}{x}$ attains its minimum value when $ax = \frac{b}{x}$, i.e., when $x = \sqrt{\frac{b}{a}}$.

Therefore, in $(0, +\infty)$, the function $f(x)$ attains its maximum value at the point $x = \sqrt{\frac{b}{a}}$, and it is equal to $\frac{1}{2\sqrt{ab}}$.

One can find the minimum value of the function $f(x)$ similarly as in the first proof.

3.34. (a) We have $y = \frac{5\sqrt{(x+2)(x+4)}+12}{x+3}$, and thus if $x \geq -2$, then $y = \frac{\sqrt{(25x+50)(x+4)}+12}{x+3} \leq \frac{\frac{25x+50+x+4}{2}+12}{x+3} = 13$, and $y = 13$ if $25x + 50 = x + 4$, i.e., if $x = -\frac{23}{12}$. If $x \leq -4$, then $y < 0$.

Therefore, the maximum value of the function y is equal to 13.

Answer. 13.

(b) We have

$$y = \frac{\sqrt[3]{(x^2+1)(x^2+1)(\frac{2}{5}x^2+\frac{6}{5})\cdot\frac{5}{2}}}{3x^2+4} = \sqrt[3]{\frac{5}{2}}\cdot\frac{\sqrt[3]{(x^2+1)(x^2+1)(\frac{2}{5}x^2+\frac{6}{5})}}{3x^2+4} \leq$$

$$\leq \sqrt[3]{\frac{5}{2}}\cdot\frac{x^2+1+x^2+1+\frac{2}{5}x^2+\frac{6}{5}}{3(3x^2+4)} = \sqrt[3]{\frac{5}{2}}\cdot\frac{12x^2+16}{15(3x^2+4)} = \frac{4}{15}\cdot\sqrt[3]{\frac{5}{2}}.$$

Note that $y = \frac{4}{15}\cdot\sqrt[3]{\frac{5}{2}}$, if $x^2+1 = \frac{2}{5}x^2+\frac{6}{5}$, i.e., if $x^2 = \frac{1}{3}$.

Therefore, the maximum value of the function y is equal to $\frac{4}{15}\cdot\sqrt[3]{\frac{5}{2}}$.

Answer. $\frac{4}{15}\cdot\sqrt[3]{\frac{5}{2}}$.

3.35. Let $PA = x$, $PB = y$, $PC = z$. Thus, it follows that $S = x+y+z+\sqrt{x^2+y^2}+\sqrt{y^2+z^2}+\sqrt{z^2+x^2}$ and $V_{PABC} = \frac{1}{6}xyz$. Since $a^2+b^2 \geq 2ab$, it follows that $S \geq x+y+z+\sqrt{2}(\sqrt{xy}+\sqrt{yz}+\sqrt{xz}) \geq 3\sqrt[3]{xyz}+3\sqrt{2}\sqrt[3]{\sqrt{xy}\cdot\sqrt{yz}\cdot\sqrt{xz}}$.

Therefore, $V \leq \frac{1}{6}\cdot\left(\frac{S}{3(1+\sqrt{2})}\right)^3$, and if $x = y = z = \frac{S}{3(1+\sqrt{2})}$, we have $V = \frac{1}{6}\cdot\left(\frac{S}{3(1+\sqrt{2})}\right)^3$. Then the volume of the pyramid with edge lengths $x = y = z = \frac{S}{3(1+\sqrt{2})}$ is maximal.

3.36. From inequality (3.2), it follows that $1 \geq z^2 + 1$, and therefore, $z = 0$, $x = y = 1$.

3.37. Note that the inequality $\frac{x^2+y^2+z^2}{3} \geq \left(\frac{x+y+z}{3}\right)^2$ becomes an equality, since $x + y + z = 3$ and $x^2 + y^2 + z^2 = 3$.
Therefore, $x = y = z = 1$ (see the proof of Problem 2.2).
Answer. (1, 1, 1).

3.38. We have
$\frac{a^2+b^2+c^2+d^2}{4} \geq \left(\frac{a+b+c+d}{4}\right)^2$ (Problem 2.2). It follows that $\frac{16-e^2}{4} \geq \left(\frac{8-e}{4}\right)^2$, whence $5e^2 - 16e \leq 0$, and therefore, $0 \leq e \leq 3.2$.
If $a = b = c = d = 1, 2$, then $e = 3.2$. Thus, it follows that the maximum value of e is equal to 3.2.
Answer. 3.2.

3.39. We have

$$\frac{x_1}{x_2} + \frac{x_3}{x_4} + \frac{x_5}{x_6} \geq \frac{1}{x_2} + \frac{x_2}{x_4} + \frac{x_4}{x_6} \geq 3\sqrt[3]{\frac{1}{x_2} \cdot \frac{x_2}{x_4} \cdot \frac{x_4}{x_6}} = 3\sqrt[3]{\frac{1}{x_6}} \geq 0.3.$$

If $x_1 = 1$, $x_2 = 10$, $x_3 = 10$, $x_4 = 10^2$, $x_5 = 10^2$, $x_6 = 10^3$, then $\frac{x_1}{x_2} + \frac{x_3}{x_4} + \frac{x_5}{x_6} = 0.3$.
Therefore, the minimum value of the expression $\frac{x_1}{x_2} + \frac{x_3}{x_4} + \frac{x_5}{x_6}$ is equal to 0.3.
Answer. 0.3.

3.40. From Problem 2.1, for the numbers x^4, y^4, 1, 1, it follows that $x^4+y^4+1+1 \geq 4\sqrt[4]{x^4 \cdot y^4 \cdot 1 \cdot 1} = 4|x| \cdot |y| \geq 4xy$, whence $x^4+y^4+2 \geq 4xy$, and therefore, $x^4 = y^4 = 1$.
Answer. (1, 1), (−1, −1).

3.41. We have $(xy)^2 + (yz)^2 + (zx)^2 = 3xyz$, whence $xyz > 0$.
On the other hand, $xyz \in Z$, and hence $xyz \geq 1$. From Problem 2.1, it follows that $3xyz = (xy)^2 + (yz)^2 + (zx)^2 \geq 3xyz \cdot \sqrt[3]{xyz} \geq 3xyz$, and hence $xyz = 1$, and the inequality becomes an equality. It follows that $(xy)^2 = (yz)^2 = (zx)^2$.
Therefore, $x^2 = y^2 = z^2 = 1$.
Answer. (1, 1, 1), (1, −1, −1), (−1, 1, −1), (−1, −1, 1).

3.42. (a) Note that for $x > 0, \alpha > \beta \geq 0$, we have $x^\alpha - x^\beta \geq x^{\alpha-\beta} - 1$, since

$$(x^{\alpha-\beta} - 1)(x^\beta - 1) \geq 0.$$

Therefore,

$$x_1^\alpha + \cdots + x_n^\alpha - (x_1^\beta + \cdots + x_n^\beta) = (x_1^\alpha - x_1^\beta) + \cdots + (x_n^\alpha - x_n^\beta) \geq$$
$$\geq (x_1^{\alpha-\beta} - 1) + \cdots + (x_n^{\alpha-\beta} - 1) = x_1^{\alpha-\beta} + \cdots + x_n^{\alpha-\beta} - n \geq n\sqrt[n]{x_1^{\alpha-\beta} \cdots x_n^{\alpha-\beta}} - n = 0.$$

It follows that $x_1^\alpha + \cdots + x_n^\alpha \geq x_1^\beta + \cdots + x_n^\beta$.

(b) If $\beta \geq 0$, then from Problem 3.42(a), it follows that $x_1^\alpha + \cdots + x_n^\alpha \geq x_1^\beta + \cdots + x_n^\beta$.
If $\beta < 0$, then

$$x_1^\beta + \cdots + x_n^\beta = \frac{1}{x_1^{-\beta}} + \cdots + \frac{1}{x_n^{-\beta}} = x_2^{-\beta} \cdots x_n^{-\beta} + \cdots + x_1^{-\beta} \cdots x_{n-1}^{-\beta} \leq$$
$$\leq \frac{x_2^{-\beta(n-1)} + \cdots + x_n^{-\beta(n-1)}}{n-1} + \cdots + \frac{x_1^{-\beta(n-1)} + \cdots + x_{n-1}^{-\beta(n-1)}}{n-1}$$
$$= x_1^{-\beta(n-1)} + \cdots + x_n^{-\beta(n-1)} \leq x_1^\alpha + \cdots + x_n^\alpha.$$

Therefore, $x_1^\alpha + \cdots + x_n^\alpha \geq x_1^\beta + \cdots + x_n^\beta$.

3.43. **Proof**. We have
$y^2+xz > \frac{4}{27}(3-3z+z^2)$, $z^2+xy > \frac{4}{27}(3-3x+x^2)$, $x^2+yz > \frac{4}{27}(3-3y+y^2)$. Summing these inequalities, we deduce that $23(x^2+y^2+z^2)+27(xy+yz+zx) > 24$, which leads to a contradiction, since $23 = 23(x+y+z)^2 \geq 23(x^2+y^2+z^2) + 27(xy+yz+zx)$.
Let $z^2 + xy \leq \frac{4}{27}(3 - 3x + x^2)$. Then

$$x^2y + y^2z + z^2x = x(z^2 + xy) + y^2z \leq \frac{4}{27}x(3 - 3x + x^2) + 4\left(\frac{0,5y + 0,5y + z}{3}\right)^3 = \frac{4}{27},$$

and therefore, $x^2y + y^2z + z^2x \leq \frac{4}{27}$.
Alternative proof. Without loss of generality one can assume that $\max(x, y, z) = x$, in which case $x^2y+y^2z+z^2x \leq x^2y+xyz+0.5z^2x+0.5zx^2 = 0.5x(x+z)(2y+z)$.
According to Problem 2.1 (AM-GM inequality), we obtain

$$x^2y + y^2z + z^2x \leq 0.5x(x+z)(2y+z)$$
$$\leq 0.5\left(\frac{x + (x+z) + (2y+z)}{3}\right)^3 = \frac{4}{27},$$

and therefore, $x^2y + y^2z + z^2x \leq \frac{4}{27}$.

3.44 (a) Note that

$$\frac{1+a}{1+ab}+\frac{1+b}{1+bc}+\frac{1+c}{1+cd}+\frac{1+d}{1+da}=\frac{cd+acd}{cd+abcd}+\frac{ad+adb}{ad+abcd}+\frac{1+c}{1+cd}+\frac{1+d}{1+da}=$$

$$=1+\frac{c(1+ad)}{1+cd}+1+\frac{d(1+ab)}{1+da}\geq 2+2\sqrt{\frac{cd(1+ab)}{1+cd}}=4,$$

whence $\frac{1+a}{1+ab}+\frac{1+b}{1+bc}+\frac{1+c}{1+cd}+\frac{1+d}{1+da}\geq 4$.

(b) Note that

$$\frac{1+ab}{1+a}+\frac{1+bc}{1+b}+\frac{1+cd}{1+c}+\frac{1+da}{1+d}=\frac{cd+abcd}{cd+acd}+\frac{1+bc}{1+b}+\frac{1+cd}{1+c}+\frac{bc+abcd}{bc+bcd}\geq$$

$$\geq(1+cd)\frac{4}{cd+acd+1+c}+(1+bc)\frac{4}{1+b+bc+bcd}=(1+cd)\frac{4b}{bcd+abcd+b+bc}+$$

$$+(1+bc)\frac{4}{1+b+bc+bcd}=\frac{4(b(1+cd)+1+bc)}{1+b+bc+bcd}=4,$$

and therefore, $\frac{1+ab}{1+a}+\frac{1+bc}{1+b}+\frac{1+cd}{1+c}+\frac{1+da}{1+d}\geq 4$.

3.45. Note that $(c-b)(c-d)+(e-f)(e-d)+(e-f)(c-b)<0$, and therefore, $(bd+df+fb)-(ac+ce+ea)<(c+e)(b+d+f-a-c-e)$, or $\tau-\sigma<s(T-S)$, where $\tau=bd+df+fb$, $\sigma=ac+ce+ea$, $s=c+e$.
We have $S\tau+T\sigma=S(\tau-\sigma)+(S+T)\sigma<Ss(T-S)+(S+T)(ce+as)\leq$ $\leq Ss(T-S)+(S+T)\left(\frac{s^2}{4}+as\right)=s\left(2ST-\frac{3}{4}(S+T)s\right)$. It follows that $\sqrt{\frac{3}{4}(S+T)(S\tau+T\sigma)}<\sqrt{\frac{3}{4}(S+T)s\cdot\left(2ST-\frac{3}{4}(S+T)s\right)}\leq$ $\frac{1}{2}\left(\frac{3}{4}(S+T)s+\left(2ST-\frac{3}{4}(S+T)s\right)\right)=ST$.
Therefore, $\sqrt{3(S+T)(S(bd+df+fb)+T(ac+ce+ea))}<2ST$.

3.46. Note that

$$\frac{a+\sqrt{ab}+\sqrt[3]{abc}+\sqrt[4]{abcd}}{\sqrt[4]{a\cdot\frac{a+b}{2}\cdot\frac{a+b+c}{3}\cdot\frac{a+b+c+d}{4}}}=$$

$$=\sqrt[4]{1\cdot\frac{2a}{a+b}\cdot\frac{3a}{a+b+c}\cdot\frac{4a}{a+b+c+d}}+\sqrt[4]{1\cdot\frac{2a}{a+b}\cdot\frac{3b}{a+b+c}\cdot\frac{4b}{a+b+c+d}}+$$

$$+\sqrt[12]{1\cdot1\cdot1\cdot\frac{2b}{a+b}\cdot\frac{2b}{a+b}\cdot\frac{2b}{a+b}\cdot\frac{3a}{a+b+c}\cdot\frac{3b}{a+b+c}\cdot\frac{3c}{a+b+c}\cdot\frac{4c}{a+b+c+d}\cdot\frac{4c}{a+b+c+d}\cdot\frac{4c}{a+b+c+d}}+$$

$$+\sqrt[4]{1\cdot\frac{2b}{a+b}\cdot\frac{3c}{a+b+c}\cdot\frac{4d}{a+b+c+d}}\leq\frac{1}{4}\left(1+\frac{2a}{a+b}+\frac{3a}{a+b+c}+\frac{4a}{a+b+c+d}\right)+$$

$$+\frac{1}{4}\left(1+\frac{2a}{a+b}+\frac{3b}{a+b+c}+\frac{4b}{a+b+c+d}\right)+$$

$$+\frac{1}{12}\left(3+\frac{6b}{a+b}+\frac{3a}{a+b+c}+\frac{3b}{a+b+c}+\frac{3c}{a+b+c}+\frac{12c}{a+b+c+d}\right)+$$

$$+\frac{1}{4}\left(1+\frac{2b}{a+b}+\frac{3c}{a+b+c}+\frac{4d}{a+b+c+d}\right)=4.$$

Therefore,

$$\frac{a+\sqrt{ab}+\sqrt[3]{abc}+\sqrt[4]{abcd}}{4}\leq\sqrt[4]{a\cdot\frac{a+b}{2}\cdot\frac{a+b+c}{3}\cdot\frac{a+b+c+d}{4}}.$$

3.47. Without loss of generality one can assume that $a\geq 0,\ b\geq 0,\ c\geq 0,\ d\geq 0$.

Let x, y, z be arbitrary positive numbers. Then according to Problem 2.1 (AM-GM inequality), we obtain

$$a^{12}+(ab)^6+(abc)^4+(abcd)^3=a^{12}+\frac{1}{x^6}(xa\cdot b)^6+\frac{1}{x^4y^8}(xya\cdot yb\cdot c)^4+$$
$$+\frac{1}{x^3y^6z^9}(xyza\cdot yzb\cdot zc\cdot d)^3\leq a^{12}+\frac{1}{2x^6}(x^{12}a^{12}+b^{12})+$$
$$+\frac{1}{3x^4y^8}(x^{12}y^{12}a^{12}+y^{12}b^{12}+c^{12})+$$
$$+\frac{1}{4x^3y^6z^9}(x^{12}y^{12}z^{12}a^{12}+y^{12}z^{12}b^{12}+z^{12}c^{12}+d^{12})=A(a^{12}+b^{12}+c^{12}+d^{12}).$$

We take the numbers x, y, z such that

$$1+\frac{x^6}{2}+\frac{x^8y^4}{3}+\frac{x^9y^6z^3}{4}=\frac{1}{2x^6}+\frac{y^4}{3x^4}+\frac{y^6z^3}{4x^3}$$
$$=\frac{1}{3x^4y^8}+\frac{z^3}{4x^3y^6}=\frac{1}{4x^3y^6z^9}=A.$$

Therefore, $x^{12}=1-\frac{1}{A}$, $y^{12}=1-\frac{1}{2\sqrt{A(A-1)}}$, $z^{12}=1-\frac{1}{3\sqrt[3]{A(\sqrt{A(A-1)}-0,5)^2}}$, and

$$\frac{256}{27}A\left(3\sqrt[3]{A\left(\sqrt{A(A-1)}-0,5\right)^2}-1\right)^3=1. \tag{1}$$

Consider the following function:

$f(t)=\frac{256}{27}t\left(3\sqrt[3]{t(\sqrt{t(t-1)}-0,5)^2}-1\right)^3-1$ in $[1,42;\ 1,43]$.

Note that

$$f(1,42)=\frac{256}{27}\cdot 1,42\left(3\sqrt[3]{1,42\left(\sqrt{1,42\cdot 0,42}-0,5\right)^2}-1\right)^3-1<$$
$$<\frac{256}{27}\cdot 1,42\left(3\sqrt[3]{1,42\cdot 0,2723^2}-1\right)^3-1<\frac{256}{27}\cdot 1,42(3\cdot 0,4723-1)^3-1<0 \text{ and}$$
$$f(1,43)=\frac{256}{27}\cdot 1,43\left(3\sqrt[3]{1,43\left(\sqrt{1,43\cdot 0,43}-0,5\right)^2}-1\right)^3-1>$$
$$>\frac{256}{27}\cdot 1,43\left(3\sqrt[3]{1,43\cdot 0,08}-1\right)^3-1>\frac{256}{27}\cdot 1,43(3\cdot 0,48-1)^3-1>0.$$

Therefore, there exists a number A such that $1,42<A<1,43$ and $f(A)=0$. Then (1) holds. It follows that $x>0$, $y>0$, $z>0$ $(A>1.42)$ and $a^{12}+(ab)^6+(abc)^4+(abcd)^3\leq A(a^{12}+b^{12}+c^{12}+d^{12})\leq 1,43(a^{12}+b^{12}+c^{12}+d^{12})$.

Problems for Independent Study

Prove the following inequalities (1–16, 20–26).

1. $ab \leq \frac{a^p}{p} + \frac{b^q}{q}$, if $\frac{1}{p} + \frac{1}{q} = 1$, $a > 0,\ b > 0,\ p > 0,\ q > 0$ where p and q are rational numbers.
2. $\left(1 + \frac{1}{n}\right)^n > 2$, where $n \in \mathbb{N}$.
3. $(1 + a_1) \cdots (1 + a_n) \leq 1 + \frac{S}{1!} + \cdots + \frac{S^n}{n!}$, where $n \geq 2$, $S = a_1 + \cdots + a_n$, $a_i > 0$, $i = 1, \ldots, n$.
4. $\left(1 + \frac{1}{a}\right)\left(1 + \frac{1}{b}\right)\left(1 + \frac{1}{c}\right) \geq 64$, where $a > 0,\ b > 0,\ c > 0$ and $a + b + c = 1$.
5. $\sqrt[n]{a^{2n-k}} + \sqrt[n]{a^{2n+k}} \geq 3a - 1$, where $n \geq 2$, $a > 0$, $n > k$, $n, k \in \mathbb{N}$.
6. $\frac{a^n - 1}{a^n(a-1)} \geq n + 1 - a^{\frac{n(n+1)}{2}}$, where $a > 0$, $a \neq 1$.
7. $na^{n+1} + 1 \geq (n + 1)a^n$, where $a > 0$.
8. $\left(\sqrt{k} + \sqrt{k+1}\right)\left(\sqrt{k+1} + \sqrt{k+2}\right) \cdots \left(\sqrt{n} + \sqrt{n+1}\right) \geq \left(\sqrt{n} - \sqrt{k}\right)\left(\sqrt{n} + \sqrt{k} - 1\right) + 2$, where $n > k$, $n, k \in \mathbb{N}$.
9. $\frac{a_1}{a_2} + \cdots + \frac{a_{n-1}}{a_n} + \frac{a_n}{a_1} \geq n$, where $a_i > 0$, $i = 1, \ldots, n$.
10. $a_{n+1} + \frac{1}{a_1(a_2-a_1)(a_3-a_2)\cdots(a_{n+1}-a_n)} \geq n + 2$, where $0 < a_k < a_{k+1}$, $k = 1, \ldots, n$.
11. $1 + \frac{x}{2} \leq \frac{1}{\sqrt{1-x}}$, where $0 \leq x < 1$.
12. $\sin(2\alpha) < \frac{2}{3\alpha - \alpha^3}$, where $0 < \alpha < \frac{\pi}{2}$.
13. $\left(\frac{a}{b}\right)^4 + \left(\frac{b}{c}\right)^4 + \left(\frac{c}{d}\right)^4 + \left(\frac{d}{e}\right)^4 + \left(\frac{e}{a}\right)^4 \geq \frac{a}{b} + \frac{b}{c} + \frac{c}{d} + \frac{d}{e} + \frac{e}{a}$, where $abcde \neq 0$.
14. $\left(\frac{a}{b}\right)^{1999} + \left(\frac{b}{c}\right)^{1999} + \left(\frac{c}{d}\right)^{1999} + \left(\frac{d}{a}\right)^{1999} \geq \frac{a}{b} + \frac{b}{c} + \frac{c}{d} + \frac{d}{a}$, where $a > 0,\ b > 0,\ c > 0, d > 0$.
15. (a) $\sqrt{\frac{a_1 + a_2}{a_3}} + \sqrt{\frac{a_2 + a_3}{a_4}} + \cdots + \sqrt{\frac{a_{n-1} + a_n}{a_1}} + \sqrt{\frac{a_n + a_1}{a_2}} \geq n\sqrt{2}$, where $n > 2$ and $a_1 > 0, \ldots, a_n > 0$.

    (b) $\frac{x}{1+x^2} + \frac{y}{1+y^2} + \frac{z}{1+z^2} \leq \frac{3\sqrt{3}}{4}$, where $x^2 + y^2 + z^2 = 1$.
16. $\left(\frac{1}{a_1^2} - 1\right) \cdots \left(\frac{1}{a_n^2} - 1\right) \geq \left(n^2 - 1\right)^n$, where $n \geq 2$, $a_1 > 0, \ldots, a_n > 0$, and $a_1 + \cdots + a_n = 1$.
17. Find the maximum and minimum values of the expression $(1 + u)(1 + v)(1 + w)$ if $0 < u \leq \frac{7}{16}$, $0 < v \leq \frac{7}{16}$, $0 < w \leq \frac{7}{16}$, and $u + v + w = 1$.
18. Find the maximum value of the expression $x^p y^q$ if $x + y = a, x > 0,\ y > 0$, and $p, q \in N$.
19. Find the maximum value of the expression $a + 2c$ if for all x, one has $ax^2 + bx + c \leq \frac{1}{\sqrt{1-x^2}}$, where $|x| < 1$.

    Hint. Take $x = \frac{1}{\sqrt{2}}$ and $x = -\frac{1}{\sqrt{2}}$. Then it follows that $a + 2c \leq 2\sqrt{2}$.
    Prove that if $a = \sqrt{2}$, $b = 0$, $c = \frac{1}{\sqrt{2}}$, then $\sqrt{2}x^2 + \frac{1}{\sqrt{2}} \leq \frac{1}{\sqrt{1-x^2}}$ for $|x| < 1$.
20. $\left(1 + \frac{a}{b}\right)\left(1 + \frac{b}{c}\right)\left(1 + \frac{c}{a}\right) \geq 2\left(1 + \frac{a+b+c}{\sqrt[3]{abc}}\right)$, where $a > 0,\ b > 0,\ c > 0$.

    Hint. $\left(1 + \frac{a}{b}\right)\left(1 + \frac{b}{c}\right)\left(1 + \frac{c}{a}\right) = 2 + \frac{a}{b} + \frac{b}{a} + \frac{b}{c} + \frac{c}{b} + \frac{a}{c} + \frac{c}{a}$ and $\frac{1}{3}\frac{a}{b} + \frac{1}{3}\frac{a}{b} + \frac{1}{3}\frac{b}{c} \geq \sqrt[3]{\frac{a^2}{bc}} = \frac{a}{\sqrt[3]{abc}}$.

21. $\frac{1+a_1}{1-a_1}\cdot\frac{1+a_2}{1-a_2}\cdots\frac{1+a_{n+1}}{1-a_{n+1}} \geq n^{n+1}$, where $-1 < a_1, a_2, \ldots, a_{n+1} < 1$ and $a_1 + a_2 + \cdots + a_{n+1} \geq n-1$.
    Hint. We have that $1+a_i \geq (1-a_1)+\cdots+(1-a_{i-1})+(1-a_{i+1})+\cdots+(1-a_{n+1})$.
22. $(a+b)^3(b+c)^3(c+d)^3(d+a)^3 \geq 16a^2b^2c^2d^2(a+b+c+d)^4$, where $a > 0,\ b > 0,\ c > 0, d > 0$.
    Hint. We have $(a+b+c+d)^2 = (a+b)(b+c)+(a+b)(d+a)+(b+c)(c+d)+(c+d)(d+a)$.
23. $\left(\left(1+\frac{a}{b}\right)^2+\left(1+\frac{b}{c}\right)^2+\left(1+\frac{c}{a}\right)^2\right)\left(\left(1+\frac{b}{a}\right)^2+\left(1+\frac{c}{b}\right)^2+\left(1+\frac{a}{c}\right)^2\right) \geq 4\left(\frac{a+b}{c}+\frac{b+c}{a}+\frac{c+a}{b}\right)^2$, where $a > 0,\ b > 0,\ c > 0$.
    Hint. We have $\left(1+\frac{a}{b}\right)^2+\left(1+\frac{b}{c}\right)^2+\left(1+\frac{c}{a}\right)^2 \geq \frac{1}{3}\left(1+\frac{a}{b}+1+\frac{b}{c}+1+\frac{c}{a}\right)^2$.
24. $(a^2+bc)^3(b^2+ac)^3(c^2+ab)^3 \geq 64(a^3+b^3)(b^3+c^3)(c^3+a^3)a^3b^3c^3$, where $a > 0,\ b > 0,\ c > 0$.
    Hint. We have $(a^2+bc)(b^2+ac) = c(a^3+b^3)+ab(ab+c^2)$.
25. (a) $a+\sqrt{ab}+\sqrt[3]{abc} \leq \frac{4}{3}(a+b+c)$, where $a > 0,\ b > 0,\ c > 0$.
    (b) $a+\sqrt{ab}+\sqrt[3]{abc} \leq 3\sqrt[3]{a\cdot\frac{a+b}{2}\cdot\frac{a+b+c}{3}}$, where $a > 0,\ b > 0,\ c > 0$.
26. $(ab)^{\frac{5}{4}}+(bc)^{\frac{5}{4}}+(ca)^{\frac{5}{4}} \leq \frac{\sqrt{3}}{9}$, where $a > 0,\ b > 0,\ c > 0$ and $a+b+c=1$.

Chapter 4
The Cauchy–Bunyakovsky–Schwarz Inequality

Historical origins. The *Cauchy–Bunyakovsky–Schwarz inequality*, also known as the *Cauchy–Schwarz inequality,* is one of the most important inequalities in mathematics. The inequality for sums was published in 1821 by the French mathematician *Augustin-Louis Cauchy,* born 21 August 1789 in Paris, France, died 23 May 1857 in Sceaux, one of the wealthy suburbs of Paris. Despite the fact that the influential French mathematician *Pierre-Simon Laplace* and the Italian–French mathematician *Joseph-Louis Lagrange* were Cauchy's father's friends and that on Lagrange's advice, Cauchy was enrolled in the École Centrale du Panthéon in Paris, France, Cauchy never obtained a doctorate in mathematics. Besides being a prolific writer, Cauchy was awarded the title of *baron* (in return for his services to *King Charles X* of France, who asked Cauchy to be tutor to his grandson, the *duke of Bordeaux*). The names of Cauchy, Laplace, and Lagrange are among the 72 names inscribed on the *Eiffel Tower*. Cauchy was an advisor of 2 doctoral students, including a well-known Russian mathematician *Viktor Bunyakovsky*.

The corresponding inequality for integrals was first proved in 1859 by Viktor Bunyakovsky, born 16 December 1804 in Bar, Russian Empire (now Bar, Ukraine), died 12 December 1889 in Saint Petersburg, Russian Empire (now Russia). In 1824, Bunyakovsky received his bachelor's degree from the University of Paris (the Sorbonne), Paris, France, and continuing his studies, he wrote three doctoral dissertations under Cauchy's supervision and obtained his doctorate in 1825.

The modern proof of the integral inequality was given in 1888 by the Prussian mathematician *Karl Hermann Amandus Schwarz,* born 25 January 1843 in Hermsdorf, Prussia (now Jerzmanowa, Poland), died 30 November 1921 in Berlin, Germany. He obtained his doctorate from the University of Berlin in 1864 under the supervision of the well-known mathematicians *Ernst Kummer* and *Karl Weierstrass*. Schwarz was the advisor to 20 doctoral students.

H. Sedrakyan and N. Sedrakyan, *Algebraic Inequalities*, Problem Books in Mathematics, https://doi.org/10.1007/978-3-319-77836-5_4

There are several generalizations of this inequality, one of which is known as *Hölder's inequality*, named after the German mathematician *Otto Ludwig Hölder,* born 22 December in Stuttgart, Germany, died 29 August 1937 in Leipzig, Germany. In 1877, Hölder began his studies at the University of Berlin, where he was a student of the well-known mathematicians *Leopold Kronecker*, *Ernst Kummer,* and *Karl Weierstrass,* and he obtained his doctorate under the supervision of *Paul du Bois-Reymond* from the University of Tübingen in 1882. Hölder was the advisor of 58 doctoral students.

In this chapter we study some inequalities whose proofs involve the Cauchy–Bunyakovsky–Schwarz inequality. But first, let us prove the simplest case of the Cauchy–Bunyakovsky–Schwarz inequality for arbitrary numbers $a_1,\ a_2,\ b_1,\ b_2$.

Consider the vectors $\vec{a}(a_1, a_2)$ and $\vec{b}(b_1, b_2)$.

It is known that $\vec{a} \cdot \vec{b} = a_1 b_1 + a_2 b_2 = |\vec{a}| \cdot \left|\vec{b}\right| \cos\left(\vec{a}, \vec{b}\right)$.

Let us estimate the absolute values of the scalar product $\vec{a} \cdot \vec{b}$. We have

$$\left|\vec{a} \cdot \vec{b}\right| = |\vec{a}| \cdot \left|\vec{b}\right| \cdot \left|\cos\left(\vec{a}, \vec{b}\right)\right| \leq |\vec{a}| \cdot \left|\vec{b}\right|.$$

On the other hand,

$$\left|\vec{a} \cdot \vec{b}\right| = |a_1 b_1 + a_2 b_2| \leq |\vec{a}| \cdot \left|\vec{b}\right| = \sqrt{a_1^2 + a_2^2} \cdot \sqrt{b_1^2 + b_2^2}, \text{ or}$$

$$(a_1 b_1 + a_2 b_2)^2 \leq \left(a_1^2 + a_2^2\right)\left(b_1^2 + b_2^2\right). \tag{4.1}$$

Inequality (4.1) is the simplest formulation of the Cauchy–Bunyakovsky–Schwarz inequality for arbitrary numbers $a_1,\ a_2,\ b_1,\ b_2$.

Note that in (4.1), equality holds if and only if $a_1 b_2 - a_2 b_1 = 0$.

The following generalization of inequality (4.1) for arbitrary numbers $a_1, \ldots, a_n, b_1, \ldots, b_n$, is called the **Cauchy–Bunyakovsky–Schwarz inequality**:

$$\left(a_1^2 + \cdots + a_n^2\right)\left(b_1^2 + \cdots + b_n^2\right) \geq (a_1 b_1 + \cdots + a_n b_n)^2. \tag{4.2}$$

First, let us prove inequality (4.2) for the numbers $a_1 \geq 0, \ldots, a_n \geq 0, b_1 \geq 0, \ldots, b_n \geq 0$.

Proof Let $x_k = \sqrt{\left(a_1^2 + \cdots + a_k^2\right)\left(b_1^2 + \cdots + b_k^2\right)}$, where $k = 1, \ldots, n$.

In this case,

$$x_{k+1} = \sqrt{\left(a_1^2 + \cdots + a_k^2 + a_{k+1}^2\right)\left(b_1^2 + \cdots + b_k^2 + b_{k+1}^2\right)} =$$

$$= \sqrt{\left(\left(\sqrt{a_1^2 + \cdots + a_k^2}\right)^2 + a_{k+1}^2\right)\left(\left(\sqrt{b_1^2 + \cdots + b_k^2}\right)^2 + b_{k+1}^2\right)} \geq$$

$$\geq \sqrt{\left(\sqrt{a_1^2 + \cdots + a_k^2} \cdot \sqrt{b_1^2 + \cdots + b_k^2} + a_{k+1} \cdot b_{k+1}\right)^2} = x_k + a_{k+1} b_{k+1}.$$

Therefore, we obtain $x_{k+1} \geq x_k + a_{k+1}b_{k+1}$, where $k = 1, \ldots, n-1$.
Summing these inequalities, we deduce that

$$\sqrt{\left(a_1^2 + \cdots + a_n^2\right)\left(b_1^2 + \cdots + b_n^2\right)} \geq a_1b_1 + \cdots + a_nb_n, \text{ or}$$
$$\left(a_1^2 + \cdots + a_n^2\right)\left(b_1^2 + \cdots + b_n^2\right) \geq (a_1b_1 + \cdots + a_nb_n)^2.$$

Now let us consider the case in which $a_1, \ldots, a_n, b_1, \ldots, b_n$ are arbitrary real numbers.

In this case,

$$\left(a_1^2 + \cdots + a_n^2\right)\left(b_1^2 + \cdots + b_n^2\right) = \left(|a_1|^2 + \cdots + |a_n|^2\right)\left(|b_1|^2 + \cdots + |b_n|^2\right) \geq$$
$$\geq (|a_1b_1| + \cdots + |a_nb_n|)^2 \geq |a_1b_1 + \cdots + a_nb_n|^2 = (a_1b_1 + \cdots + a_nb_n)^2.$$

This ends the proof.

Alternative proof.

$$\left(a_1^2 + \cdots + a_n^2\right)\left(b_1^2 + \cdots + b_n^2\right) - (a_1b_1 + \cdots + a_nb_n)^2 = \sum_{\substack{i;\, j=1 \\ i \geq j}}^{n} \left(a_ib_j - b_ia_j\right)^2 \geq 0.$$

This ends the proof.

Let us consider the following examples.

Example 4.1 Prove that $\sin\alpha \cdot \sin\beta + \cos\alpha + \cos\beta \leq 2$.

Proof According to the Cauchy–Bunyakovsky–Schwarz inequality, we have

$$\sin\alpha \cdot \sin\beta + \cos\alpha \cdot 1 + 1 \cdot \cos\beta \leq \sqrt{\sin^2\alpha + \cos^2\alpha + 1^2} \cdot \sqrt{\sin^2\beta + \cos^2\beta + 1^2} = 2.$$

This ends the proof.

Example 4.2 Prove that $a^3 + b^3 > a^2 + b^2$ if $a > 0, \quad b > 0$, and $a^2 + b^2 > a + b$.

Proof According to the Cauchy–Bunyakovsky–Schwarz inequality, we have

$$(a^3+b^3)(a+b) = \left(\left(a^{\frac{3}{2}}\right)^2 + \left(b^{\frac{3}{2}}\right)^2\right)\left(\left(a^{\frac{1}{2}}\right)^2 + \left(b^{\frac{1}{2}}\right)^2\right) \geq \left(a^{\frac{3}{2}} \cdot a^{\frac{1}{2}} + b^{\frac{3}{2}} \cdot b^{\frac{1}{2}}\right)^2 = (a^2+b^2)^2.$$

Since $\frac{a^3+b^3}{a^2+b^2} \geq \frac{a^2+b^2}{a+b} > 1$, it follows that $a^3 + b^3 > a^2 + b^2$.

This ends the proof.

Problems

Prove the following inequalities (4.1–4.19, 4.22, 4.23).

4.1. $a^2 + b^2 + c^2 \geq 14$ if $a + 2b + 3c \geq 14$.

4.2. $ab + \sqrt{(1 - a^2)(1 - b^2)} \leq 1$ if $|a| \leq 1$, $|b| \leq 1$.

4.3. $\sqrt{c(a - c)} + \sqrt{c(b - c)} \leq \sqrt{ab}$ if $a > c$, $b > c$, $c > 0$.

4.4. $a\sqrt{a^2 + c^2} + b\sqrt{b^2 + c^2} \leq a^2 + b^2 + c^2$.

4.5. $\frac{1}{\sqrt{ab}} + \frac{1}{\sqrt{bc}} + \frac{1}{\sqrt{ca}} \leq \frac{1}{a} + \frac{1}{b} + \frac{1}{c}$, where $a > 0$, $b > 0$, $c > 0$.

4.6. $\sqrt{a}(a+c-b)+\sqrt{b}(a+b-c)+\sqrt{c}(b+c-a) \leq \sqrt{(a^2 + b^2 + c^2)(a + b + c)}$, where a, b, c are the side lengths of a triangle.

4.7. $(a_1 + \cdots + a_n)\left(\frac{1}{a_1} + \cdots + \frac{1}{a_n}\right) \geq n^2$, where $a_1 > 0, \ldots, a_n > 0$.

4.8. $\frac{a_1^2+\cdots+a_n^2}{n} \geq \left(\frac{a_1+\cdots+a_n}{n}\right)^2$.

4.9. $a_1a_2 + a_2a_3 + \cdots + a_9a_{10} + a_{10}a_1 \geq -1$ if $a_1^2 + \cdots + a_{10}^2 = 1$.

4.10. $x^4 + y^4 \geq x^3y + xy^3$.

4.11. $\left(|a_1|^3 + \cdots + |a_n|^3\right)^2 \leq \left(a_1^2 + \cdots + a_n^2\right)^3$.

4.12.
$$3\left(a^2 + b^2 + c^2 + x^2 + y^2 + z^2\right) + 6\sqrt{(a^2 + b^2 + c^2)(x^2 + y^2 + z^2)} \geq (a + b + c + x + y + z)^2.$$

4.13. $a^2 + b^2 + c^2 \geq ab + bc + ca$.

4.14. $(a_1 + \cdots + a_n)\left(a_1^7 + \cdots + a_n^7\right) \geq \left(a_1^3 + \cdots + a_n^3\right)\left(a_1^5 + \cdots + a_n^5\right)$, where $a_1 > 0, \ldots, a_n > 0$.

4.15. (a) $\sqrt{a + 1} + \sqrt{2a - 3} + \sqrt{50 - 3a} \leq 12$, where $\frac{3}{2} \leq a \leq \frac{50}{3}$,
(b) $a + b + c \leq abc + 2$, where $a^2 + b^2 + c^2 = 2$,
(c) $2(a + b + c) - abc \leq 10$, where $a^2 + b^2 + c^2 = 9$,
(d) $1 + abc \geq 3\min(a, b, c)$, where $a^2 + b^2 + c^2 = 9$.

4.16. $\left(\sum_{i=1}^{n} a_i^{k+1}\right)\left(\sum_{i=1}^{n} a_i^{-1}\right) \geq n \sum_{i=1}^{n} a_i^k$, where $k, n \in \mathbb{N}$ and $a_1 > 0, \ldots, a_n > 0$.

4.17. $\frac{a+b+c}{3} \geq \sqrt[3]{abc}$, where $a > 0$, $b > 0$, $c > 0$.

4.18. $\frac{a_1^k+\cdots+a_n^k}{n} \geq \left(\frac{a_1+\cdots+a_n}{n}\right)^k$, where $k, n \in \mathbb{N}$ and $a_1 \geq 0, \ldots, a_n \geq 0$.

4.19. $\left(1 + \frac{1}{\sin\alpha}\right)\left(1 + \frac{1}{\cos\alpha}\right) > 5$, where $0 < \alpha < \frac{\pi}{2}$.

4.20. Find the distance from a point $A(x_0, y_0)$ to the line defined by the equation $ax + by + c = 0$ $(a^2 + b^2 \neq 0)$.

4.21. Find the smallest possible value of the expression $(u - v)^2 + \left(\sqrt{2 - u^2} - \frac{9}{v}\right)^2$ if $0 < u < \sqrt{2}$, $v > 0$.

4.22. $x_1^2 + \left(\frac{x_1+x_2}{2}\right)^2 + \cdots + \left(\frac{x_1+\cdots+x_n}{n}\right)^2 \leq 4\left(x_1^2 + \cdots + x_n^2\right)$.
This inequality is a particular case of *Hardy's inequality*
$\sum_{k=1}^{n} \left(\frac{a_1+\cdots+a_k}{k}\right)^p \leq \left(\frac{p}{p-1}\right)^p \cdot \sum_{k=1}^{n} a_k^p$, where $p > 1$, $a_i \geq 0$, $i = 1, \ldots, n$.

4.23. (a) $\frac{1}{a_1} + \frac{2}{a_1+a_2} + \ldots + \frac{n}{a_1+\ldots+a_n} < 4\left(\frac{1}{a_1} + \ldots + \frac{1}{a_n}\right)$, where $a_1 > 0, \ldots, a_n > 0$,

(b) $\frac{1}{a_1}+\frac{2}{a_1+a_2}+\cdots+\frac{n}{a_1+\cdots+a_n}<2\left(\frac{1}{a_1}+\cdots+\frac{1}{a_n}\right)$, where $a_1>0,\ldots,a_n>0$.

Proofs

4.1. From (4.2) we have that $\left(a^2+b^2+c^2\right)\left(1^2+2^2+3^2\right)\geq(a+2b+3c)^2\geq 14^2$, and therefore, $a^2+b^2+c^2\geq 14$.

4.2. We have $a\cdot b+\sqrt{1-a^2}\cdot\sqrt{1-b^2}\leq\sqrt{a^2+\left(\sqrt{1-a^2}\right)^2}\cdot\sqrt{b^2+\left(\sqrt{1-b^2}\right)^2}=1.$

4.3. We have

$$\sqrt{c(a-c)}+\sqrt{c(b-c)}=\sqrt{c}\cdot\sqrt{a-c}+\sqrt{b-c}\cdot\sqrt{c}\leq$$
$$\leq\sqrt{\left(\sqrt{c}\right)^2+\left(\sqrt{b-c}\right)^2}\cdot\sqrt{\left(\sqrt{a-c}\right)^2+\left(\sqrt{c}\right)^2}=\sqrt{ab}.$$

4.4. Note that

$$a\cdot\sqrt{a^2+c^2}+\sqrt{b^2+c^2}\cdot b\leq\sqrt{a^2+\left(\sqrt{b^2+c^2}\right)^2}\cdot\sqrt{\left(\sqrt{a^2+c^2}\right)^2+b^2}$$
$$=a^2+b^2+c^2.$$

4.5. $\frac{1}{\sqrt{a}}\cdot\frac{1}{\sqrt{b}}+\frac{1}{\sqrt{b}}\cdot\frac{1}{\sqrt{c}}+\frac{1}{\sqrt{c}}\cdot\frac{1}{\sqrt{a}}\leq\sqrt{\left(\frac{1}{a}+\frac{1}{b}+\frac{1}{c}\right)\left(\frac{1}{b}+\frac{1}{c}+\frac{1}{a}\right)}=\frac{1}{a}+\frac{1}{b}+\frac{1}{c}.$

4.6. Note that

$$\begin{aligned}&\sqrt{a}(a+c-b)+\sqrt{b}(a+b-c)+\sqrt{c}(b+c-a)=\\&=\sqrt{a(a+c-b)}\sqrt{(a+c-b)}+\sqrt{b(a+b-c)}\sqrt{(a+b-c)}+\sqrt{c(b+c-a)}\sqrt{(b+c-a)}\\&\leq\sqrt{(a(a+c-b)+b(a+b-c)+c(b+c-a))}\cdot\sqrt{(a+c-b)+(a+b-c)+(b+c-a)}\\&=\sqrt{\left(a^2+b^2+c^2\right)(a+b+c)}.\end{aligned}$$

4.7. We have

$$\left((\sqrt{a_1})^2+\cdots+(\sqrt{a_n})^2\right)\left(\left(\frac{1}{\sqrt{a_1}}\right)^2+\cdots+\left(\frac{1}{\sqrt{a_n}}\right)^2\right)\geq\left(\sqrt{a_1}\cdot\frac{1}{\sqrt{a_1}}+\cdots+\sqrt{a_n}\cdot\frac{1}{\sqrt{a_n}}\right)^2=n^2.$$

4.8. We have $\left(a_1^2+\cdots+a_n^2\right)\left(\underbrace{1^2+\cdots+1^2}_{n}\right)\geq(a_1+\cdots+a_n)^2$, whence

$$\frac{a_1^2+\cdots+a_n^2}{n}\geq\left(\frac{a_1+\cdots+a_n}{n}\right)^2.$$

4.9. We have
$|a_1a_2 + \ldots + a_9a_{10} + a_{10}a_1| \le \sqrt{(a_1^2 + \ldots + a_9^2 + a_{10}^2)(a_2^2 + \ldots + a_{10}^2 + a_1^2)} =$ 1, hence $a_1a_2 + \cdots + a_9a_{10} + a_{10}a_1 \ge -1$.

4.10. We have

$$\begin{aligned} x^4 + y^4 &= \sqrt{x^4+y^4}\cdot\sqrt{x^4+y^4} \ge \sqrt{x^4+y^4}\cdot\sqrt{2x^2y^2} \\ &= \sqrt{(x^2)^2+(y^2)^2}\cdot\sqrt{(xy)^2+(xy)^2} \ge x^3y + xy^3. \end{aligned}$$

4.11. We have

$$\begin{aligned} \left(\left|a_1^3\right| + \cdots + \left|a_n^3\right|\right)^2 &= \left(|a_1|\cdot a_1^2 + \cdots + |a_n|\cdot a_n^2\right)^2 \le \left(a_1^2+\cdots+a_n^2\right)\cdot\left(a_1^4+\cdots+a_n^4\right) \\ &\le \left(a_1^2+\cdots+a_n^2\right)\left(a_1^2+\cdots+a_n^2\right)^2 = \left(a_1^2+\cdots+a_n^2\right)^3. \end{aligned}$$

4.12. Note that the left-hand side of the inequality is equal to $3\left(\sqrt{a^2+b^2+c^2} + \sqrt{x^2+y^2+z^2}\right)^2 = B$.
According to the inequality of Problem 4.8, we have

$$B \ge 3\left(\sqrt{3\left(\frac{a+b+c}{3}\right)^2} + \sqrt{3\left(\frac{x+y+z}{3}\right)^2}\right)^2 = (a+b+c+x+y+z)^2.$$

4.13. $ab + bc + ca \le \sqrt{(a^2+b^2+c^2)(b^2+c^2+a^2)} = a^2+b^2+c^2$.

4.14. We have

$$\begin{aligned} &(a_1+\cdots+a_n)\left(a_1^5+\cdots+a_n^5\right) = \left((\sqrt{a_1})^2+\cdots+(\sqrt{a_n})^2\right)\left(\left(\sqrt{a_1^5}\right)^2+\cdots+\left(\sqrt{a_n^5}\right)^2\right) \\ &\ge \left(a_1^3+\cdots+a_n^3\right)^2. \end{aligned}$$

In a similar way, we deduce that $\left(a_1^7+\cdots+a_n^7\right)\left(a_1^3+\cdots+a_n^3\right) \ge \left(a_1^5+\cdots+a_n^5\right)^2$.
Multiplying these inequalities, we obtain $(a_1+\cdots+a_n)\left(a_1^7+\cdots+a_n^7\right) \ge \left(a_1^3+\cdots+a_n^3\right)\left(a_1^5+\cdots+a_n^5\right)$.

4.15. (a) We have

$$\begin{aligned} &1\cdot\sqrt{\alpha+1} + 1\cdot\sqrt{2\alpha-3} + 1\cdot\sqrt{50-3\alpha} \le \\ &\le \sqrt{(1^2+1^2+1^2)\left(\left(\sqrt{\alpha+1}\right)^2 + \left(\sqrt{2\alpha-3}\right)^2 + \left(\sqrt{50-3\alpha}\right)^2\right)} = 12. \end{aligned}$$

(b) We have

$$a\cdot(1-bc) + (b+c)\cdot 1 \le \sqrt{(a^2+(b+c)^2)((1-bc)^2+1^2)} = \sqrt{4(1-b^2c^2)+2(1+bc)b^2c^2} \le 2,$$

since $bc \leq \frac{b^2+c^2}{2} \leq 1$.

(c) Let $a^2 = \max(a^2, b^2, c^2)$; hence $a^2 \geq 3$ and $x = bc \leq \frac{b^2+c^2}{2} \leq 3$, and therefore,

$$a \cdot (2 - bc) + (b + c) \cdot 2 \leq \sqrt{(a^2 + (b + c)^2)((2 - bc)^2 + 2^2)}$$
$$= \sqrt{100 + (2x - 7)(x + 2)^2} \leq 10.$$

(d) Let $a \leq b \leq c$.
If $0 \leq a \leq 1$, then $a(b - a)(c - a) \geq 0$ and $9 = a^2 + b^2 + c^2 \leq bc + b^2 + c^2 \leq (b + c)^2$.
Hence $1 + abc \geq 1 - a^3 + a^2(b + c) \geq (1 - a) \cdot 3a + a^2 \cdot 3 = 3a$, whence $1 + abc \geq 3a$.
If $a > 1$ and $c \geq 2$, then $1 + abc \geq 1 + 2a^2 > 3a$, and thus $1 + abc > 3a$.
If $a > 1$ and $c < 2$, then $a^2 + b^2 > 5$, whence $b^2 > 2,5$ and $1 + abc \geq 1 + ab^2 > 1 + 2,5a > 0,5c + 2,5a \geq 3a$, and therefore, $1 + abc > 3a$.
If $a < 0$ and $bc \leq 3$, then $1 > 0 \geq a(3 - bc)$, and therefore, $1 + abc > 3a$.
If $a < 0$ and $bc > 3$, since $9 = a^2 + b^2 + c^2 \geq a^2 + 2bc$, or $bc \leq \frac{9-a^2}{2}$, it follows that

$$-a(bc - 3) = |a|(bc - 3) \leq |a|\frac{3 - a^2}{2} = 1 - \frac{(|a| - 1)^2(|a| + 2)}{2} \leq 1,$$

whence $1 + abc \geq 3a$.

4.16. Define $\sum_{i=1}^{n} a_i^S = A_S$. From inequality (4.2), it follows that

$$A_{k+1} \cdot A_{k-1} = \left(\sum_{i=1}^{n} \left(a_i^{\frac{k+1}{2}}\right)^2\right)\left(\sum_{i=1}^{n} \left(a_i^{\frac{k-1}{2}}\right)^2\right) \geq \left(\sum_{i=1}^{n} a_i^{\frac{k+1}{2}} \cdot a_i^{\frac{k-1}{2}}\right)^2 = A_k^2.$$

In a similar way, we obtain $A_k \cdot A_{k-2} \geq A_{k-1}^2$, $A_{k-1} \cdot A_{k-3} \geq A_{k-2}^2, \ldots, A_2 \cdot A_0 \geq A_1^2$, $A_1 \cdot A_{-1} \geq A_0^2$.
Multiplying these inequalities, we deduce that $A_{k+1} \cdot A_{-1} \geq A_k \cdot A_0$, or

$$\left(\sum_{i=1}^{n} a_i^{k+1}\right)\left(\sum_{i=1}^{n} a_i^{-1}\right) \geq \sum_{i=1}^{n} a_i^k.$$

4.17. We have

$$(a + b + c)(b + c + a) = \left((\sqrt{a})^2 + \left(\sqrt{b}\right)^2 + (\sqrt{c})^2\right)\left(\left(\sqrt{b}\right)^2 + (\sqrt{c})^2 + (\sqrt{a})^2\right) \geq$$
$$\geq \left(\sqrt{ab} + \sqrt{bc} + \sqrt{ca}\right)^2,$$
$$\left(\sqrt{ab} + \sqrt{bc} + \sqrt{ca}\right)^2\left(\sqrt{c} + \sqrt{a} + \sqrt{b}\right)^2 \geq \left(\sqrt[4]{abc} + \sqrt[4]{abc} + \sqrt[4]{abc}\right)^4,$$
$$(a + b + c)(1 + 1 + 1) \geq \left(\sqrt{a} + \sqrt{b} + \sqrt{c}\right)^2.$$

Multiplying these inequalities, we obtain that $(a + b + c)^3 \geq 27abc$, or

$$\frac{a+b+c}{3} \geq \sqrt[3]{abc}.$$

4.18. We have that $A_k \cdot A_{k-2} \geq A_{k-1}^2$, $A_{k-1} \cdot A_{k-3} \geq A_{k-2}^2, \ldots, A_2 \cdot A_0 \geq A_1^2$, therefore multiplying these inequalities, we obtain

$$A_k \cdot A_0 \geq A_{k-1} \cdot A_1, A_{k-1} \cdot A_0 \geq A_{k-2} \cdot A_1, \ldots, A_2 \cdot A_0 \geq A_1 \cdot A_1, A_1 \cdot A_0 \geq A_0 \cdot A_1.$$

Multiplying these inequalities, we deduce that

$$A_k \cdot A_0^k \geq A_0 \cdot A_1^k, \text{ or } \frac{a_1^k + \ldots + a_n^k}{n} \geq \left(\frac{a_1 + \ldots + a_n}{n}\right)^k.$$

Here $A_k = \sum_{i=1}^{n} a_i^k$.

4.19. We have

$$\left(1^2 + \left(\frac{1}{\sqrt{\sin\alpha}}\right)^2\right)\left(1^2 + \left(\frac{1}{\sqrt{\cos\alpha}}\right)^2\right) \geq \left(1 + \frac{1}{\sqrt{\sin\alpha\, \cos\alpha}}\right)^2 =$$

$$= \left(1 + \frac{1}{\sqrt{\frac{\sin 2\alpha}{2}}}\right)^2 \geq \left(1 + \frac{1}{\sqrt{\frac{1}{2}}}\right)^2 = \left(1 + \sqrt{2}\right)^2 > 5.$$

4.20. Let $M(x, y)$ be a point on the line defined by the equation $ax + by + c = 0$. Let us estimate the minimum value of the distance $AM = \sqrt{(x - x_0)^2 + (y - y_0)^2}$.
According to inequality (4.1), we have $(a^2 + b^2)\left((x - x_0)^2 + (y - y_0)^2\right) \geq (a(x - x_0) + b(y - y_0))^2$,
whence $\sqrt{a^2 + b^2} \cdot \sqrt{(x - x_0)^2 + (y - y_0)^2} \geq |ax + by - ax_0 - by_0|$, or $\sqrt{(x - x_0)^2 + (y - y_0)^2} \geq \frac{|ax_0+by_0+c|}{\sqrt{a^2+b^2}}$.
In the last inequality, equality holds if $\frac{x-x_0}{a} = \frac{y-y_0}{b}$.
Solving the following system:

$$\begin{cases} ax + by + c = 0, \\ \dfrac{x - x_0}{a} = \dfrac{y - y_0}{b}. \end{cases}$$

We obtain $x = -a\frac{ax_0+by_0+c}{a^2+b^2} + x_0, \quad y = -b\frac{ax_0+by_0+c}{a^2+b^2} + y_0$.
Therefore, the distance from point A to the given line is equal to $\frac{|ax_0+by_0+c|}{\sqrt{a^2+b^2}}$.

4.21. We have

$$(u - v)^2 + \left(\sqrt{2 - u^2} - \frac{9}{v}\right)^2 = 2 - 2\left(uv + \frac{9}{v}\sqrt{2 - u^2}\right) + v^2 + \left(\frac{9}{v}\right)^2 \geq$$

$$\geq 2 - 2\sqrt{\left(u^2 + \left(\sqrt{2 - u^2}\right)^2\right)\left(v^2 + \left(\frac{9}{v}\right)^2\right)} + v^2 + \left(\frac{9}{v}\right)^2 = \left(\sqrt{v^2 + \frac{81}{v^2}} - \sqrt{2}\right)^2 \geq \left(\sqrt{2 \cdot 9} - \sqrt{2}\right)^2 = 8.$$

If $u = 1,\ v = 3$, then $(u - v)^2 + \left(\sqrt{2 - u^2} - \frac{9}{v}\right)^2 = 8$; hence the minimum value of the given expression is equal to 8.

4.22. We have

$$a_1^2 \geq a_1^2\left(1 - \frac{1}{\sqrt{2}}\right) + a_1^2\left(\frac{1}{\sqrt{2}} - \frac{1}{\sqrt{3}}\right) + \cdots + a_1^2\left(\frac{1}{\sqrt{n}} - \frac{1}{\sqrt{n+1}}\right),$$

$$a_2^2 \geq a_2^2\sqrt{2}\left(\frac{1}{\sqrt{2}} - \frac{1}{\sqrt{3}}\right) + a_2^2\sqrt{2}\left(\frac{1}{\sqrt{3}} - \frac{1}{\sqrt{4}}\right) + \cdots + a_2^2\sqrt{2}\left(\frac{1}{\sqrt{n}} - \frac{1}{\sqrt{n+1}}\right),$$

....

$$a_n^2 \geq a_n^2\sqrt{n}\left(\frac{1}{\sqrt{n}} - \frac{1}{\sqrt{n+1}}\right).$$

Therefore,

$$a_1^2 + a_2^2 + \cdots + a_n^2 \geq a_1^2\left(1 - \frac{1}{\sqrt{2}}\right) + \left(a_1^2 + a_2^2\sqrt{2}\right)\left(\frac{1}{\sqrt{2}} - \frac{1}{\sqrt{3}}\right) + \cdots +$$
$$+ \left(a_1^2 + a_2^2\sqrt{2} + \cdots + a_n^2\sqrt{n}\right)\left(\frac{1}{\sqrt{n}} - \frac{1}{\sqrt{n+1}}\right).$$

Let us prove that

$$\frac{\sqrt{k+1} - \sqrt{k}}{\sqrt{k} \cdot \sqrt{k+1}}\left(a_1^2 + \cdots + a_k^2\sqrt{k}\right) \geq \frac{1}{4}\left(\frac{a_1 + \cdots + a_k}{k}\right)^2. \qquad (4.3)$$

According to inequality (4.2), it follows that

$$\left(a_1^2 + \cdots + a_k^2\sqrt{k}\right)\left(1 + \cdots + \frac{1}{\sqrt{k}}\right) \geq (a_1 + \cdots + a_k)^2.$$

In order to prove inequality (4.3), it is sufficient to prove that $\frac{4(\sqrt{k+1} - \sqrt{k})}{\sqrt{k}\sqrt{k+1}}k^2 > 1 + \cdots + \frac{1}{\sqrt{k}}$. We have

$$1 + \frac{1}{\sqrt{2}} + \cdots + \frac{1}{\sqrt{k}} \leq 1 + \frac{2}{\sqrt{2} + 1} + \cdots + \frac{2}{\sqrt{k} + \sqrt{k-1}} =$$
$$= 1 + 2\left(\sqrt{2} - 1\right) + \cdots + 2\left(\sqrt{k} - \sqrt{k-1}\right) = 2\sqrt{k} - 1.$$

Let us prove that

$$2\sqrt{k} - 1 < \frac{4k^2\left(\sqrt{k+1} - \sqrt{k}\right)}{\sqrt{k}\sqrt{k+1}} = \frac{4k\sqrt{k}}{k + 1 + \sqrt{k^2 + k}}.$$

Since $\frac{2k}{\sqrt{k+1} + \sqrt{k}} < k \leq k + \left(\sqrt{k} - 1\right)^2 < k + \sqrt{k^2 + k} - 2\sqrt{k} + 1$, it follows that

$$\frac{2k}{\sqrt{k+1}+\sqrt{k}} < k+\sqrt{k^2+k}-2\sqrt{k}+1, \text{ or}$$
$$2k\left(\sqrt{k+1}-\sqrt{k}\right) < k+\sqrt{k^2+k}-2\sqrt{k}+1, \text{ or}$$
$$2\sqrt{k}\left(k+1+\sqrt{k^2+k}\right)-\left(k+1+\sqrt{k^2+k}\right) < 4k\sqrt{k}.$$

Hence we deduce that

$$2\sqrt{k}-1 < \frac{4k\sqrt{k}}{k+1+\sqrt{k^2+k}}.$$

Now let us prove that on the right-hand side, it is impossible to choose a multiplier smaller than 4.
Suppose the inequality $x_1^2+\left(\frac{x_1+x_2}{2}\right)^2+\cdots+\left(\frac{x_1+\cdots+x_n}{n}\right)^2 \le c\left(x_1^2+x_2^2+\cdots+x_n^2\right)$ holds for all $x_1,\ldots,x_n$.
Taking $x_i=\frac{1}{\sqrt{i}}$, $i=1,\ldots,n$, we deduce that

$$1+\left(\frac{1+\frac{1}{\sqrt{2}}}{2}\right)^2+\cdots+\left(\frac{1+\frac{1}{\sqrt{2}}+\cdots+\frac{1}{\sqrt{n}}}{n}\right)^2 \le c\left(1+\frac{1}{2}+\cdots+\frac{1}{n}\right).$$

According to the inequality of Problem 14.16, we have
$\sum\limits_{k=1}^{n}\left(\frac{2}{\sqrt{k}}-\frac{2}{k}\right)^2 < c\left(1+\frac{1}{2}+\cdots+\frac{1}{n}\right)$, whence $4-8\frac{1/1\sqrt{1}+\cdots+1/n\sqrt{n}}{1+1/2+\cdots+1/n} < c$.
Passing to the limit, we obtain
$\lim\limits_{n\to\infty}\left(4-8\frac{\frac{1}{1\sqrt{1}}+\cdots+\frac{1}{n\sqrt{n}}}{1+\frac{1}{2}+\cdots+\frac{1}{n}}\right) \le \lim\limits_{n\to\infty} c$. Since
$\lim\limits_{n\to\infty}\frac{\frac{1}{1\sqrt{1}}+\cdots+\frac{1}{n\sqrt{n}}}{1+\frac{1}{2}+\cdots+\frac{1}{n}}=0$, we must have $4\le c$.

4.23. (a) The proof follows straightforwardly from the proof of part (b).
(b) We have

$$\frac{1}{a_1} > \frac{1}{a_1}\left(1-\frac{1}{2^2}\right)+\frac{1}{a_1}\left(\frac{1}{2^2}-\frac{1}{3^2}\right)+\cdots+\frac{1}{a_1}\left(\frac{1}{n^2}-\frac{1}{(n+1)^2}\right),$$
$$\frac{1}{a_2} > \frac{2^2}{a_2}\left(\frac{1}{2^2}-\frac{1}{3^2}\right)+\frac{2^2}{a_2}\left(\frac{1}{3^2}-\frac{1}{4^2}\right)+\cdots+\frac{2^2}{a_2}\left(\frac{1}{n^2}-\frac{1}{(n+1)^2}\right),\ldots,\frac{1}{a_n} > \frac{n^2}{a_n}\left(\frac{1}{n^2}-\frac{1}{(n+1)^2}\right).$$

Therefore,

$$2\left(\frac{1}{a_1}+\frac{1}{a_2}+\cdots+\frac{1}{a_n}\right) > 2\frac{1}{a_1}\left(1-\frac{1}{2^2}\right)+2\left(\frac{1}{a_1}+\frac{2^2}{a_2}\right)\left(\frac{1}{2^2}-\frac{1}{3^2}\right)+\cdots$$
$$+2\left(\frac{1}{a_1}+\cdots+\frac{n^2}{a_n}\right)\left(\frac{1}{n^2}-\frac{1}{(n+1)^2}\right).$$

Let us prove that

$$2\left(\frac{1}{a_1}+\cdots+\frac{k^2}{a_k}\right)\left(\frac{1}{k^2}-\frac{1}{(k+1)^2}\right) > \frac{k}{a_1+\cdots+a_k},$$

or equivalently,

$$(a_1+\cdots+a_k)\left(\frac{1}{a_1}+\cdots+\frac{k^2}{a_k}\right)>\frac{k^3(k+1)^2}{2(2k+1)}.$$

We have

$$(a_1+\cdots+a_k)\left(\frac{1}{a_1}+\cdots+\frac{k^2}{a_k}\right)\geq\left(\sqrt{a_1\cdot\frac{1}{a_1}}+\cdots+\sqrt{\frac{k^2}{a_k}\cdot a_k}\right)^2=\frac{k^2(k+1)^2}{4}>\frac{k^3(k+1)^2}{2(2k+1)}.$$

It follows that

$$2\left(\frac{1}{a_1}+\cdots+\frac{1}{a_n}\right)>\sum_{k=1}^{n}2\left(\frac{1}{a_1}+\cdots+\frac{k^2}{a_k}\right)\left(\frac{1}{k^2}-\frac{1}{(k+1)^2}\right)$$
$$>\sum_{k=1}^{n}\frac{k}{a_1+\cdots+a_k}.$$

Now let us prove that on the right-hand side, it is impossible to choose a multiplier smaller than 2. Indeed, let $\frac{1}{a_1}+\cdots+\frac{n}{a_1+\cdots+a_n}\leq c\left(\frac{1}{a_1}+\cdots+\frac{1}{a_n}\right)$.

Assume that $a_k=k,\quad k=1,\ldots,n$, and thus $\frac{2x_{n+1}-2}{x_n}\leq c$, where $x_n=1+\frac{1}{2}+\cdots+\frac{1}{n}$. Therefore, $2\lim\limits_{n\to\infty}\left(\frac{x_{n+1}}{x_n}-\frac{1}{x_n}\right)\leq c$, or $2\leq c$. This ends the proof.

Problems for Independent Study

Prove the following inequalities (1–16).

1. $(\sin\alpha_1+\cdots+\sin\alpha_n)^2+(\cos\alpha_1+\cdots+\cos\alpha_n)^2\leq n^2$.
2. $\frac{a_1+\cdots+a_n}{n}\geq\sqrt[n]{a_1\cdots a_n}$, where $n\geq 2,\quad a_1>0,\ldots,a_n>0$.
3. $\sqrt{a_1b_1}+\cdots+\sqrt{a_nb_n}\leq\sqrt{a_1+\cdots+a_n}\cdot\sqrt{b_1+\cdots+b_n}$, where $a_i\geq 0,\quad b_i\geq 0,\quad i=1,\ldots,n$.
4. $(x_1y_2-x_2y_1)^2+(x_2y_3-x_3y_2)^2+(x_1y_3-x_3y_1)^2\leq(x_1^2+x_2^2+x_3^2)(y_1^2+y_2^2+y_3^2)$.
5. $\left(\sum\limits_{i=1}^{n}\sqrt{a_ib_i}\right)^2\leq\left(\sum\limits_{i=1}^{n}a_ix_i\right)\left(\sum\limits_{i=1}^{n}\frac{b_i}{x_i}\right)$, where $x_i>0,\quad a_i>0,\quad b_i>0\quad i=1,\ldots,n$.
6. $\left(\sum\limits_{i=1}^{n}x_iy_i\right)\left(\sum\limits_{i=1}^{n}\frac{x_i}{y_i}\right)\geq\left(\sum\limits_{i=1}^{n}x_i\right)^2$, where $x_i>0,\quad y_i>0\quad i=1,\ldots,n$.
7. $ax+by+cz+\sqrt{(a^2+b^2+c^2)(x^2+y^2+z^2)}\geq\frac{2}{3}(a+b+c)(x+y+z)$.
8. $(p_1q_1-p_2q_2-\cdots-p_nq_n)^2\geq(p_1^2-p_2^2-\cdots-p_n^2)(q_1^2-q_2^2\quad\cdots-q_n^2)$, if $p_1^2\geq p_2^2+\cdots+p_n^2,q_1^2\geq q_2^2+\cdots+q_n^2$.

9. $\sqrt{x^2+xy+y^2}\sqrt{y^2+yz+z^2}+\sqrt{y^2+yz+z^2}\sqrt{z^2+zx+x^2}+$ $+\sqrt{z^2+zx+x^2}\sqrt{x^2+xy+y^2}\geq(x+y+z)^2$.
Hint. We have
$$\sqrt{x^2+xy+y^2}\sqrt{y^2+yz+z^2}=\sqrt{\left(y+\frac{x}{2}\right)^2+\left(\frac{\sqrt{3}x}{2}\right)^2}\sqrt{\left(y+\frac{z}{2}\right)^2+\left(\frac{\sqrt{3}z}{2}\right)^2}\geq$$
$$\geq\left(y+\frac{x}{2}\right)\left(y+\frac{z}{2}\right)+\frac{3xz}{4}.$$
10. $a_1(b_1+a_2)+a_2(b_2+a_3)+\cdots+a_n(b_n+a_1)<1$, where $n\geq 3, a_1,\ldots,a_n>0$ and $a_1+\cdots+a_n=1$, $b_1^2+\cdots+b_n^2=1$.
11. $\sqrt{1-\left(\frac{x+y}{2}\right)^2}+\sqrt{1-\left(\frac{y+z}{2}\right)^2}+\sqrt{1-\left(\frac{z+x}{2}\right)^2}\geq\sqrt{6}$, where $x,y,z\geq 0$, $x^2+y^2+z^2=1$.
12. $\sqrt{\frac{a}{b+c}}+\sqrt{\frac{b}{c+a}}+\sqrt{\frac{c}{a+b}}\geq 2\sqrt{1+\frac{abc}{(a+b)(b+c)(c+a)}}$, where $a,b,c>0$.
13. $\sqrt{a+(b-c)^2}+\sqrt{b+(c-a)^2}+\sqrt{c+(a-b)^2}\geq\sqrt{3}$, where $a,b,c\geq 0$ and $a+b+c=1$.
14. $\sqrt{\frac{a+b}{2}-ab}+\sqrt{\frac{b+c}{2}-bc}+\sqrt{\frac{c+a}{2}-ca}\geq\sqrt{2}$, where $a,b,c\geq 0$ and $a+b+c=2$.
15. $\sqrt{1-xy}\sqrt{1-yz}+\sqrt{1-yz}\sqrt{1-zx}+\sqrt{1-zx}\sqrt{1-xy}\geq 2$, where $x,y,z\geq 0$ and $x^2+y^2+z^2=1$.
16. $x\sqrt{1-yz}+y\sqrt{1-zx}+z\sqrt{1-xy}\geq\frac{2\sqrt{2}}{3}$, where $x,y,z\geq 0$ and $x+y+z=1$.
17. Find the distance from a point $A(x_0,y_0,z_0)$ to the plane defined by the equation $ax+by+cz+d=0\quad\left(a^2+b^2+c^2\neq 0\right)$.
18. Prove

(a) the following identity:
$$(a_1c_1+\cdots+a_nc_n)(b_1d_1+\cdots+b_nd_n)-(a_1d_1+\cdots+a_nd_n)(b_1c_1+\cdots+b_nc_n)$$
$$=\sum_{1\leq i<k\leq n}(a_ib_k-a_kb_i)(c_id_k-c_kd_i).$$

(b) $(a_1c_1+\cdots+a_nc_n)(b_1d_1+\cdots+b_nd_n)\geq(a_1d_1+\cdots+a_nd_n)(b_1c_1+\cdots+b_nc_n)$, where $b_id_i>0\ (i=1,\ldots,n)$ or $b_id_i<0\ (i=1,\ldots,n)$ and $\frac{a_1}{b_1}\leq\frac{a_2}{b_2}\leq\ldots\leq\frac{a_n}{b_n}$, $\frac{c_1}{d_1}\leq\frac{c_2}{d_2}\leq\ldots\leq\frac{c_n}{d_n}$.

19. Find the maximum and minimum values of the expression
$$\frac{\sqrt{x^2+y^2}+\sqrt{(x-2)^2+(y-1)^2}}{\sqrt{x^2+(y-1)^2}+\sqrt{(x-2)^2+y^2}}.$$

20. Find the minimum value of the expression

$$\left(\frac{1}{x^n}+\frac{1}{a^n}-1\right)\left(\frac{1}{y^n}+\frac{1}{b^n}-1\right),$$

where $x, y, a, b > 0, x + y = 1, a + b = 1$.

Hint. We have

$$\left(\frac{1}{x^n}+\frac{1}{a^n}-1\right)\left(\frac{1}{y^n}+\frac{1}{b^n}-1\right)=\left(\frac{1}{x^n}+\frac{b(1+a+\cdots+a^{n-1})}{a^n}\right)\left(\frac{1}{y^n}+\frac{a(1+b+\cdots+b^{n-1})}{b^n}\right)\geq$$

$$\geq\left(\left(\frac{1}{\sqrt{xy}}\right)^n+\sqrt{\frac{(1+a+\cdots+a^{n-1})(1+b+\cdots+b^{n-1})}{a^{n-1}b^{n-1}}}\right)^2\geq\left(\left(\frac{1}{\sqrt{xy}}\right)^n+\frac{1+\sqrt{ab}+\cdots+\left(\sqrt{ab}\right)^{n-1}}{\left(\sqrt{ab}\right)^{n-1}}\right)^2$$

$$\geq(2^{n+1}-1)^2.$$

Chapter 5
Change of Variables Method

The *change of variables method* is a basic technique used to simplify problems in which the original variable or variables are replaced with other variables in order to express the given problem in new variables in such a way that the problem becomes simpler or easier to prove.

In order to prove inequalities, one of the most useful techniques is the change of variables method. Depending on the problem, one can deal with the *single-variable* case or the *multiple-variables* case. Both cases can be treated using this method.

Let us explain this method by considering the following examples.

Example 5.1 Prove that $-\frac{1}{2} \leq \frac{(x+y)(1-xy)}{(1+x^2)(1+y^2)} \leq \frac{1}{2}$.

Proof Let us perform the following change of variables: $x = \tan\alpha, \quad y = \tan\beta$.

It then follows that $\frac{(x+y)(1-xy)}{(1+x^2)(1+y^2)} = \frac{(\tan\alpha+\tan\beta)(1-\tan\alpha\tan\beta)}{(1+\tan^2\alpha)(1+\tan^2\beta)} = \frac{\frac{\sin(\alpha+\beta)}{\cos\alpha\cos\beta}\cdot\frac{\cos(\alpha+\beta)}{\cos\alpha\cos\beta}}{\frac{1}{\cos^2\alpha}\cdot\frac{1}{\cos^2\beta}} =$
$= \sin(\alpha+\beta)\cos(\alpha+\beta) = \frac{1}{2}\sin 2(\alpha+\beta)$, therefore $-\frac{1}{2} \leq \frac{1}{2}\sin 2(\alpha+\beta) \leq \frac{1}{2}$. This completes the proof.

Example 5.2 Prove that $a^2\left(1+b^4\right)+b^2\left(1+a^4\right) \leq \left(1+a^4\right)\left(1+b^4\right)$.

Proof Let us perform the following change of variables: $a^2 = \tan\alpha, \quad b^2 = \tan\beta$.
Then $\tan\alpha\left(1+\tan^2\beta\right)+\tan\beta\left(1+\tan^2\alpha\right) \leq \left(1+\tan^2\alpha\right)\left(1+\tan^2\beta\right)$.
This inequality is equivalent to the following inequality: $\sin 2\alpha + \sin 2\beta \leq 2$.
This ends the proof.

Example 5.3 Prove that if $a > 0, \ b > 0, \ c > 0, a+b+c = 1$, then $a^3+b^3+c^3 \geq \frac{1}{9}$.

H. Sedrakyan and N. Sedrakyan, *Algebraic Inequalities*, Problem Books in Mathematics, https://doi.org/10.1007/978-3-319-77836-5_5

Proof Let us perform the following change of variables: $a = x + \frac{1}{3}$, $b = y + \frac{1}{3}$, $c = z + \frac{1}{3}$. Then $a^3 = x^3 + x^2 + \frac{x}{3} + \frac{1}{27}$, $b^3 = y^3 + y^2 + \frac{y}{3} + \frac{1}{27}$, $c^3 = z^3 + z^2 + \frac{z}{3} + \frac{1}{27}$, and therefore, $a^3 + b^3 + c^3 = x^2(1+x) + y^2(1+y) + z^2(1+z) + \frac{x+y+z}{3} + \frac{1}{9} = x^2(1+x) + y^2(1+y) + z^2(1+z) + \frac{1}{9} \geq \frac{1}{9}$, as $x > -\frac{1}{3}$, $y > -\frac{1}{3}$, $z > -\frac{1}{3}$.
This ends the proof.

Example 5.4 Prove that if $a > 0$, $b > 0$, $c > 0$, and $abc = 1$, then
$\frac{1}{1+a+b} + \frac{1}{1+b+c} + \frac{1}{1+a+c} \leq 1.$

Proof Let us perform the following change of variables: $\sqrt[3]{a} = x$, $\sqrt[3]{b} = y$, $\sqrt[3]{c} = z$. Then $xyz = 1$ and $\frac{1}{1+a+b} = \frac{xyz}{xyz+x^3+y^3} \leq \frac{xyz}{xyz+xy(x+y)} = \frac{z}{x+y+z}$. In a similar way, we obtain that $\frac{1}{1+b+c} \leq \frac{x}{x+y+z}$ and $\frac{1}{1+a+c} \leq \frac{y}{x+y+z}$. Therefore, we deduce that
$\frac{1}{1+a+b} + \frac{1}{1+b+c} + \frac{1}{1+a+c} \leq \frac{x}{x+y+z} + \frac{y}{x+y+z} + \frac{z}{x+y+z} = 1.$
This ends the proof.

Example 5.5 Prove that if $x_1 > 0, \ldots, x_n > 0$ and $\frac{1}{x_1} + \cdots + \frac{1}{x_n} = n$, then $x_1 + \frac{x_2^2}{2} + \cdots + \frac{x_n^n}{n} \geq 1 + \frac{1}{2} + \cdots + \frac{1}{n}$.

Proof Let us perform the following change of variables: $x_i = 1 + y_i$, $i = 1, \ldots, n$, where $y_i > -1$. We have $(x_1 + \cdots + x_n)\left(\frac{1}{x_1} + \cdots + \frac{1}{x_n}\right) \geq n^2$, and hence $y_1 + \cdots + y_n \geq 0$.

According to Bernoulli's inequality (Problem 3.27a), we deduce that
$x_1 + \frac{x_2^2}{2} + \cdots + \frac{x_n^n}{n} \geq 1 + \frac{1}{2} + \cdots + \frac{1}{n} + y_1 + y_2 + \cdots + y_n \geq 1 + \frac{1}{2} + \cdots + \frac{1}{n}$. Therefore, $x_1 + \frac{x_2^2}{2} + \cdots + \frac{x_n^n}{n} \geq 1 + \frac{1}{2} + \cdots + \frac{1}{n}$. This ends the proof.

Example 5.6 Prove that if $a_1 \geq 1, \ldots, a_n \geq 1$ and $n \geq 2$, $k \geq 2$, $k \in \mathbb{N}$, then
$\sqrt[k]{a_1 - 1} + \cdots + \sqrt[k]{a_n - 1} \leq \frac{(n-1)^{\frac{n-1}{k}}}{n^{\frac{n-k}{k}}} \sqrt[k]{a_1 \cdot \cdots \cdot a_n}.$

Proof Let us perform the following change of variables: $b_i = \sqrt[k]{(n-1)(a_i - 1)}$, where $i = 1, 2, \ldots, n$ and $b_1^k < 1, \ldots, b_l^k < 1, b_{l+1}^k \geq 1, \ldots, b_n^k \geq 1$. Then from Problems 10.6, 11.14, it follows that

$$\left(1 + \frac{b_1^k - 1}{n}\right) \cdot \cdots \cdot \left(1 + \frac{b_l^k - 1}{n}\right)\left(1 + \frac{b_{l+1}^k - 1}{n}\right) \cdot \cdots \cdot \left(1 + \frac{b_n^k - 1}{n}\right) \geq$$
$$\geq \left(1 + \frac{b_1^k + \cdots + b_l^k - l}{n}\right)\left(1 + \frac{b_{l+1}^k + \cdots + b_n^k - (n-l)}{n}\right) =$$
$$= \left(\frac{b_1^k + \ldots + b_l^k + \overbrace{1^k + \cdots + 1^k}^{n-l}}{n}\right)\left(\frac{\overbrace{1^k + \cdots + 1^k}^{l} + b_{l+1}^k + \cdots + b_n^k}{n}\right) \geq \left(\frac{b_1 + b_2 + \cdots + b_n}{n}\right)^k.$$

Note that $1 + \frac{b_i^k - 1}{n} = \frac{(n-1)a_i}{n}$, $i = 1, 2, \ldots, n$, and thus it follows that
$\left(\frac{n-1}{n}\right)^n a_1 \cdot \cdots \cdot a_n \geq \left(\frac{\sqrt[k]{n-1}(\sqrt[k]{a_1 - 1} + \cdots + \sqrt[k]{a_n - 1})}{n}\right)^k$, whence
$\frac{(n-1)^{\frac{n-1}{k}}}{n^{\frac{n-k}{k}}} \sqrt[k]{a_1 \ldots a_n} \geq \sqrt[k]{a_1 - 1} + \cdots + \sqrt[k]{a_n - 1}.$
This ends the proof.

Problems

Prove the following inequalities (5.1–5.5, 5.8–5.12, 5.16–5.20).

5.1. $4 \le a^2+b^2+ab+\sqrt{4-a^2}\cdot\sqrt{9-b^2} \le 19$, where $0 \le a \le 2$ and $0 \le b \le 3$.

5.2. $n\sqrt{m-1}+m\sqrt{n-1} \le mn$, where $m \ge 1,\ \ n \ge 1$.

5.3. $\sqrt{m^2-n^2}+\sqrt{2mn-n^2} \ge m$, where $m > n > 0$.

5.4. $x > \sqrt{x-1}+\sqrt{x\left(\sqrt{x}-1\right)}$, where $x \ge 1$.

5.5. Prove that $1+\frac{1}{\sqrt{2}}+\ldots+\frac{1}{\sqrt{n}} \ge n\sqrt{\frac{2}{n+1}}$, *where* $n \in \mathbb{N}$

5.6. Prove that among seven arbitrary numbers one can find two numbers x and y such that $0 \le \frac{x-y}{1+xy} < \frac{\sqrt{3}}{3}$.

5.7. Prove the inequality of Problem 4.3.

5.8. $\frac{|a-b|}{\sqrt{1+a^2}\cdot\sqrt{1+b^2}} \le \frac{|a-c|}{\sqrt{1+a^2}\cdot\sqrt{1+c^2}}+\frac{|b-c|}{\sqrt{1+b^2}\cdot\sqrt{1+c^2}}$.

5.9. (a) *Huygens's inequality*: $\sqrt[n]{(a_1+b_1)\cdots(a_n+b_n)} \ge \sqrt[n]{a_1\cdot\cdots\cdot a_n}+\sqrt[n]{b_1\cdots b_n}$, where $a_i > 0,\ \ b_i > 0,\ \ i=1,\ldots,n$,

(b) *Milne's inequality*: $\frac{a_1b_1}{a_1+b_1}+\cdots+\frac{a_nb_n}{a_n+b_n} \le \frac{(a_1+\cdots+a_n)(b_1+\cdots+b_n)}{(a_1+\cdots+a_n)+(b_1+\cdots+b_n)}$, where $a_i > 0,\ \ b_i > 0,\ \ i=1,\ldots,n$.

5.10. $\frac{8}{(x_1+x_2)(y_1+y_2)-(z_1+z_2)^2} \le \frac{1}{x_1y_1-z_1^2}+\frac{1}{x_2y_2-z_2^2}$, where $x_1 > 0,\ \ x_2 > 0$, and $x_1y_1-z_1^2 > 0$, $x_2y_2-z_2^2 > 0$.

5.11. (a) $\sqrt{a-1}+\sqrt{b-1}+\sqrt{c-1} \le \frac{2}{\sqrt{3}}\sqrt{abc}$, where $a \ge 1,\ \ b \ge 1,\ \ c \ge 1$,

(b) $\sqrt{a-1}+\sqrt{b-1}+\sqrt{c-1}+\sqrt{d-1} \le \frac{3\sqrt{3}}{4}\sqrt{abcd}$, where $a \ge 1,\ \ b \ge 1,\ \ c \ge 1,\ \ d \ge 1$.

5.12. $\frac{a_1^k+\cdots+a_n^k}{n} \ge \left(\frac{a_1+\cdots+a_n}{n}\right)^k$, where $k \in \mathbb{N}$ and $a_1 > 0,\ldots,a_n > 0$.

5.13. Let $f(x)$ be a given differentiable function on $[a, a+4]$. Prove that there exists a point x_0 $(x_0 \in [a, a+4])$ such that $f'(x_0) < 1+f^2(x_0)$.

5.14. Let $f(x)$ be a function defined in $[a, b]$ such that $f(x) > 0,\ \ x \in (a, b)$, and $f(a) = f(b) = 0,\ \ f(x)+f^n(x) > 0$ for all $x \in [a, b]$. Prove that $b-a > \pi$.

5.15. Prove that if $a \ge \frac{1}{2},\ b \ge \frac{1}{2}$, then $\left(\frac{a^2-b^2}{2}\right)^2 \ge \sqrt{\frac{a^2+b^2}{2}}-\frac{a+b}{2}$.

5.16. (a) $x_1+\cdots+x_n \le \frac{n}{3}$, where $x_1^3+\cdots+x_n^3 = 0$ and $x_i \in [-1, 1],\ \ i=1,\ldots,n$,

(b) $\left|x_1^3+\cdots+x_n^3\right| \le 2n$, where $x_1+\cdots+x_n = 0$ and $x_i \in [-2.2], i = 1, 2, \ldots, n$.

5.17. $\frac{x}{1+x^2}+\frac{y}{1+y^2}+\frac{z}{1+z^2} \le \frac{3\sqrt{3}}{4}$, where $x^2+y^2+z^2 = 1$.

5.18. $1 < \frac{a}{\sqrt{a^2+b^2}}+\frac{b}{\sqrt{b^2+c^2}}+\frac{c}{\sqrt{c^2+a^2}} \le \frac{3\sqrt{2}}{2}$, where $a, b, c > 0$.

5.19. $\sqrt{1-a}+\sqrt{1-b}+\sqrt{1-c}+\sqrt{1-d} \ge \sqrt{a}+\sqrt{b}+\sqrt{c}+\sqrt{d}$, where $a, b, c, d > 0,\ \ a^2+b^2+c^2+d^2 = 1$.

5.20. $\frac{a+b+c}{3}-\sqrt[3]{abc} \le \max\left(\left(\sqrt{a}-\sqrt{b}\right)^2, \left(\sqrt{b}-\sqrt{c}\right)^2, \left(\sqrt{c}-\sqrt{a}\right)^2\right)$, where $a > 0, b > 0, c > 0$.

Proofs

5.1. Let $a = 2\cos\alpha$, $b = 3\cos\beta$, where $\alpha, \beta \in \left[0, \frac{\pi}{2}\right]$. It follows that

$a^2 + b^2 + ab + \sqrt{4 - a^2} \cdot \sqrt{9 - b^2} = 4\cos^2\alpha + 9\cos^2\beta + 6\cos\alpha\cos\beta + 6\sin\alpha\sin\beta =$

$= 4\cos^2\alpha + 9\cos^2\beta + 6\cos(\alpha - \beta) \leq 4 + 9 + 6 = 19.$

On the other hand, $6\sin\alpha\sin\beta \geq 6\sin^2\alpha$ or $6\sin\alpha\sin\beta \geq 6\sin^2\beta$.
If $6\sin\alpha\sin\beta \geq 6\sin^2\alpha$, then we have that
$4\cos^2\alpha + 9\cos^2\beta + 6\cos\alpha\cos\beta + 6\sin\alpha\sin\beta \geq 4\cos^2\alpha + 6\sin^2\alpha + 9\cos^2\beta + 6\cos\alpha\cos\beta \geq 4$.
On the other hand, if $6\sin\alpha\sin\beta \geq 6\sin^2\beta$, then
$4\cos^2\alpha + 9\cos^2\beta + 6\cos\alpha\cos\beta + 6\sin\alpha\sin\beta \geq 6$.
Therefore, in both cases we obtain $a^2 + b^2 + ab + \sqrt{4 - a^2} \cdot \sqrt{9 - b^2} \geq 4$.

5.2. Let $m = \frac{1}{\cos^2\alpha}$, $n = \frac{1}{\cos^2\beta}$, where $\alpha, \beta \in \left[0, \frac{\pi}{2}\right)$. Then

$\frac{n\sqrt{m-1}+m\sqrt{n-1}}{mn} = \frac{\frac{1}{\cos^2\beta}\cdot\frac{\sin\alpha}{\cos\alpha}+\frac{1}{\cos^2\alpha}\cdot\frac{\sin\beta}{\cos\beta}}{\frac{1}{\cos^2\alpha\cos^2\beta}} = \frac{\sin 2\alpha+\sin 2\beta}{2} \leq 1.$

Therefore, $n\sqrt{m-1} + m\sqrt{n-1} \leq mn$.

5.3. Let $\frac{n}{m} = \sin\alpha$, where $\alpha \in \left(0; \frac{\pi}{2}\right)$. It follows that

$\frac{\sqrt{m^2-n^2}+\sqrt{2mn-n^2}}{m} = \frac{m\cos\alpha+m\sqrt{2\sin\alpha-\sin^2\alpha}}{m} = \cos\alpha + \sqrt{2\sin\alpha - \sin^2\alpha} =$

$= \cos\alpha + \sqrt{\sin\alpha + \sin\alpha(1 - \sin\alpha)} \geq \cos\alpha + \sqrt{\sin\alpha} \geq \cos^2\alpha + \sin^2\alpha = 1.$

Therefore, $\sqrt{m^2 - n^2} + \sqrt{2mn - n^2} \geq m$.

5.4. Let $x = \frac{1}{\cos^2\alpha}$, where $\alpha \in \left[0, \frac{\pi}{2}\right)$. Then the given inequality can be rewritten as

$$\frac{1}{\cos^2\alpha} > \tan\alpha + \frac{1}{\cos\alpha}\sqrt{\frac{1-\cos\alpha}{\cos\alpha}}, \text{ or } 1 > \sin\alpha\cos\alpha + \sqrt{\cos\alpha(1-\cos\alpha)}. \quad (5.1)$$

From inequality (3.2), it follows that $\sqrt{\cos\alpha(1-\cos\alpha)} \leq \frac{1}{2}$ and $\sin\alpha\cos\alpha \leq \frac{1}{2}$, where equality does not hold simultaneously in both inequalities. Summing these inequalities, we obtain inequality (5.1).

5.5. First, let us prove that $\frac{1}{\sqrt{k}} + \frac{1}{\sqrt{n+1-k}} \geq 2\sqrt{\frac{2}{n+1}}$, where $k = 1, \ldots, n$.
Let $k = (n+1)\sin^2\alpha$, where $0 < \alpha < \frac{\pi}{2}$. It follows that

$\frac{1}{\sqrt{n+1}\cdot\sin\alpha} + \frac{1}{\sqrt{n+1}\cdot\cos\alpha} = \frac{1}{\sqrt{n+1}}\left(\frac{1}{\sin\alpha} + \frac{1}{\cos\alpha}\right) \geq \frac{2}{\sqrt{n+1}}\sqrt{\frac{2}{\sin 2\alpha}} \geq 2\sqrt{\frac{2}{n+1}}.$

Since $1 + \frac{1}{\sqrt{2}} + \cdots + \frac{1}{\sqrt{n}} = \frac{1}{2}\left(\frac{1}{\sqrt{1}} + \frac{1}{\sqrt{n}}\right) + \cdots + \frac{1}{2}\left(\frac{1}{\sqrt{n}} + \frac{1}{\sqrt{1}}\right)$, we have $1 + \frac{1}{\sqrt{2}} + \cdots + \frac{1}{\sqrt{n}} \geq n\sqrt{\frac{2}{n+1}}$.

5.6. Let $x_k = \tan\alpha_k$, where $-\frac{\pi}{2} < \alpha_k < \frac{\pi}{2}$, $k = 1, \ldots, 7$. Let us divide $\left[-\frac{\pi}{2}, \frac{\pi}{2}\right)$ into six equal parts. According to Dirichlet's principle, at least two among the numbers $\alpha_1, \dots, \alpha_7$ belong to the same interval, and therefore, $0 \le \alpha_i - \alpha_j < \frac{\pi}{6}$. It follows that

$$0 \le \frac{x_i - x_j}{1 + x_i x_j} = \tan\left(\alpha_i - \alpha_j\right) < \tan\frac{\pi}{6} = \frac{\sqrt{3}}{3}.$$

5.7. Define $a = \frac{c}{\cos^2\alpha}$, $b = \frac{c}{\cos^2\beta}$, where $\alpha, \beta \in \left(0, \frac{\pi}{2}\right)$. After several similar transformations, we get the inequality $\sin(\alpha+\beta) \le 1$. The last inequality obviously holds.

5.8. Set $a = \tan\alpha$, $b = \tan\beta$, $c = \tan\gamma$. Then the given inequality can be rewritten as $|\sin(\alpha-\beta)| \le |\sin(\alpha-\gamma)| + |\sin(\beta-\gamma)|$.
That can be proved in the following way:
$|\sin(\alpha-\beta)| = |\sin(\alpha-\gamma)\cos(\gamma-\beta) + \sin(\gamma-\beta)\cos(\alpha-\gamma)| \le$
$\le |\sin(\alpha-\gamma)\cos(\gamma-\beta)| + |\sin(\gamma-\beta)\cos(\alpha-\gamma)| \le |\sin(\alpha-\gamma)| + |\sin(\gamma-\beta)|$, therefore
$|\sin(\alpha-\beta)| \le |\sin(\alpha-\gamma)| + |\sin(\beta-\gamma)|$.

5.9. (a) Let $\sqrt{\frac{a_i}{b_i}} = \tan\alpha_i$, $i = 1, \dots, n$. After several similar transformations, we obtain
$\sqrt[n]{\sin^2\alpha_1 \cdot \dots \cdot \sin^2\alpha_n} + \sqrt[n]{\cos^2\alpha_1 \cdot \dots \cdot \cos^2\alpha_n} \le 1$. From Problem 2.1, it follows that

$$\sqrt[n]{\sin^2\alpha_1 \cdots \sin^2\alpha_n} + \sqrt[n]{\cos^2\alpha_1 \cdots \cos^2\alpha_n} \le \frac{\sin^2\alpha_1 + \dots + \sin^2\alpha_n}{n} + \frac{\cos^2\alpha_1 + \dots + \cos^2\alpha_n}{n} =$$
$$= \frac{1}{n}\sum_{i=1}^{n}\left(\sin^2\alpha_i + \cos^2\alpha_i\right) = 1.$$

(b) Let $\sqrt{\frac{a_i}{b_i}} = \tan\alpha_i$, $i = 1, \dots, n$. Hence, we obtain the inequality
$\frac{b_1^2\tan^2\alpha_1}{b_1\tan^2\alpha_1 + b_1} + \dots + \frac{b_n^2\tan^2\alpha_n}{b_n\tan^2\alpha_n + b_n} \le \frac{(b_1\tan^2\alpha_1 + \dots + b_n\tan^2\alpha_n)(b_1 + \dots + b_n)}{(b_1\tan^2\alpha_1 + \dots + b_n\tan^2\alpha_n) + (b_1 + \dots + b_n)}$, which is equivalent to

$$\sum_{i=1}^{n} b_i\sin^2\alpha_i \le \frac{\left(\sum\limits_{i=1}^{n} b_i\tan^2\alpha_i\right)\left(\sum\limits_{i=1}^{n} b_i\right)}{\sum\limits_{i=1}^{n}\frac{b_i}{\cos^2\alpha_i}}, \text{ or}$$

$$\left(\sum_{i=1}^{n} b_i\sin^2\alpha_i\right)\left(\sum_{i=1}^{n}\frac{b_i}{\cos^2\alpha_i}\right) \le \left(\sum_{i=1}^{n} b_i\tan^2\alpha_i\right)\left(\sum_{i=1}^{n} b_i\right).$$

The factor $b_i b_j\left(\frac{\sin^2\alpha_i}{\cos^2\alpha_j} + \frac{\sin^2\alpha_j}{\cos^2\alpha_i}\right)$ on the left-hand side is not greater than the factor $b_i b_j\left(\tan^2\alpha_i + \tan^2\alpha_j\right)$ on the right-hand side, that is, $\frac{\sin^2\alpha_i}{\cos^2\alpha_j} + \frac{\sin^2\alpha_j}{\cos^2\alpha_i} \le \tan^2\alpha_i + \tan^2\alpha_j$. This holds because we have that
$\tan^2\alpha_i + \tan^2\alpha_j - \left(\frac{\sin^2\alpha_i}{\cos^2\alpha_j} + \frac{\sin^2\alpha_j}{\cos^2\alpha_i}\right) = \frac{\left(\sin^2\alpha_i - \sin^2\alpha_j\right)^2}{\cos^2\alpha_i \cdot \cos^2\alpha_j} \ge 0.$

5.10. Let $z_1 = \sqrt{x_1 y_1} \cdot \sin\alpha$, $z_2 = \sqrt{x_2 y_2} \cdot \sin\beta$. Then the given inequality can be rewritten as

$$A = \frac{8}{x_1 y_1 \cos^2\alpha + x_2 y_2 \cos^2\beta + x_1 y_2 \cos^2\alpha + x_2 y_1 \cos^2\beta + \left(\sqrt{x_1 y_2}\sin\alpha - \sqrt{y_1 x_2}\sin\beta\right)^2} \le$$
$$\le \frac{1}{x_1 y_1 \cos^2\alpha} + \frac{1}{x_2 y_2 \cos^2\beta}.$$

Note that

$$A \le \frac{8}{x_1 y_1 \cos^2\alpha + x_2 y_2 \cos^2\beta + x_1 y_2 \cos^2\alpha + x_2 y_1 \cos^2\beta}. \tag{5.2}$$

On the other hand,
$\left(x_1 y_1 \cos^2\alpha + x_2 y_2 \cos^2\beta + x_1 y_2 \cos^2\alpha + x_2 y_1 \cos^2\beta\right)\left(\frac{1}{x_1 y_1 \cos^2\alpha} + \frac{1}{x_2 y_2 \cos^2\beta}\right) =$
$= 2 + \left(\frac{x_1 y_1 \cos^2\alpha}{x_2 y_2 \cos^2\beta} + \frac{x_2 y_2 \cos^2\beta}{x_1 y_1 \cos^2\alpha}\right) + \left(\frac{y_2}{y_1} + \frac{y_1}{y_2}\right) + \left(\frac{x_2 \cos^2\beta}{x_1 \cos^2\alpha} + \frac{x_1 \cos^2\alpha}{x_2 \cos^2\beta}\right) \ge 8$, and therefore, we obtain

$$\frac{8}{x_1 y_1 \cos^2\alpha + x_2 y_2 \cos^2\beta + x_1 y_2 \cos^2\alpha + x_2 y_1 \cos^2\beta}$$
$$\le \frac{1}{x_1 y_1 \cos^2\alpha} + \frac{1}{x_2 y_2 \cos^2\beta}. \tag{5.3}$$

From (5.2) and (5.3) follows the required inequality.

5.11. Let $a = \frac{1}{\cos^2\alpha}$, $b = \frac{1}{\cos^2\beta}$, $c = \frac{1}{\cos^2\gamma}$, $d = \frac{1}{\cos^2\varphi}$, where $\alpha, \beta, \gamma, \varphi \in \left[0; \frac{\pi}{2}\right)$.

(a) The given inequality can be rewritten as
$\tan\alpha + \tan\beta + \tan\gamma \le \frac{2}{\sqrt{3}\cos\alpha\,\cos\beta\,\cos\gamma}$, or
$\cos\alpha\,\cos\beta\,\cos\gamma\,(\tan\alpha + \tan\beta + \tan\gamma) \le \frac{2}{\sqrt{3}}$.
Since $\cos\alpha\,\cos\beta\,\cos\gamma\,(\tan\alpha + \tan\beta + \tan\gamma) = \sin(\alpha+\beta)\cos\gamma + \cos\alpha\,\cos\beta\,\cos\gamma = B$, it follows from inequality (4.1) that $B \le \sqrt{\left(\sin^2(\alpha+\beta) + \cos^2\alpha\,\cos^2\beta\right)\left(\cos^2\gamma + \sin^2\gamma\right)} = \sqrt{\sin^2(\alpha+\beta) + \cos^2\alpha\,\cos^2\beta}$.
Now let us prove $\sin^2(\alpha+\beta) + \cos^2\alpha\,\cos^2\beta \le \frac{4}{3}$.
Indeed,

$$\sin^2(\alpha+\beta) + \left(\frac{1}{2}(\cos(\alpha-\beta) + \cos(\alpha+\beta))\right)^2 = \sin^2(\alpha+\beta) +$$
$$+\frac{1}{4}\cos^2(\alpha-\beta) + \frac{1}{2}\cos(\alpha-\beta)\cos(\alpha+\beta) + \frac{1}{4}\cos^2(\alpha+\beta) \le \sin^2(\alpha+\beta) + \frac{1}{4}+$$
$$+\frac{1}{2}|\cos(\alpha+\beta)| + \frac{1}{4}\cos^2(\alpha+\beta) = -\frac{3}{4}\cos^2(\alpha+\beta) + \frac{5}{4} + \frac{1}{2}|\cos(\alpha+\beta)| =$$
$$= -\frac{3}{4}\left(|\cos(\alpha+\beta)| - \frac{1}{3}\right)^2 + \frac{4}{3} \le \frac{4}{3}$$

(b) The given inequality can be rewritten as $M = \cos\alpha\,\cos\beta\,\cos\gamma\,\cos\varphi\,(\tan\alpha+\tan\beta+\tan\gamma+\tan\varphi)\le\frac{3\sqrt{3}}{4}$.
One can easily prove that
$M=\sin(\alpha+\beta)\cos\gamma\,\cos\varphi+\sin(\gamma+\varphi)\cos\alpha\,\cos\beta$.
According to inequality (3.2), we have that $M\le\sin(\alpha+\beta)\left(\frac{\cos\gamma+\cos\varphi}{2}\right)^2+\sin(\gamma+\varphi)\left(\frac{\cos\alpha+\cos\beta}{2}\right)^2$.
Since $\frac{\cos x+\cos y}{2}=\cos\frac{x+y}{2}\cos\frac{x-y}{2}\le\cos\frac{x+y}{2}$, where $x,y\in\left[0;\frac{\pi}{2}\right]$, we must have
$M\le\sin(\alpha+\beta)\cos^2\frac{\gamma+\varphi}{2}+\sin(\gamma+\varphi)\cos^2\frac{\alpha+\beta}{2}=$
$=2\cos\frac{\alpha+\beta}{2}\cos\frac{\gamma+\varphi}{2}\left(\sin\frac{\alpha+\beta}{2}\cos\frac{\gamma+\varphi}{2}+\sin\frac{\gamma+\varphi}{2}\cos\frac{\alpha+\beta}{2}\right)=$
$=2\cos\frac{\alpha+\beta}{2}\cos\frac{\gamma+\varphi}{2}\sin\frac{\alpha+\beta+\gamma+\varphi}{2}\le2\cos^2\frac{\alpha+\beta+\gamma+\varphi}{4}\sin\frac{\alpha+\beta+\gamma+\varphi}{2}=$
$=4\cos^3\frac{\alpha+\beta+\gamma+\varphi}{4}\sin\frac{\alpha+\beta+\gamma+\varphi}{4}$.
Since $\cos^3 t\,\sin t\le\frac{3\sqrt{3}}{16}$ (see Problem 3.28), we have $M\le\frac{3\sqrt{3}}{4}$.

5.12. Let $\frac{a_1+\ldots+a_n}{n}=a$, $a_i=a+x_i$, $i=1,\ldots,n$. Then from the condition $a+x_i>0$, it follows that $\frac{x_i}{a}>-1$. Hence, according to Bernoulli's inequality, we have $\left(1+\frac{x_i}{a}\right)^k\ge1+k\frac{x_i}{a}$, $i=1,\ldots,n$, and therefore, $(a+x_i)^k\ge a^k+ka^{k-1}x_i$.
Summing these inequalities, we obtain
$(a+x_1)^k+\cdots+(a+x_n)^k\ge na^k+ka^{k-1}(x_1+\ldots+x_n)=na^k$, since $x_1+\ldots+x_n=a_1+\ldots+a_n-na=0$. Thus, we deduce that $\frac{a_1^k+\ldots+a_n^k}{n}\ge\left(\frac{a_1+\ldots+a_n}{n}\right)^k$.

5.13. Proof by contradiction. Assume that for all $x\in[a,a+4]$,
$f'(x)\ge1+f^2(x)$.
Let $\arctan f(x)=g(x)$. Then $-\frac{\pi}{2}<g(x)<\frac{\pi}{2}$ and $f(x)=\tan(g(x))$.
It follows that $f'(x)=\frac{g'(x)}{\cos^2 g(x)}\ge1+f^2(x)=\frac{1}{\cos^2 g(x)}$, and hence $g'(x)\ge1$.
Therefore, $g(x)-x$ is a nondecreasing function.
Hence,

$$g(b)-b\ge g(a)-a,\text{ where }b=a+4\text{ or }g(b)-g(a)\ge b-a. \quad (5.4)$$

Since

$$g(b),g(a)\in\left(-\frac{\pi}{2};\frac{\pi}{2}\right),\text{ we have }g(b)-g(a)<\pi. \quad (5.5)$$

From (5.4)–(5.5) and the condition $b-a=4$ it follows that $\pi>4$; this leads to a contradiction.
Therefore, there exists x_0 such that $x_0\in[a,a+4]$ and $f'(x_0)<1+f^2(x_0)$.

5.14. Without loss of generality one can assume that $a=0$ (otherwise, consider the function $f(x-a)$). Then we need to prove that $b>\pi$.
We proceed with a proof by contradiction argument. Assume that $b\le\pi$.
Consider the function $F(x)=-\frac{\pi}{b}f(x)\cos\left(\frac{\pi}{b}x\right)+f'(x)\sin\left(\frac{\pi}{b}x\right)$ in $[0,b]$.
Note that $F(0)=F(b)=0$; Hence by Rolle's theorem, there exists $c\in(0,b)$ such that $F'(c)=0$.

On the other hand, we have
$F'(c) = \frac{\pi^2}{b^2} f(c) \sin\left(\frac{\pi}{b}c\right) + f''(c) \sin\left(\frac{\pi}{b}c\right) =$
$= \left(\frac{\pi^2}{b^2} f(c) + f''(c)\right) \sin\left(\frac{\pi}{b}c\right) \geq \left(f(c) + f''(c)\right) \sin\left(\frac{\pi}{b}c\right) > 0.$
It follows that $F'(c) > 0$, which leads to a contradiction with $F'(c) = 0$.
Therefore, $b > \pi$.

5.15. Without loss of generality one can assume that $a \geq b$. Let $a = b \tan\alpha$, where $\frac{\pi}{4} \leq \alpha < \frac{\pi}{2}$. Then the given inequality can be rewritten as
$b^4 \frac{\cos^2 2\alpha}{4\cos^4\alpha} \geq b \frac{1}{\sqrt{2}\cos\alpha} - b \frac{\sin\alpha+\cos\alpha}{2\cos\alpha}$, or
$b^3 \cos^2 2\alpha \geq 2\cos^3\alpha \left(\sqrt{2} - \sin\alpha - \cos\alpha\right).$
In order to finish the proof, it is sufficient to prove that for α in $\left[\frac{\pi}{4}, \frac{\pi}{2}\right)$, one has $\frac{1}{8}\cos^2 2\alpha \geq 2\cos^3\alpha\left(\sqrt{2} - \sin\alpha - \cos\alpha\right)$, which is equivalent to the following inequality: $\frac{1}{2}\cos^2(\alpha + \pi/4)\sin^2(\alpha + \pi/4) \geq 2\sqrt{2}\cos^3\alpha\,(1 - \sin(\alpha + \pi/4))$, or
$\frac{1}{2}\left(1 - \sin^2(\alpha + \pi/4)\right)\sin^2(\alpha + \pi/4) \geq 2\sqrt{2}\cos^3\alpha\,(1 - \sin(\alpha + \pi/4)).$
This inequality holds because for the difference between its left- and right-hand sides we have $\frac{1}{2}(1 - \sin(\alpha + \pi/4))\left(\sin^2(\alpha + \pi/4) + \sin^3(\alpha + \pi/4) - 4\sqrt{2}\cos^3\alpha\right) \geq 0$, since $1 \geq \sin\left(\alpha + \frac{\pi}{4}\right)$ and $\sin^2\left(\alpha + \frac{\pi}{4}\right) \geq 2\cos^2\alpha$, $1 + \sin\left(\alpha + \frac{\pi}{4}\right) \geq 2\sqrt{2} \cdot \cos\alpha \left(\frac{\pi}{4} \leq \alpha < \frac{\pi}{2}\right)$.

5.16. (a) Let $x_i = \sin\alpha_i$, $i = 1, \ldots, n$. We have
$\sum_{i=1}^{n}\left(3\sin\alpha_i - 4\sin^3\alpha_i\right) = \sum_{i=1}^{n}\sin 3\alpha_i$, and therefore,
$\sum_{i=1}^{n} x_i = \sum_{i=1}^{n}\sin\alpha_i = \frac{1}{3}\sum_{i=1}^{n}\sin 3\alpha_i \leq \frac{n}{3}.$

(b) Let $x_i = 2\sin\alpha_i, i = 1, 2, \ldots, n$. We have $\sum_{i=1}^{n}(3\sin\alpha_i - 4\sin^3\alpha_i) =$
$\sum_{i=1}^{n}\sin 3\alpha_i$, and therefore,
$$\left|x_1^3 + x_2^3 + \ldots + x_n^3\right| = 2\left|\sum_{i=1}^{n}\sin 3\alpha_i\right| \leq 2\sum_{i=1}^{n}|\sin 3\alpha_i| \leq 2n.$$
It follows that $\left|x_1^3 + x_2^3 + \ldots + x_n^3\right| \leq 2n$.

5.17. Without loss of generality one can assume that $x, y, z \geq 0$. Indeed, note that $\frac{x}{1+x^2} + \frac{y}{1+y^2} + \frac{z}{1+z^2} \leq \frac{|x|}{1+|x|^2} + \frac{|y|}{1+|y|^2} + \frac{|z|}{1+|z|^2}$ and $|x|^2+|y|^2+|z|^2= 1$.
Let $x = \tan\alpha$, $y = \tan\beta$, $z = \tan\gamma$, where $\alpha, \beta, \gamma \in \left[0, \frac{\pi}{4}\right]$.
We have $xy + yz + xz \leq x^2 + y^2 + z^2$ (Problem 4.13), whence $\tan\alpha\,\tan\beta + \tan\beta\,\tan\gamma + \tan\gamma\,\tan\alpha \leq 1$.
Hence, $\tan\gamma \cdot \frac{\tan\alpha+\tan\beta}{1-\tan\alpha\,\tan\beta} \leq 1$, or $\tan(\alpha + \beta) \leq \tan\left(\frac{\pi}{2} - \gamma\right)$.

It follows that $\alpha+\beta \leq \frac{\pi}{2}-\gamma$, that is, $\alpha+\beta+\gamma \leq \frac{\pi}{2}$. Note that $\frac{x}{1+x^2}+\frac{y}{1+y^2}+\frac{z}{1+z^2}$ $= \frac{1}{2}(\sin 2\alpha + \sin 2\beta + \sin 2\gamma) \leq \frac{1}{2}\cdot 3 \sin\frac{2\alpha+2\beta+2\gamma}{3} \leq \frac{3}{2}\sin\frac{\pi}{3} = \frac{3\sqrt{3}}{4}$.

5.18. Note that
$\frac{a}{\sqrt{a^2+b^2}}+\frac{b}{\sqrt{b^2+c^2}}+\frac{c}{\sqrt{c^2+a^2}} > \frac{a}{a+b}+\frac{b}{b+c}+\frac{c}{c+a} > \frac{a}{a+b+c}+\frac{b}{a+b+c}+\frac{c}{a+b+c} = 1$,
and therefore, $\frac{a}{\sqrt{a^2+b^2}}+\frac{b}{\sqrt{b^2+c^2}}+\frac{c}{\sqrt{c^2+a^2}} > 1$.
Set $a^2+b^2=m^2$, $b^2+c^2=n^2$, $c^2+a^2=k^2$, where $m, n, k > 0$.
We have $(m+n)^2 > m^2+n^2 > k^2$, whence $m+n > k$, and in a similar way, we obtain $n+k > m$, $m+k > n$.
Let an acute triangle with side lengths m, n, k have angles α, β, γ. Then using the law of sines and the law of cosines, this inequality can be rewritten as
$\sin\alpha\sqrt{\sin 2\gamma}+\sin\beta\sqrt{\sin 2\alpha}+\sin\gamma\sqrt{\sin 2\beta} \leq 3\sqrt{\sin\alpha\ \sin\beta\ \sin\gamma}$.
Therefore, from inequality (4.2), it follows that

$$\begin{aligned}
&\sin\alpha\sqrt{\sin 2\gamma}+\sin\beta\sqrt{\sin 2\alpha}+\sin\gamma\sqrt{\sin 2\beta}\\
&\quad\leq \sqrt{(\sin^2\alpha+\sin^2\beta+\sin^2\gamma)(\sin 2\alpha+\sin 2\beta+\sin 2\gamma)}\\
&\quad= \sqrt{4\sin\alpha\ \sin\beta\ \sin\gamma(2+0,25\ \cos^2(\alpha-\beta)-(\cos\gamma-0,5\ \cos(\alpha-\beta))^2}\\
&\quad\leq 3\sqrt{\sin\alpha\ \sin\beta\ \sin\gamma}.
\end{aligned}$$

It follows that
$\sin\alpha\sqrt{\sin 2\gamma}+\sin\beta\sqrt{\sin 2\alpha}+\sin\gamma\sqrt{\sin 2\beta} \leq 3\sqrt{\sin\alpha\ \sin\beta\ \sin\gamma}$.

5.19. Let us prove that for an arbitrary number a, where $0 \leq a \leq 1$, the following inequality holds:

$$\sqrt{1-a} \geq \sqrt{a}+\frac{\sqrt{2}}{2}(1-4a^2).$$

Let $a=\sin^2\alpha$, where $\alpha \in \left[0, \frac{\pi}{2}\right]$. Then the given inequality can be rewritten as

$$\cos\alpha-\sin\alpha \geq \frac{\sqrt{2}}{4}\cos 2\alpha(2-\cos 2\alpha). \tag{1}$$

If $0 \leq \alpha \leq \frac{\pi}{4}$, then $\cos\alpha-\sin\alpha \geq 0$, $\cos\alpha+\sin\alpha \leq \sqrt{2}$, and $2-\cos 2\alpha \leq 2$. Therefore, inequality (1) holds.
If $\frac{\pi}{4} \leq \alpha \leq \frac{\pi}{2}$, then on setting $\alpha = \frac{\pi}{4}+\beta$, the inequality (1) can be rewritten as $\cos^2\beta\ \cos\frac{\beta}{2} \geq \sin\frac{\beta}{2}$. In order to prove this inequality it is sufficient to prove that
$\cos^2\beta(1+\cos\beta) \geq \sin\beta$. Since $0 \leq \beta \leq \frac{\pi}{4}$, we have $\cos^2\beta(1+\cos\beta) \geq \frac{1}{2}(1+\frac{\sqrt{2}}{2}) > \frac{\sqrt{2}}{2} \geq \sin\beta$.
It follows that $\cos^2\beta(1+\cos\beta) > \sin\beta$. We have
$\sqrt{1-a}+\sqrt{1-b}+\sqrt{1-c}+\sqrt{1-d} \geq \sqrt{a}+\sqrt{b}+\sqrt{c}+\sqrt{d}+\frac{\sqrt{2}}{2}(1-4a^2+1-4b^2+1-4c^2+1-4d^2)$,

whence

$$\sqrt{1-a}+\sqrt{1-b}+\sqrt{1-c}+\sqrt{1-d} \geq \sqrt{a}+\sqrt{b}+\sqrt{c}+\sqrt{d}.$$

5.20. Without loss of generality one can assume that $a \geq b \geq c$.
Let $a = x^3,\ b = y^3,\ c = z^3$. Therefore, $x \geq y \geq z > 0$.
Thus, the given inequality can be rewritten as $\frac{x^3+y^3+z^3}{3} - xyz \leq \left(\sqrt{x^3}-\sqrt{z^3}\right)^2$.
We have

$$\frac{x^3+y^3+z^3}{3} - xyz = \frac{1}{6}(x+y+z)\left((x-z)^2+(x-y)^2+(y-z)^2\right) \leq$$
$$\leq \frac{x}{2}\left((x-z)^2+(x-y+y-z)^2\right) = (\sqrt{x}-\sqrt{z})^2\,(x+\sqrt{xz})^2 \leq (\sqrt{x}-\sqrt{z})^2\,(x+\sqrt{xz}+z)^2 =$$
$$= \left(\sqrt{x^3}-\sqrt{z^3}\right)^2, \text{ whence } \frac{x^3+y^3+z^3}{3} - xyz \leq \left(\sqrt{x^3}-\sqrt{z^3}\right)^2.$$

This ends the proof.

Problems for Independent Study

Prove the following inequalities (2–27).

1. Given that $a^2+b^2=1$. Prove that
(a) $|a+b| \leq \sqrt{2}$, (b) $|a-b| \leq \sqrt{2}$, (c) $|ab| \leq \frac{1}{2}$, (d) $|a^2b+ab^2| \leq \frac{\sqrt{2}}{2}$.
2. $\left|xy-\sqrt{(1-x^2)(1-y^2)}\right| \leq 1$, where $|x| \leq 1,\ \ |y| \leq 1$.
3. $\sqrt{1-x^2}+\sqrt{1-y^2} \leq 2\sqrt{1-\left(\frac{x+y}{2}\right)^2}$, where $|x| \leq 1,\ \ |y| \leq 1$.
4. $\left(x+\frac{1}{x}\right)\operatorname{arccot} x > 1$, where $x > 0$.
5. $\frac{2}{\cos\alpha+\cos\beta} - 1 \leq \sqrt{\left(\frac{1}{\cos\alpha}-1\right)\left(\frac{1}{\cos\beta}-1\right)}$, where $\frac{\pi}{3} \leq \alpha < \frac{\pi}{2},\ \ \frac{\pi}{3} \leq \beta < \frac{\pi}{2}$.
6. $\frac{a_1}{1-a_1}+\cdots+\frac{a_n}{1-a_n} \geq \frac{n(a_1+\cdots+a_n)}{n-(a_1+\cdots+a_n)}$, where $0 \leq a_1 < 1, \ldots, 0 \leq a_n < 1$.
7. $\frac{1}{\sqrt{1+a^2}}+\frac{1}{\sqrt{1+b^2}}+\frac{1}{\sqrt{1+c^2}} \leq \frac{3}{2}$, where $a, b, c > 0$ and $a+b+c = abc$.
8. $\frac{|x-y|}{1+a|x-y|}+\frac{|y-z|}{1+a|y-z|} \geq \frac{|x-z|}{1+a|x-z|}$, where $a > 0$.
9. $\frac{2x(1-x^2)}{(1+x^2)^2}+\frac{2y(1-y^2)}{(1+y^2)^2}+\frac{2z(1-z^2)}{(1+z^2)^2} \leq \frac{x}{1+x^2}+\frac{y}{1+y^2}+\frac{z}{1+z^2}$, where $x > 0,\ \ y > 0,\ \ z > 0$ and $xy+yz+zx = 1$.
10. $\sqrt{a_1+a_2+\cdots+a_n} \leq \sqrt{1}\left(\sqrt{a_1}-\sqrt{a_2}\right)+\sqrt{2}\left(\sqrt{a_2}-\sqrt{a_3}\right)+\cdots+\sqrt{n}\left(\sqrt{a_n}-\sqrt{a_{n+1}}\right)$, where $a_1 \geq \ldots \geq a_n \geq a_{n+1} = 0$.
11. $\frac{1}{\frac{1}{1+a_1}+\cdots+\frac{1}{1+a_n}} - \frac{1}{\frac{1}{a_1}+\cdots+\frac{1}{a_n}} \geq \frac{1}{n}$, where $a_1 > 0, \ldots, a_n > 0$.
12. $a+b+c-2\sqrt{abc} \geq ab+bc+ca-2abc$, where $0 \leq a \leq 1,\ \ 0 \leq b \leq 1,\ \ 0 \leq c \leq 1$.
13. $\sqrt{a(1-b)(1-c)}+\sqrt{b(1-a)(1-c)}+\sqrt{c(1-a)(1-b)} \leq 1+\sqrt{abc}$, where $0 \leq a \leq 1, 0 \leq b \leq 1,\ \ 0 \leq c \leq 1$.

14. $((x+y)(y+z)(x+z))^2 \geq xyz(2x+y+z)(2y+z+x)(2z+x+y)$, where $x, y, z \geq 0$.
Hint. If $x^2+y^2+z^2 \neq 0$, then without loss of generality one can assume that $x+y+z=1$.
Let $x = \tan\alpha \tan\beta$, $y = \tan\beta \tan\gamma$, $z = \tan\alpha \tan\gamma$, where $\alpha+\beta+\gamma = \frac{\pi}{2}$.
Then one needs to prove that $((1-x)(1-y)(1-z))^2 \geq xyz(1+x)(1+y)(1+z)$.
15. $\frac{ab(1-a)(1-b)}{(1-ab)^2} < \frac{1}{4}$, where $0 < a < 1, 0 < b < 1$.
Hint. Let $a = \sin^2\alpha$, $b = \sin^2\beta$, where $\alpha, \beta \in \left(0, \frac{\pi}{2}\right)$.
16. $\max(a_1, \ldots, a_n) \geq 2$, where $n > 3$, $a_1 + \cdots + a_n \geq n$, $a_1^2 + \cdots + a_n^2 \geq n^2$.
Hint. Let $a_i = 2 - b_i$ and $b_i > 0, i = 1, \ldots, n$. We have that $b_1 + \cdots + b_n \leq n$, $b_1^2 + \cdots + b_n^2 - 4(b_1 + \cdots + b_n) \geq n(n-4)$, and therefore, $(b-4)(b_1 + \cdots + b_n) \geq n(n-4)$, where $b = \max(b_1, \ldots, b_n)$.
17. $\sqrt{a_1 + \frac{(a_n - a_{n-1})^2}{4(n-2)}} + \cdots + \sqrt{a_{n-2} + \frac{(a_n - a_{n-1})^2}{4(n-2)}} + \sqrt{a_{n-1}} + \sqrt{a_n} \leq \sqrt{n}$,
where $n \geq 3$, $a_1, \ldots, a_n \geq 0$, and $a_1 + a_2 + \cdots + a_n = 1$.
Hint. $\sqrt{a_{n-1}} = x + t$, $\sqrt{a_n} = x - t$. Then $x \leq \frac{1}{\sqrt{2}}$, and the left-hand side of the inequality is not greater than $\sqrt{(n-2)(1-2x^2)} + 2x$.
18. $2\sqrt{(x^2-1)(y^2-1)} \leq 2(x-1)(y-1) + 1$, where $0 \leq x, y \leq 1$.
19. $a^3 + b^3 + c^3 - 3abc \leq \sqrt{(a^2+b^2+c^2)^3}$.
20. $\frac{1}{n-1+x_1} + \cdots + \frac{1}{n-1+x_n} \leq 1$, where $x_1, \ldots, x_n > 0$ and $x_1 \cdot \cdots \cdot x_n = 1$.
Hint. Let $x_i = y_i^n, i = 1, \ldots, n$, where $y_i > 0$. Then $y_1 \cdot \cdots \cdot y_n = 1$ and $n - 1 + x_i = n - 1 + \frac{y_i^n}{y_1 \cdot \cdots \cdot y_n} \geq n - 1 + \frac{(n-1)y_i^{n-1}}{y_1^{n-1} + \cdots + y_n^{n-1} - y_i^{n-1}}$.
21. $\frac{x}{\sqrt{1-x}} + \frac{y}{\sqrt{1-y}} \geq \frac{x+y}{\sqrt{1-\frac{x+y}{2}}}$, where $0 \leq x < 1, 0 \leq y < 1$.
22. $\frac{x_1}{\sqrt{1-x_1}} + \cdots + \frac{x_n}{\sqrt{1-x_n}} \geq \frac{\sqrt{x_1} + \cdots + \sqrt{x_n}}{\sqrt{n-1}}$, where $n \geq 2, n \in \mathbb{N}, x_1, \ldots, x_n > 0$ and $x_1 + \ldots + x_n = 1$.
Hint. Take $1 - x_i = a_i, i = 1, \ldots, n$.
23. $\frac{x}{\sqrt{4y^2+1}} + \frac{y}{\sqrt{4x^2+1}} \leq \frac{\sqrt{2}}{2}$, where $0 \leq x, y \leq 0.5$.
Hint. Let $2x = \tan\alpha$, $2y = \tan\beta$, where $0 \leq \alpha, \beta \leq \frac{\pi}{4}$.
24. $0 \leq ab+bc+ca-abc \leq 2$, where $a > 0, b > 0, c > 0$ and $a^2+b^2+c^2+abc = 4$.
25. $a + b + c \leq 3$, where $a > 0, b > 0, c > 0$, and $a^2 + b^2 + c^2 + abc = 4$.
26. $(x-1)(y-1)(z-1) \leq 6\sqrt{3} - 10$, where $x > 0$, $y > 0$, $z > 0$, and $x + y + z = xyz$.
Hint. Let $x = \tan\alpha$, $y = \tan\beta$, $z = \tan\gamma$, where $\alpha, \beta, \gamma \in \left(0, \frac{\pi}{2}\right)$. Then $\alpha + \beta + \gamma = \pi$ and $\beta > \frac{\pi}{4}$. Note that $(\tan\alpha - 1)(\tan\gamma - 1) = 2 + \frac{\cos\beta - \sin\beta}{\cos\alpha \cos\gamma} \leq 2 + \frac{\cos\beta - \sin\beta}{\cos\frac{\pi}{3}\cos\left(\alpha+\gamma-\frac{\pi}{3}\right)}$.

27. $\sqrt[3]{\frac{x+y}{2z}}+\sqrt[3]{\frac{y+z}{2x}}+\sqrt[3]{\frac{z+x}{2y}} \leq \frac{5(x+y+z)+9}{8}$, where $x>0,\ y>0,\ z>0$, and $xyz=1$.

    Hint. $\sqrt[3]{\frac{1}{2}a^3b^3\left(a^3+b^3\right)} \leq \sqrt[3]{\frac{1}{2}(a+b)\left(\frac{ab+ab+ab+a^2-ab+b^2}{4}\right)^4} = \frac{(a+b)^3}{8}$,
    where $a>0,\ b>0$.
28. Prove that among four arbitrary numbers there are two numbers a and b such that $\frac{1+ab}{\sqrt{1+a^2}\cdot\sqrt{1+b^2}} > \frac{1}{2}$.
29. Given that $x+y+z=0$ and $x^2+y^2+z^2=6$. Find all possible values of the expression $x^2y+y^2z+z^2x$.
30. Let (h_n) be a sequence such that $h_1=\frac{1}{2}$ and $h_{n+1}=\sqrt{\frac{1-\sqrt{1-h_n^2}}{2}}$, $n=1,2,\ldots$. Prove that $h_1+\cdots+h_n \leq 1.03$.

Chapter 6
Using Symmetry and Homogeneity

In order to prove certain inequalities, one often needs to use the symmetry and homogeneity of mathematical expressions.

Definition 1 A mathematical expression is called *symmetric with respect to the set of variables* $x_1, x_2, \ldots, x_n$, $n \in \mathbb{N}$, if its value remains unchanged under every permutation of variables $x_1, x_2, \ldots, x_n$.

Remark For purposes of simplicity and brevity, if a mathematical expression is symmetric with respect to the set of variables $x_1, x_2, \ldots, x_n$, then hereinafter we shall call it *symmetric*.

Definition 2 A mathematical expression is called *homogeneous of degree k with respect to the set of variables* $x_1, x_2, \ldots, x_n$, $n \in \mathbb{N}$ if for every positive λ, on replacing the variables $x_1, x_2, \ldots, x_n$ by the variables $\lambda x_1, \lambda x_2, \ldots, \lambda x_n$, the value of the given mathematical expression is multiplied by λ^k.

Remark For purposes of simplicity and brevity, if a mathematical expression is homogeneous of degree k with respect to the set of variables $x_1, x_2, \ldots, x_n$, then hereinafter we shall call it *homogeneous*.

The symmetry and homogeneity of mathematical expressions allow us to deduce additional conditions for variables. For example, if a mathematical expression is symmetric, then an additional condition for variables can be assuming without loss of generality that $x_1 \leq x_2 \leq \cdots \leq x_n$, and if it is homogeneous, then such a condition can be assuming without loss of generality that $x_1 \cdot x_2 \cdots x_n = 1$ or $x_1^k + x_2^k + \cdots + x_n^k = 1$, where $k \in \mathbb{N}$.

In order to demonstrate how these additional conditions can be obtained, let us consider the following examples of symmetric or/and homogeneous mathematical expressions.

H. Sedrakyan and N. Sedrakyan, *Algebraic Inequalities*, Problem Books in Mathematics, https://doi.org/10.1007/978-3-319-77836-5_6

Example 6.1 Prove that

$$x(x-z)^2 + y(y-z)^2 \geq (x-z)(y-z)(x+y-z), \tag{6.1}$$

where $x \geq 0,\ y \geq 0,\ z \geq 0.$

Proof Note that (6.1) can be rewritten as
$x^3 + y^3 + z^3 - x^2y - x^2z - y^2x - y^2z - z^2x - z^2y + 3xyz \geq 0$, whence we obtain a symmetric inequality with respect to the variables $x,\ y,\ z$. Therefore, without loss of generality, one can assume that $x \geq z \geq y$. It follows that

$$x(x-z)^2 + y(y-z)^2 \geq 0 \geq (x-z)(y-z)(x+y-z), \text{ whence}$$
$$x(x-z)^2 + y(y-z)^2 \geq (x-z)(y-z)(x+y-z).$$

This ends the proof.

Example 6.2 Hölder's inequality (particular case):
Prove that $(a_1b_1c_1 + \cdots + a_nb_nc_n)^3 \leq \left(a_1^3 + \cdots + a_n^3\right)\left(b_1^3 + \cdots + b_n^3\right)\left(c_1^3 + \cdots + c_n^3\right)$, where $a_i > 0, b_i > 0, c_i > 0, \quad i = 1, \ldots, n$.

Proof Note (that if we substitute the variables $a_1, \ldots, a_n$ by the variables $\lambda a_1, \ldots, \lambda a_n$, where λ is an arbitrary positive number, then we obtain an equivalent inequality.

Therefore, we can choose λ such that $(\lambda a_1)^3 + \cdots + (\lambda a_n)^3 = 1$.

Without loss of generality one can assume that $a_1^3 + \cdots + a_n^3 = b_1^3 + \cdots + b_n^3 = c_1^3 + \cdots + c_n^3 = 1$. Hence in order to complete the proof, it is sufficient to prove that $a_1b_1c_1 + \cdots + a_nb_nc_n \leq 1$.

We have $a_ib_ic_i \leq \frac{a_i^3+b_i^3+c_i^3}{3}$, $i = 1, \ldots, n$ (Problem 2.1).

Summing these inequalities, we obtain $a_1b_1c_1 + \cdots + a_nb_nc_n \leq 1$.

This ends the proof.

Example 6.3 Prove that $\sqrt{\frac{a}{b+c}} + \sqrt{\frac{b}{a+c}} + \sqrt{\frac{c}{a+b}} > 2$, where $a > 0,\ b > 0,\ c > 0$.

Proof Without loss of generality one can assume that $a + b + c = 1$.
Note that

$$\sqrt{\frac{a}{1-a}} \geq 2a, \tag{1}$$

and the equality in (1) holds if $a = \frac{1}{2}$. It follows that

$$\sqrt{\frac{a}{b+c}} + \sqrt{\frac{b}{a+c}} + \sqrt{\frac{c}{a+b}} = \sqrt{\frac{a}{1-a}} + \sqrt{\frac{b}{1-b}} + \sqrt{\frac{c}{1-c}} > 2a + 2b + 2c = 2.$$

This ends the proof.

Problems

Prove the following inequalities.

6.1. $abc \geq (a+b-c)(b+c-a)(a+c-b)$, where $a > 0,\ b > 0,\ c > 0$.

6.2. $(a_1b_1 + \cdots + a_nb_n)^2 \leq \left(a_1^2 + \cdots + a_n^2\right)\left(b_1^2 + \cdots + b_n^2\right)$.

6.3. $(a+b)^2 \cdot \ldots \cdot (a^n + b^n)^2 \geq \left(a^{n+1} + b^{n+1}\right)^n$, where $a > 0,\ b > 0$.

6.4. $\left(a_1^\alpha + \cdots + a_n^\alpha\right)^\beta \leq \left(a_1^\beta + \cdots + a_n^\beta\right)^\alpha$, where $0 < \beta < \alpha,\ \ a_1 > 0, \ldots, a_n > 0$.

6.5. *Nesbitt's inequality*: $\frac{a}{b+c} + \frac{b}{c+a} + \frac{c}{a+b} \geq \frac{3}{2}$, where $a > 0,\ b > 0,\ c > 0$.

6.6. $\sqrt{\frac{a}{b+c+d}} + \sqrt{\frac{b}{a+c+d}} + \sqrt{\frac{c}{a+b+d}} + \sqrt{\frac{d}{a+b+c}} > 2$, where $a > 0,\ b > 0,\ c > 0,\ d > 0$.

6.7. $\sqrt[3]{\frac{abc+abd+acd+bcd}{4}} \leq \sqrt{\frac{ab+ac+ad+bc+bd+cd}{6}}$, where $a > 0,\ b > 0,\ c > 0,\ d > 0$.

6.8. $2\sqrt{ab+bc+ac} \leq \sqrt{3}\sqrt[3]{(b+c)(c+a)(a+b)}$, where $a > 0,\ b > 0,\ c > 0$.

6.9. $8\left(x^3 + y^3 + z^3\right)^2 \geq 9\left(x^2 + yz\right)\left(y^2 + xz\right)\left(z^2 + xy\right)$, where $x > 0,\ y > 0,\ z > 0$.

6.10. (a) $4a^3 + 4b^3 + 4c^3 + 15abc \geq 1$, where $a \geq 0,\ b \geq 0,\ c \geq 0$ and $a+b+c = 1$,

(b) $a^3 + b^3 + c^3 + abcd \geq \min\left(\frac{1}{4}, \frac{1}{9} + \frac{d}{27}\right)$, where $a \geq 0,\ b \geq 0,\ c \geq 0$ and $a+b+c = 1$.

6.11. $\frac{a_1+\cdots+a_n}{n} \geq \frac{1}{n}\sqrt{\frac{a_1^2+\cdots+a_n^2}{n}} + \left(1 - \frac{1}{n}\right)\sqrt[n]{a_1 \cdots a_n}$, where $n \geq 2,\ \ a_i > 0,\ \ i = 1, \ldots, n$.

6.12. *Turkevici's inequality*:

$$a^4 + b^4 + c^4 + d^4 + 2abcd \geq a^2b^2 + a^2c^2 + a^2d^2 + b^2c^2 + b^2d^2 + c^2d^2,$$

where $a \geq 0,\ b \geq 0,\ c \geq 0, d \geq 0$.

6.13. $\frac{a_1^3}{b_1} + \cdots + \frac{a_n^3}{b_n} \geq 1$, where $a_i > 0, b_i > 0,\ \ i = 1, \ldots, n$, and $(a_1^2 + \cdots + a_n^2)^3 = b_1^2 + \cdots + b_n^2$.

6.14. $\frac{a}{b} + \frac{b}{c} + \frac{c}{a} \geq \frac{a+c}{b+c} + \frac{b+a}{c+a} + \frac{c+b}{a+b}$, where $a > 0,\ b > 0,\ c > 0$.

6.15. $\sqrt{\frac{a_1^n}{a_1^n+\lambda a_1\cdots a_n}} + \cdots + \sqrt{\frac{a_n^n}{a_n^n+\lambda a_1\cdots a_n}} \geq \frac{n}{\sqrt{1+\lambda}}$, where $n \geq 2,\ \ a_1 > 0, \ldots, a_n > 0$, and $\lambda \geq n^2 - 1$.

6.16. $(\sqrt[k]{2} - 1)(a_1 + \cdots + a_n) < \sqrt[k]{2a_1^k + \cdots + 2^n a_n^k}$, where $k \in \mathbb{N},\ \ k \geq 2, a_1 > 0, \ldots, a_n > 0$.

6.17. $3(x^2y + y^2z + z^2x)(xy^2 + yz^2 + zx^2) \geq xyz(x+y+z)^3$, where $x > 0, y > 0, z > 0$.

6.18. $(x_1 + \cdots + x_n + y_1 + \cdots + y_n)^2 \geq 4n(x_1y_1 + \cdots + x_ny_n)$, where $x_1 \leq \cdots \leq x_n \leq y_1 \leq \cdots \leq y_n$.

Proofs

6.1. Without loss of generality one can assume that $a \le b \le c$. Note that the given inequality is equivalent to the following inequality: $(b-c)^2(b+c-a) \ge a(a-b)(c-a)$.
The last inequality can be proved in the following way: $(b-c)^2(b+c-a) \ge 0 \ge a(a-b)(c-a)$.

6.2. If $a_1^2 + \cdots + a_n^2 \neq 0$ and $b_1^2 + \cdots + b_n^2 \neq 0$, then without loss of generality one can assume that $a_1^2 + \cdots + a_n^2 = 1, \quad b_1^2 + \cdots + b_n^2 = 1$.
We have that $-\frac{a_i^2+b_i^2}{2} \le a_i b_i \le \frac{a_i^2+b_i^2}{2}, \quad i = 1, \ldots, n$.
Summing these inequalities we obtain
$-1 \le a_1 b_1 + \cdots + a_n b_n \le 1$; hence

$$(a_1 b_1 + \cdots + a_n b_n)^2 \le 1 = \left(a_1^2 + \cdots + a_n^2\right)\left(b_1^2 + \cdots + b_n^2\right).$$

If $a_1^2 + \cdots + a_n^2 = 0$ or $b_1^2 + \cdots + b_n^2 = 0$, then the proof is obvious.

6.3. Without loss of generality one can assume that $a^{n+1} + b^{n+1} = 1$.
Therefore, $0 < a < 1, \quad 0 < b < 1$ and $a^i + b^i \ge a^{n+1} + b^{n+1} = 1, \quad i = 1, \ldots, n$.
Thus, it follows that $(a+b)^2 \cdots (a^n + b^n)^2 \ge 1 = \left(a^{n+1} + b^{n+1}\right)^n$.

6.4. Without loss of generality one can assume that $a_1^\beta + \cdots + a_n^\beta = 1$, and therefore, $0 < a_i \le 1, \quad i = 1, \ldots, n$. Since $\alpha > \beta$, we obtain that $a_i^\alpha \le a_i^\beta, \quad i = 1, \ldots, n$.
It follows that $a_1^\alpha + \cdots + a_n^\alpha < a_1^\beta + \cdots + a_n^\beta = 1$, and hence $\left(a_1^\alpha + \cdots + a_n^\alpha\right)^\beta \le 1 = \left(a_1^\beta + \cdots + a_n^\beta\right)^\alpha$.

6.5. Without loss of generality one can assume that $a + b + c = 1$.
Note that if $0 < x < 1$, then $\frac{x}{1-x} \ge \frac{9x-1}{4}$. Therefore,

$$\frac{a}{b+c} + \frac{b}{c+a} + \frac{c}{a+b} = \frac{a}{1-a} + \frac{b}{1-b} + \frac{c}{1-c} \ge \frac{9a-1}{4} + \frac{9b-1}{4} + \frac{9c-1}{4} = \frac{3}{2}.$$

6.6. Without loss of generality one can assume that $a + b + c + d = 1$.
We have

$$\sqrt{\frac{a}{b+c+d}} + \sqrt{\frac{b}{a+c+d}} + \sqrt{\frac{c}{a+b+d}} + \sqrt{\frac{d}{a+b+c}}$$
$$= \sqrt{\frac{a}{1-a}} + \sqrt{\frac{b}{1-b}} + \sqrt{\frac{c}{1-c}} + \sqrt{\frac{d}{1-d}} > 2a + 2b + 2c + 2d = 2.$$

6.7. Without loss of generality one can assume that $d = 1$. From Problems 2.1 and 2.2, it follows that

$$\sqrt[3]{\frac{abc+ab+bc+ac}{4}} \le \sqrt[3]{\frac{\sqrt{\left(\frac{ab+bc+ac}{3}\right)^3}+ab+bc+ac}{4}} = A \text{ and}$$

$$\sqrt{\frac{ab+bc+ac+a+b+c}{6}} \ge \sqrt{\frac{ab+bc+ac+\sqrt{3(ab+bc+ac)}}{6}} = B.$$

Let us prove that $B \ge A$.

Let $ab + bc + ac = 3t^2$, where $t > 0$. Then the inequality $B \ge A$ is equivalent to the inequality $(t-1)^2(t+2) \ge 0$.

6.8. Without loss of generality one can assume that

$$ab + bc + ac = 1. \tag{1}$$

We need to prove that

$$(b+c)(a+b)(a+c) \ge \frac{8}{3\sqrt{3}}.$$

Let $a = \tan\alpha, b = \tan\beta, c = \tan\gamma$, where $\alpha, \beta, \gamma \in \left(0, \frac{\pi}{2}\right)$. From (1) it follows that $\alpha+\beta+\gamma = \frac{\pi}{2}$, and therefore, $(b+c)(a+b)(a+c) = \frac{1}{\cos\alpha\,\cos\beta\,\cos\gamma}$. In order to end the proof, it is sufficient to prove that

$$\cos\alpha\,\cos\beta\,\cos\gamma \le \frac{3\sqrt{3}}{8}, \text{ if } \alpha, \beta, \gamma \in \left(0, \frac{\pi}{2}\right) \text{ and } \alpha+\beta+\gamma = \frac{\pi}{2}.$$

Note that

$$\cos\alpha\,\cos\beta\,\cos\gamma = \frac{1}{2}(\cos(\alpha+\beta)+\cos(\alpha-\beta))\cos\gamma \le \frac{1}{2}(1+\sin\gamma)\cos\gamma$$
$$= \frac{1}{2}\sqrt{(1+\sin\gamma)^3(1-\sin\gamma)} = \frac{1}{2\sqrt{3}}\sqrt{(1+\sin\gamma)^3(3-3\sin\gamma)}$$
$$\le \frac{1}{2\sqrt{3}}\left(\frac{1+\sin\gamma+1+\sin\gamma+1+\sin\gamma+3-3\sin\gamma}{4}\right)^2 = \frac{3\sqrt{3}}{8}.$$

6.9. Without loss of generality one can assume that

$$x^3 + y^3 + z^3 = 1. \tag{1}$$

Then we have

$$A = \left(x^2+yz\right)\left(y^2+xz\right)\left(z^2+xy\right) = 2(xyz)^2 + xyz + x^3y^3 + y^3z^3 + z^3x^3.$$

Since $x^3 + y^3 + z^3 = 1$, it follows that $xyz \le \frac{1}{3}$ and $x^3y^3 + y^3z^3 + z^3x^3 \le \frac{(x^3+y^3+z^3)^2}{3} = \frac{1}{3}$, and therefore, $A \le \frac{2}{9} + \frac{1}{3} + \frac{1}{3} = \frac{8}{9}$.

6.10. (a) Note that the inequality $4a^3+4b^3+4c^3+15abc \geq (a+b+c)^3$ is equivalent to the inequality $a^3+b^3+c^3+3abc \geq a^2b+a^2c+b^2a+b^2c+c^2a+c^2b$ (see Example 6.1).

(b) If $d \geq \frac{15}{4}$, then

$$a^3+b^3+c^3+abcd \geq a^3+b^3+c^3+\frac{15}{4}abc = \frac{1}{4}\left(4a^3+4b^3+4c^3+15abc\right) \geq \frac{1}{4}.$$

If $d < \frac{15}{4}$, then according to 6.10(a) and the inequality $abc \leq \left(\frac{a+b+c}{3}\right)^3 = \frac{1}{27}$, we have $27(a^3+b^3+c^3) - 3 \geq \frac{15}{4}(1-27abc)$, and therefore, $27(a^3+b^3+c^3)-3 \geq d(1-27abc)$, or $a^3+b^3+c^3+abcd \geq \frac{1}{9}+\frac{d}{27}$.

6.11. Without loss of generality one can assume that $a_1a_2\cdots a_n = 1$. We need to prove that

$$\left(1-\frac{1}{n}\right)\left(a_1^2+\cdots+a_n^2\right) - \frac{2}{\sqrt{n}}(n-1)\sqrt{a_1^2+\cdots+a_n^2} + 2a_1a_2+\cdots+2a_1a_{n-1}+2a_1a_n+\cdots+2a_{n-1}a_n \geq (n-1)^2, \text{ or}$$

$$(n-1)\left(\frac{1}{\sqrt{n}}\sqrt{a_1^2+\cdots+a_n^2}-1\right)^2 + 2a_1a_2+\cdots+2a_1a_{n-1}+2a_1a_n+\cdots+2a_{n-1}a_n \geq n(n-1).$$

The proof of this inequality follows from Problem 4.1:

$$a_1a_2+a_1a_3+\cdots+a_1a_{n-1}+a_1a_n+\cdots+a_{n-1}a_n \geq C_n^2\sqrt[C_n^2]{(a_1a_2\ldots a_n)^{n-1}} = C_n^2.$$

6.12. Without loss of generality one can assume that $a \geq b \geq c \geq d \geq 0$. We have

$$\begin{aligned} A &= a^4+b^4+c^4+d^4+2abcd-a^2b^2-a^2c^2-a^2d^2-b^2c^2-b^2d^2-c^2d^2 \\ &= (a^2-b^2)(a^2-c^2-d^2)+2bcd(a-b)+(b^2+cd-c^2-d^2)^2+2cd(c-d)^2 \\ &= (a-b)\left(a(a^2-d^2)-b(c-d)^2\right)+a(a-b)(ab-c^2) \\ &\quad +(b^2+cd-c^2-d^2)^2+2cd(c-d)^2. \end{aligned}$$

Since $a-b \geq 0$, $a(a^2-d^2) \geq b(a-d)^2 \geq b(c-d)^2$ and $ab \geq c^2$, it follows that $A \geq 0$.

6.13. Let $\sqrt[3]{b_i} = c_i$, $i = 1, \ldots, n$. Then we have $c_1^6+\cdots+c_n^6 = (a_1^2+\cdots+a_n^2)^3$. We need to prove that $\frac{a_1^3}{c_1^3}+\cdots+\frac{a_n^3}{c_n^3} \geq 1$.
Without loss of generality one can assume that $a_1^2+\cdots+a_n^2 = 1$, and hence $c_1^6+\cdots+c_n^6 = 1$.
From inequality (8.4) from Chapter 8, it follows that

$$\frac{a_1^3}{c_1^3}+\cdots+\frac{a_n^3}{c_n^3} = \frac{a_1^4}{a_1c_1^3}+\cdots+\frac{a_n^4}{a_nc_n^3} \geq \frac{(a_1^2+\cdots+a_n^2)^2}{a_1c_1^3+\cdots+a_nc_n^3} = \frac{1}{a_1c_1^3+\cdots+a_nc_n^3} \geq 1.$$

Then from inequality (4.2) from Chapter 4, it follows that $1 = (a_1^2+\cdots+a_n^2)(c_1^6+\cdots+c_n^6) \geq (a_1c_1^3+\cdots+a_nc_n^3)^2$, whence
$1 \geq a_1c_1^3+\cdots+a_nc_n^3$, and hence $\frac{a_1^3}{c_1^3}+\cdots+\frac{a_n^3}{c_n^3} \geq 1$.

6.14. Without loss of generality one can assume that $\max(a,b,c) = a$. Then $a \geq b \geq c$ or $a \geq c \geq b$.

If $a \geq b \geq c$, then we have

$$\frac{a}{b}+\frac{b}{c}+\frac{c}{a}-\frac{a+c}{b+c}-\frac{b+a}{c+a}-\frac{c+b}{a+b}=\frac{c(a-b)}{b(b+c)}+\frac{a(b-c)}{c(a+c)}-\frac{b(a-c)}{a(b+a)}$$
$$\geq \frac{c(a-b)}{a(b+a)}+\frac{a(b-c)}{a(a+b)}-\frac{b(a-c)}{a(b+a)}=0,$$

whence

$$\frac{a}{b}+\frac{b}{c}+\frac{c}{a}\geq \frac{a+c}{b+c}+\frac{b+a}{c+a}+\frac{c+b}{a+b}.$$

If $a \geq c \geq b$, then we have

$$\frac{a}{b}+\frac{b}{c}+\frac{c}{a}-\frac{a+c}{b+c}-\frac{b+a}{c+a}-\frac{c+b}{a+b}=\frac{c(a-b)}{b(b+c)}-\frac{a(c-b)}{c(a+c)}-\frac{b(a-c)}{a(b+a)}$$
$$\geq \frac{c(a-b)}{b(b+c)}-\frac{a(c-b)}{b(b+c)}-\frac{b(a-c)}{b(b+c)}=0,$$

and therefore,

$$\frac{a}{b}+\frac{b}{c}+\frac{c}{a}\geq \frac{a+c}{b+c}+\frac{b+a}{c+a}+\frac{c+b}{a+b}.$$

6.15. Without loss of generality one can assume that $a_1 \cdot \ldots \cdot a_n = 1$. Then the expression on the left-hand side of the given inequality is equal to $\sqrt{\frac{a_1^n}{a_1^n+\lambda}} + \cdots + \sqrt{\frac{a_n^n}{a_n^n+\lambda}}$.

By the Cauchy–Bunyakovsky–Schwarz inequality and inequality (8.4) from Chapter 8, we have

$$\sqrt{\frac{a_1^n}{a_1^n+\lambda}}+\cdots+\sqrt{\frac{a_n^n}{a_n^n+\lambda}}=\frac{\left(a_1^{\frac{n}{2}}\right)^2}{\sqrt{a_1^{\frac{n}{2}}(a_1^n+\lambda)}}+\cdots+\frac{\left(a_n^{\frac{n}{2}}\right)^2}{\sqrt{a_n^{\frac{n}{2}}(a_n^n+\lambda)}}$$
$$\geq \frac{\left(a_1^{\frac{n}{2}}+\cdots+a_n^{\frac{n}{2}}\right)^2}{\sqrt{a_1^{\frac{n}{2}}(a_1^n+\lambda)}+\cdots+\sqrt{a_n^{\frac{n}{2}}(a_n^n+\lambda)}}\geq \frac{\left(a_1^{\frac{n}{2}}+\cdots+a_n^{\frac{n}{2}}\right)^2}{\sqrt{\left(a_1^{\frac{n}{2}}+\cdots+a_n^{\frac{n}{2}}\right)(a_1^n+\cdots+a_n^n+n\lambda)}}.$$

In order to complete the proof, it is sufficient to prove that $(1+\lambda)(b_1+\cdots+b_n)^3 \geq n^2(b_1^3+\cdots+b_n^3)+n^3\lambda$, where $b_i = a_i^{\frac{n}{3}} > 0,\quad i = 1,\ldots,n$, and $b_1 \cdot \ldots \cdot b_n = 1$.

Note that

$$(b_1+\cdots+b_n)^3 = (b_1^2+\cdots+b_n^2+2b_1b_2+\cdots+2b_1b_n+\cdots+2b_{n-1}b_n)(b_1+\cdots+b_n)$$

$$\geq (b_1^2+\dots+b_n^2+n(n-1)((b_1\cdot\ldots\cdot b_n)^{n-1})^{\frac{2}{n(n-1)}})(b_1+\dots+b_n)$$
$$=(b_1^2+\dots+b_n^2+n(n-1))(b_1+\dots+b_n)\geq b_1^3+\dots+b_n^3+b_1^2b_2+\dots+b_1^2b_n+\dots+b_{n-1}^2b_n$$
$$+n^2(n-1)\geq b_1^3+\dots+b_n^3+n(n-1)+n^2(n-1)=b_1^3+\dots+b_n^3+n^3-n,$$

whence $(b_1+\dots+b_n)^3 \geq b_1^3+\dots+b_n^3+n^3-n$. Thus

$$(1+\lambda)(b_1+\dots+b_n)^3-n^2(b_1^3+\dots+b_n^3)-n^3\lambda\geq(1+\lambda)(b_1^3+\dots+b_n^3+n^3-n)$$
$$-n^2(b_1^3+\dots+b_n^3)-n^3\lambda\geq(\lambda-(n^2-1))(b_1^3+\dots+b_n^3)+(1+\lambda)(n^3-n)-n^3\lambda$$
$$\geq(\lambda-(n^2-1))n+(1+\lambda)(n^3-n)-n^3\lambda=0,$$

and therefore, $(1+\lambda)(b_1+\dots+b_n)^3 \geq n^2(b_1^3+\dots+b_n^3)+n^3\lambda$.

6.16. Without loss of generality one can assume that $2a_1^k+\dots+2^na_n^k=1$, whence $2^ia_i^k<1$.
Therefore, $a_1+\ldots+a_n < \frac{1}{\sqrt[k]{2}}+\ldots+\left(\frac{1}{\sqrt[k]{2}}\right)^n < \frac{1}{\sqrt[k]{2}}+\ldots+\left(\frac{1}{\sqrt[k]{2}}\right)^n+\ldots=\frac{1}{\sqrt[k]{2}-1}$, and thus

$$(\sqrt[k]{2}-1)(a_1+\dots+a_n)<1=\sqrt[k]{2a_1^k+\dots+2^na_n^k}.$$

6.17. According to Example 6.2 from Chapter 6, we have

$$3(x^2y+y^2z+z^2x)(xy^2+yz^2+zx^2)\geq xyz(x+y+z)^3,$$
$$3(x^2y+y^2z+z^2x)(xy^2+yz^2+zx^2)=(1^3+1^3+1^3)\left(\left(\sqrt[3]{x^2y}\right)^3+\left(\sqrt[3]{y^2z}\right)^3+\left(\sqrt[3]{z^2x}\right)^3\right)$$
$$\cdot\left(\left(\sqrt[3]{zx^2}\right)^3+\left(\sqrt[3]{xy^2}\right)^3+\left(\sqrt[3]{yz^2}\right)^3\right)\geq\left(\sqrt[3]{xyz}(x+y+z)\right)^3=xyz(x+y+z)^3,$$

and therefore, $3(x^2y+y^2z+z^2x)(xy^2+yz^2+zx^2)\geq xyz(x+y+z)^3$.

6.18. Note that if we substitute the numbers $x_1,\dots,x_n,y_1,\dots,y_n$ by the numbers $x_1+x,\dots,x_n+x,y_1+x,\dots,y_n+x$, where x is an arbitrary number, then we obtain an equivalent inequality. Hence, we can choose the number x such that

$$x_1+x+\dots+x_n+x+y_1+x+\dots+y_n+x=0.$$

Therefore, without loss of generality one can assume that $x_1+\dots+x_n+y_1+\dots+y_n=0$.
Note that the given inequality is equivalent to the inequality

$$x_1y_1+\dots+x_ny_n\leq 0. \qquad (1)$$

If $x_1\leq\dots\leq x_n\leq 0\leq y_1\leq\dots\leq y_n$, then $x_1y_1\leq 0,\dots,x_ny_n\leq 0$, and hence (1) holds.

In order to complete the proof, it is sufficient to prove (1) if $x_1 \leq \dots \leq x_k \leq 0 \leq x_{k+1} \leq \dots \leq x_n \leq y_1 \leq \dots \leq y_n$ (such a number k exists, for otherwise, we could substitute the numbers $x_1, \dots, x_n, y_1, \dots, y_n$ by the numbers $-y_n, \dots, -y_1, -x_n, \dots, -x_1$).
Hence, we have that

$$x_1y_1 + \dots + x_ny_n = y_1(x_1 - x_n) + \dots + y_k(x_k - x_n) + x_{k+1}y_{k+1} + \dots + x_{n-1}y_{n-1} + x_n(-x_1 - \dots - x_n - y_{k+1} - \dots - y_{n-1}) \leq x_n(x_1 - x_n) + \dots + x_n(x_k - x_n) + x_{k+1}y_{k+1} + \dots + x_{n-1}y_{n-1} + x_n(-x_1 - \dots - x_n - y_{k+1} - \dots - y_{n-1}) = -kx_n^2 - x_n(x_{k+1} + \dots + x_n) + y_{k+1}(x_{k+1} - x_n) + \dots + y_{n-1}(x_{n-1} - x_n) \leq 0,$$

whence $x_1y_1 + \dots + x_ny_n \leq 0$.

Problems for Independent Study

Prove the following inequalities.

1. $\frac{\ln z - \ln y}{z-y} < \frac{\ln z - \ln x}{z-x} < \frac{\ln y - \ln x}{y-x}$, where $0 < x < y < z$.
2. $a^b b^c c^d d^a \geq b^a c^b d^c a^d$, where $0 < a \leq b \leq c \leq d$.
3. $\frac{x_1}{S-x_1} + \dots + \frac{x_n}{S-x_n} \geq \frac{n}{n-1}$, where $n \geq 2$, $S = x_1 + \dots + x_n$, $x_1 > 0, \dots, x_n > 0$.
4. $a^3 + b^3 + c^3 + 6abc \geq \frac{1}{4}(a+b+c)^3$, where $a \geq 0$, $b \geq 0$, $c \geq 0$.
5. $a^2(2b + 2c - a) + b^2(2a + 2c - b) + c^2(2a + 2b - c) \geq 9abc$, where a, b, c are the side lengths of some triangle.
6. (a) $\sqrt[n]{a_1 \cdot \ldots \cdot a_n} + \sqrt[n]{b_1 \cdot \ldots \cdot b_n} \leq \sqrt[n]{(a_1 + b_1) \cdot \ldots \cdot (a_n + b_n)}$, where $n \geq 2$, $a_i > 0, b_i > 0$, $i = 1, \dots, n$,

   (b) $\sqrt[n]{(n+1)!} - \sqrt[n]{n!} \geq 1$, where $n \geq 2$, $n \in \mathbb{N}$,
   (c) $\sqrt[n]{F_{n+1}} > 1 + \frac{1}{\sqrt[n]{F_n}}$, where $n \geq 2$, $n \in \mathbb{N}$, $F_1 = 1, F_2 = 2, F_{k+2} = F_{k+1} + F_k$, $k = 1, 2, \dots$,
   (d) $\sqrt[n]{C_{2n+1}^n} > 2\left(1 + \frac{1}{\sqrt[n]{n+1}}\right)$, where $n = 2, 3, \dots$,
   (e) $(1 + a_1) \cdot \ldots \cdot (n + a_n) \geq n^{\frac{n}{2}}$, where $n \geq 2, n \in \mathbb{N}$, $a_1 > 0, \dots, a_n > 0$ and $a_1 \cdot \ldots \cdot a_n = 1$,
   (f) $\sqrt[n]{\frac{(a_1+b_1)\cdot\ldots\cdot(a_n+b_n)}{(a_1-c_1)\cdot\ldots\cdot(a_n-c_n)}} \geq \frac{\sqrt[n]{a_1\cdot\ldots\cdot a_n} + \sqrt[n]{b_1\cdot\ldots\cdot b_n}}{\sqrt[n]{a_1\cdot\ldots\cdot a_n} - \sqrt[n]{c_1\cdot\ldots\cdot c_n}}$, where $n \geq 2$, $n \in \mathbb{N}$, $b_i > 0$,

   $$a_i > c_i > 0, \quad i = 1, \dots, n,$$

   Hint. Prove that $\frac{\sqrt[n]{(a_1+b_1)\cdot\ldots\cdot(a_n+b_n)}}{\sqrt[n]{a_1\cdot\ldots\cdot a_n} + \sqrt[n]{b_1\cdot\ldots\cdot b_n}} \geq 1 \geq \frac{\sqrt[n]{(a_1-c_1)\cdot\ldots\cdot(a_n-c_n)}}{\sqrt[n]{a_1\cdot\ldots\cdot a_n} - \sqrt[n]{c_1\cdot\ldots\cdot c_n}}$.
   (g) $\sqrt[3]{ab} + \sqrt[3]{cd} \leq \sqrt[3]{(a+c+d)(a+c+b)}$, where $a > 0$, $b > 0$, $c > 0$, $d > 0$.
   Hint. Note that

   $$\sqrt[3]{a\sqrt{b}\sqrt{b}} + \sqrt[3]{c\sqrt{d}\sqrt{d}} \leq \sqrt[3]{(a+c)(\sqrt{b} + \sqrt{d})^2}$$

$$\leq \sqrt[3]{(a+c)(b+d)+2(a+c)\sqrt{bd}} \leq \sqrt[3]{(a+c)(b+d)+(a+c)^2+bd}.$$

7. $x^2(x^2-1)^2 + y^2(y^2-1)^2 \geq (x^2-1)(y^2-1)(x^2+y^2-1)$.
8. $x^2(x-1)^2 + y^2(y-1)^2 + z^2(z-1)^2 \geq 2xyz(2-x-y-z)$.
   Hint. Prove that

$$x^2(x-1)^2 + y^2(y-1)^2 + z^2(z-1)^2 - 2xyz(2-x-y-z) = x^2(x-1)^2$$
$$+ (y(y-1) - z(z-1))^2 + 2xyz(x+y-1)(x+z-1).$$

9. $(x_1-x_2)(x_1-x_3)(x_1-x_4)(x_1-x_5) + (x_2-x_1)(x_2-x_3)(x_2-x_4)(x_2-x_5) + \cdots + (x_5-x_1)(x_5-x_2)(x_5-x_3)(x_5-x_4) \geq 0$.
10. $0 \leq ab+bc+ca-abc \leq 2$, where $a \geq 0$, $b \geq 0$, $c \geq 0$ and $a^2+b^2+c^2+abc = 4$.
    Hint. Note that $\min(a, b, c) \leq 1$, and without loss of generality one can assume that $(a-1)(b-1) \geq 0$.
11. $x^\lambda(x-y)(x-z) + y^\lambda(y-z)(y-x) + z^\lambda(z-x)(z-y) \geq 0$, where $x, y, z > 0$.
    Hint. Let $x \geq y \geq z$.
    Note that if $\lambda \geq 0$, then we have

$$x^\lambda(x-y)(x-z) + y^\lambda(y-z)(y-x) + z^\lambda(z-x)(z-y) = (x-y)(x^\lambda(x-z)$$
$$- y^\lambda(y-z)) + z^\lambda(z-x)(z-y) \geq 0.$$

Otherwise, if $\lambda < 0$, then we have

$$x^\lambda(x-y)(x-z) + y^\lambda(y-z)(y-x) + z^\lambda(z-x)(z-y) = x^\lambda(x-y)(x-z)$$
$$+ (y-z)(z^\lambda(x-z) - y^\lambda(x-y)) \geq 0.$$

12. $\sqrt[3]{\left(\frac{a}{b+c}\right)^2} + \sqrt[3]{\left(\frac{b}{a+c}\right)^2} + \sqrt[3]{\left(\frac{c}{a+b}\right)^2} \geq \frac{3}{\sqrt[3]{4}}$, where $a > 0,\ b > 0,\ c > 0$.
13. $(a^5-a^2+3)(b^5-b^2+3)(c^5-c^2+3) \geq (a+b+c)^3$, where $a > 0,\ b > 0,\ c > 0$.
    Hint. Note that $a^5 - a^2 + 3 = a^2(a^3-1) + 3 \geq 1 \cdot (a^3-1) + 3 = a^3 + 1^3 + 1^3$.
14. $abc + abd + bcd + acd - abcd \leq 3$, where $a > 0,\ b > 0,\ c > 0,\ d > 0$, and $a^3 + b^3 + c^3 + d^3 + abcd = 5$.
    Hint. Prove that
    $3(a^3+b^3+c^3+d^3)(a+b+c+d) + 12abcd$
    $\geq 5((abc+abd+bcd+acd)(a+b+c+d) - 4abcd)$.
15. $0 \leq ab+bc+ca-abc \leq 2$, where $a \geq 0,\ b \geq 0,\ c \geq 0$ and $a^2+b^2+c^2+abc = 4$.
16. $a^2+b^2+c^2+2abc+1 \geq 2(ab+bc+ca)$, where $a \geq 0,\ b \geq 0,\ c \geq 0$.
17. $\frac{x+y+z}{xy+yz+zx} \leq 1 + \frac{1}{48}\left((x-y)^2+(y-z)^2+(z-x)^2\right)$, where $x > 0, y > 0, z > 0$, and $xy+yz+zx+xyz = 4$.

Chapter 7
The Principle of Mathematical Induction

A large number of inequalities that at first glance appear difficult can be easily proved by a classical mathematical proof technique called the *principal of mathematical induction*.

This chapter is devoted to some applications of the principal of mathematical induction in proving algebraic inequalities. Note that there are different kinds of mathematical induction. In this chapter we mainly consider applications of the classical form of the principle of mathematical induction called the (first) principle of mathematical induction, since this variety is applied most often. Nevertheless, we also formulate two other relatively important variants of the principle of mathematical induction.

Let S_n be a statement about an arbitrary positive integer n. The (first) principle of mathematical induction is used to prove that the statement S_n holds for all positive integral values of n.

The principle of mathematical induction implies that in order to prove that statement S_n holds for all such values of n, one needs to prove first that statement S_1 holds and then that from the assumption that statement S_k holds for an arbitrary positive integer k, it follows that statement S_{k+1} holds as well.

Putting together these explanations, we deduce the following formulation.

(First) principle of mathematical induction.

Let S_n be a statement about an arbitrary positive integer n.

(i). The **basis (first step/base case)**: S_1 is true.

(ii). The **inductive step**: Whenever S_k is true for $k \in \mathbb{N}$, then S_{k+1} is also true. The assumption that S_k holds for some $k \in \mathbb{N}$ is called the **induction hypothesis (inductive hypothesis)**.

(iii). **Conclusion**: If (i) and (ii) hold, then S_n is true for all $n \in \mathbb{N}$.

Remark Note that sometimes instead of $n \in \mathbb{N}$, one can consider $n \in \mathbb{N}_0 = \{0, 1, 2, \ldots\}$, in which case one needs to verify (the basis) that S_0 is true.

H. Sedrakyan and N. Sedrakyan, *Algebraic Inequalities*, Problem Books in Mathematics, https://doi.org/10.1007/978-3-319-77836-5_7

So, the (first) principle of mathematical induction implies that if the basis and the inductive step are proved, then S_n is true for all positive integers n.

Let us consider another useful variant of the principle of mathematical induction called the second principle of mathematical induction. (It is also called "strong induction," because a stronger induction hypothesis is used.)

Second principle of mathematical induction.

Let S_n be a statement about an arbitrary positive integer n.

(i). The **basis**: S_1 is true.
(ii). The **inductive step**: Whenever S_i is true for all $i \leq k$, $k \in \mathbb{N}$, it follows that S_{k+1} is true.
(iii). **Conclusion**: If (i) and (ii) hold, then S_n is true for all $n \in \mathbb{N}$.

Now let us consider the following variant of the principle of mathematical induction that is used to prove that statement S_n holds not for all positive integers n, but only for all positive integers greater than or equal to a certain positive integer k.

Different starting point (induction basis other than 1 or 0).

Let m be a given positive integer and let S_n be a statement about an arbitrary positive integer n, where $n \geq m$.

(i). The **basis:** S_m is true.
(ii). The **inductive step**: Whenever S_i is true for all $m \leq i \leq k$, it follows that S_{k+1} is true.
(iii). **Conclusion**: If (i) and (ii) hold, then S_n is true for all $n \in \mathbb{N}$, $n \geq m$.

Below we provide examples of inequalities whose proofs make use of the (first) principle of mathematical induction.

Example 7.1 Prove that $\frac{1}{2n+1} + \frac{1}{2n+2} + \cdots + \frac{1}{2n+n} \geq \frac{11}{30}$, where $n > 1$, $n \in \mathbb{N}$.

Proof If $n = 2$, then we have $\frac{11}{30} \geq \frac{11}{30}$.

Assume that the given inequality is true for $n = k$, that is,

$$\frac{1}{2k+1} + \frac{1}{2k+2} + \cdots + \frac{1}{2k+k} \geq \frac{11}{30}. \tag{7.1}$$

Let us prove that the given inequality is true for $n = k + 1$.

For $n = k + 1$ we need to prove that

$$\frac{1}{2(k+1)+1} + \frac{1}{2(k+1)+2} + \cdots + \frac{1}{2(k+1)+(k+1)} \geq \frac{11}{30}.$$

Adding to both sides of inequality (7.1) the expression $\frac{1}{3k+1} + \frac{1}{3k+2} + \frac{1}{3k+3} - \frac{1}{2k+1} - \frac{1}{2k+2}$, we obtain

$$\frac{1}{2k+3} + \cdots + \frac{1}{3k+3} \geq \frac{11}{30} + \frac{1}{3k+1} + \frac{1}{3k+2} + \frac{1}{3k+3} - \frac{1}{2k+1} - \frac{1}{2k+2}.$$

Since $\frac{1}{3k+1} + \frac{1}{3k+2} + \frac{1}{3k+3} - \frac{1}{2k+1} - \frac{1}{2k+2} > 0$ (Problem 1.19), it follows that

$$\frac{1}{2k+3} + \cdots + \frac{1}{3k+3} \geq \frac{11}{30}.$$

Therefore, the given inequality is true for every positive integer n.
This ends the proof.

Example 7.2 Prove that

$$\frac{1}{2^2} + \cdots + \frac{1}{n^2} < 1, \tag{7.2}$$

where $n = 3, 4, \ldots$.

Proof Note that it is impossible to prove this inequality similarly to how we proved Example 7.1. In the proof of this inequality and in the proof of some other inequalities, the method of induction is applied to another inequality (a more general one), from which follows the validity of the considered inequality.

In order to prove inequality (7.2), let us prove the following inequality:

$$\frac{1}{2^2} + \cdots + \frac{1}{n^2} < 1 - \frac{1}{n}. \tag{7.3}$$

We carry out the proof by mathematical induction on n.

For $n = 3$, we have $\frac{1}{2^2} + \frac{1}{3^2} < 1 - \frac{1}{3}$.

Assume that (7.3) is true for $n = k \quad (k \geq 3)$, and let us prove that it is true for $n = k + 1$. For $n = k$, we have $\frac{1}{2^2} + \cdots + \frac{1}{k^2} < 1 - \frac{1}{k}$.

Let us add to both sides of this inequality the expression $\frac{1}{(k+1)^2}$. Then we obtain $\frac{1}{2^2} + \cdots + \frac{1}{k^2} + \frac{1}{(k+1)^2} < 1 - \frac{1}{k} + \frac{1}{(k+1)^2}$.

Since $1 - \frac{1}{k} + \frac{1}{(k+1)^2} < 1 - \frac{1}{k+1}$, we deduce that the inequality $\frac{1}{2^2} + \cdots + \frac{1}{k^2} + \frac{1}{(k+1)^2} < 1 - \frac{1}{k+1}$ coincides with the given inequality for $n = k + 1$.

Therefore, the given inequality holds for every positive integer $n \geq 3$.

This ends the proof.

In some inequalities the principle of mathematical induction is used in the following way:

(a) One first proves that the considered inequality holds for values of the positive integer n equal to $n_1, \ldots, n_k, \ldots$, where $n_1 < \cdots < n_k < \cdots$.
(b) From the validity of the considered inequality for arbitrary $n = k \quad (k \geq 2)$, follows that it holds for $n = k - 1$.

Example 7.3 Ky Fan inequality: Prove that $\frac{x_1 \cdots x_n}{(x_1 + \cdots + x_n)^n} \leq \frac{(1-x_1)\cdots(1-x_n)}{((1-x_1)+\cdots+(1-x_n))^n}$, where $n \geq 2$, $0 < x_1 \leq \frac{1}{2}, \ldots, 0 < x_n \leq \frac{1}{2}$.

Proof Let us first prove the given inequality for n equal to $2^1, \ldots, 2^k, \ldots$.

For $n = 2$, we have $\frac{x_1 x_2}{(x_1+x_2)^2} \leq \frac{(1-x_1)(1-x_2)}{((1-x_1)+(1-x_2))^2}$ (Problem 1.20).

Assume that the given inequality holds for $n = 2^k \quad (k \in \mathbb{N})$, and let us prove that it holds for $n = 2^{k+1}$.

We have

$$\frac{x_1 \cdots x_{2p}}{(x_1 + \cdots + x_{2p})^{2p}} = \frac{x_1 \cdots x_p}{(x_1 + \cdots + x_p)^p} \cdot \frac{x_{p+1} \cdots x_{2p}}{(x_{p+1} + \cdots + x_{2p})^p} \cdot \frac{(x_1 + \cdots + x_p)^p (x_{p+1} + \cdots + x_{2p})^p}{(x_1 + \cdots + x_{2p})^{2p}}$$

$$\leq \frac{(1 - x_1) \cdots (1 - x_p)}{((1 - x_1) + \cdots + (1 - x_p))^p} \cdot \frac{(1 - x_{p+1}) \cdots (1 - x_{2p})}{((1 - x_{p+1}) + \cdots + (1 - x_{2p}))^p} \cdot \left(\frac{\left(\frac{x_1 + \cdots + x_p}{p}\right)\left(\frac{x_{p+1} + \cdots + x_{2p}}{p}\right)}{\left(\frac{x_1 + \cdots + x_p}{p} + \frac{x_{p+1} + \cdots + x_{2p}}{p}\right)^2} \right)^p$$

$$\leq \frac{(1 - x_1) \cdots (1 - x_{2p})}{((1 - x_1) + \cdots + (1 - x_p))^p ((1 - x_{p+1}) + \cdots + (1 - x_{2p}))^p}$$

$$\times \left(\frac{\left(1 - \frac{x_1 + \cdots + x_p}{p}\right)\left(1 - \frac{x_{p+1} + \cdots + x_{2p}}{p}\right)}{\left(\left(1 - \frac{x_1 + \cdots + x_p}{p}\right) + \left(1 - \frac{x_{p+1} + \cdots + x_{2p}}{p}\right)\right)^2} \right)^p = \frac{(1 - x_1) \cdots (1 - x_{2p})}{((1 - x_1) + \cdots + (1 - x_{2p}))^{2p}},$$

where $p = 2^k$.

We obtain $\frac{x_1 \cdots x_{2p}}{(x_1 + \cdots + x_{2p})^{2p}} \leq \frac{(1-x_1)\cdots(1-x_{2p})}{((1-x_1)+\cdots+(1-x_{2p}))^{2p}}$, whence the given inequality holds for $n = 2^{k+1}$.

Now let us prove that if the given inequality holds for m, then it holds true for $m - 1$, where $m \geq 3 \quad (m \in \mathbb{N})$.

We have $\frac{x_1 \cdots x_m}{(x_1 + \cdots + x_m)^m} \leq \frac{(1-x_1)\cdots(1-x_m)}{((1-x_1)+\cdots+(1-x_m))^m}$.

Taking $x_m = \frac{x_1 + \cdots + x_{m-1}}{m-1}$, where by Problem 1.10, $0 < x_m \leq \frac{1}{2}$, we obtain

$$\frac{x_1 \cdots x_{m-1} \cdot \frac{x_1 + \cdots + x_{m-1}}{m-1}}{\left(x_1 + \cdots + x_{m-1} + \frac{x_1 + \cdots + x_{m-1}}{m-1}\right)^m} \leq \frac{(1 - x_1) \cdots (1 - x_{m-1}) \cdot \left(1 - \frac{x_1 + \cdots + x_{m-1}}{m-1}\right)}{\left((1 - x_1) + \cdots + (1 - x_{m-1}) + \left(1 - \frac{x_1 + \cdots + x_{m-1}}{m-1}\right)\right)^m}.$$

Therefore, we have

$$\frac{x_1 \cdots x_{m-1}}{(x_1 + \cdots + x_{m-1})^{m-1}} \leq \frac{(1 - x_1) \cdots (1 - x_{m-1})}{((1 - x_1) + \cdots + (1 - x_{m-1}))^{m-1}}.$$

Problems

Prove the following inequalities (7.1–7.4, 7.6, 7.7, 7.10, 7.11, 7.13–7.20, 7.24, 7.25).

7.1. (a) $\frac{1}{2} \cdots \frac{2n-1}{2n} \leq \frac{1}{\sqrt{3n+1}}$, where $n \in \mathbb{N}$.

(b) $\frac{1}{n+1} + \frac{1}{n+2} + \cdots + \frac{1}{2n} < \frac{25}{36}$, where $n \geq 2, \quad n \in \mathbb{N}$.

7.2. (a) $\underbrace{\sqrt{a + \sqrt{a + \cdots + \sqrt{a}}}}_{n} \leq \frac{1+\sqrt{4a+1}}{2}$, where $a \geq 0$,

(b) $\sqrt{2\sqrt{3\sqrt{4\ldots\sqrt{n}}}} < 3$, where $n \geq 2, \quad n \in \mathbb{N}$.

7.3. $x_1^2 + \cdots + (2n-1)x_n^2 \leq (x_1 + \cdots + x_n)^2$, where $x_1 \geq \cdots \geq x_n \geq 0$.

7.4. (a) $|\sin(x_1 + \cdots + x_n)| \leq |\sin x_1| + \cdots + |\sin x_n|$,
(b) $\sin(x_1 + \cdots + x_n) \leq \sin x_1 + \cdots + \sin x_n$, where $x_1, \ldots, x_n \in [0, \pi]$,
(c) $|\cos x_1| + |\cos x_2| + |\cos x_3| + |\cos x_4| + |\cos x_5| \geq 1$, where $x_1 + x_2 + x_3 + x_4 + x_5 = 0$.

7.5. Prove
(a) *Bellman's inequality*: if a function $f(x)$ is defined in $[0, a)$ (or $[0, +\infty)$) and for arbitrary numbers $x \geq y \geq z$ from that interval we have $f(x) - f(y) + f(x) \geq f(x - y + z)$ and $f(0) \leq 0$, then for all numbers $a > x_1 \geq \cdots \geq x_n \geq 0$, the following inequality holds: $f(x_1) - f(x_2) + \cdots + (-1)^{n-1} f(x_n) \geq f\big(x_1 - x_2 + \cdots + (-1)^{n-1} x_n\big)$,
(b) $\tan x_1 - \tan x_2 + \cdots + (-1)^{n-1} \tan x_n \geq \tan\big(x_1 - x_2 + \cdots + (-1)^{n-1} x_n\big)$, where $\frac{\pi}{2} > x_1 \geq \cdots \geq x_n \geq 0$,
(c) $a_1^r - a_2^r + \cdots + (-1)^{n-1} a_n^r \geq \big(a_1 - a_2 + \cdots + (-1)^{n-1} a_n\big)^r$, where $a_1 \geq \cdots \geq a_n \geq 0, \quad r \geq 1$.

7.6. (a) $(x_1 + \cdots + x_5)^2 \geq 4(x_1x_2 + x_2x_3 + x_3x_4 + x_4x_5 + x_5x_1)$, where $x_1 > 0, \ldots, x_5 > 0$.
(b) $x_1\sqrt{x_n^2 + x_2^2} + x_2\sqrt{x_1^2 + x_3^2} + \cdots + x_{n-1}\sqrt{x_{n-2}^2 + x_n^2} + x_n\sqrt{x_{n-1}^2 + x_1^2} \leq \frac{1}{2}(x_1 + \cdots + x_n)^2$, where $n \geq 3$ and $x_1 > 0, \ldots, x_n > 0$.

7.7. $\frac{1}{2}(x_1 + \cdots + x_n)^2 \leq (x_1 + \cdots + nx_n) \cdot \max(x_1, \ldots, x_n)$, where $x_1 \geq 0, \ldots, x_n \geq 0$.

7.8. Prove Theorem 11.1 (Chapter 11).

7.9. Prove Theorem 11.2 (Chapter 11).

7.10. (a) $a_1 + \cdots + a_n^n \leq na_1 \cdot \ldots \cdot a_n$, where $a_1 \geq \cdots \geq a_n \geq 1$;
(b) $a_1 + \cdots + a_n^n \geq na_1 \cdot \ldots \cdot a_n$, where $0 \leq a_1 \leq \cdots \leq a_n \leq 1$.

7.11. (a) $\frac{a_2^2}{a_1} + \cdots + \frac{a_n^2}{a_{n-1}} \geq 4(a_n - a_1)$, where $a_1 > 0, \ldots, a_n > 0$;
(b) $\frac{a_1^3}{b_1c_1} + \cdots + \frac{a_n^3}{b_nc_n} \geq \frac{(a_1 + \cdots + a_n)^3}{(b_1 + \cdots + b_n)(c_1 + \cdots + c_n)}$, where $a_i > 0, b_i > 0, c_i > 0, \quad i = 1, \ldots, n$.

7.12. Prove that
(a) if $f(x)$ is defined in I[1] and is a convex function,[2] then

$$(x_2 + x_1)(f(x_2) - f(x_1)) + \cdots + (x_n + x_{n-1})(f(x_n) - f(x_{n-1})) \geq (x_n + x_1)(f(x_n) - f(x_1)),$$

where $n \geq 2, \quad x_1 < \cdots < x_n, \quad x_1, \ldots, x_n \in I$.
(b) $a\sqrt{b} + b\sqrt{c} + c\sqrt{a} \geq a\sqrt{c} + b\sqrt{a} + c\sqrt{b}$, where $a \geq b \geq c \geq 0$;
(c) $x_1^{x_2} \cdot x_2^{x_3} \cdot \ldots \cdot x_n^{x_1} \geq x_2^{x_1} \cdot x_3^{x_2} \cdot \ldots \cdot x_n^{x_{n-1}} \cdot x_1^{x_n}$, where $x_n \geq \cdots \geq x_1 > 0, \quad n \geq 3$;
(d) $\frac{a(c-b)}{(c+b)(2a+b+c)} + \frac{b(a-c)}{(a+c)(2b+a+c)} + \frac{c(b-a)}{(b+a)(2c+b+a)} \leq 0$,
where $a \geq b \geq c > 0$.

---

[1] I is defined as the domain of function f, so $I = D(f)$

[2] See Chapter 11.

7.13. $\frac{a_1+\dots+a_{n-1}}{n-1}+\frac{a_1+\dots+a_{n+1}}{n+1}\geq 2\frac{a_1+\dots+a_n}{n}$, where $n\geq 2$, $\frac{a_k+a_{k+2}}{2}\geq a_{k+1}$, $k=1,\dots,n-1$.

7.14. (a) $\frac{a_1+a_3+\dots+a_{2n-1}}{n}\geq\frac{a_0+a_2+\dots+a_{2n}}{n+1}$, where $a_k\geq\frac{a_{k-1}+a_{k+1}}{2}$, $k=1,2,\dots,2n-1$, $n\in\mathbb{N}$,
(b) $\frac{a^0+a^2+\dots+a^{2n}}{a+a^3+\dots+a^{2n-1}}\geq\frac{n+1}{n}$, where $a>0$, $n\in\mathbb{N}$.

7.15. $1+\dots+\frac{1}{n}>\ln(n+1)$, where $n\in\mathbb{N}$.

7.16. $1+\dots+\frac{1}{n\sqrt{n}}\leq 3-\frac{2}{\sqrt{n}}$, where $n\in\mathbb{N}$.

7.17. $(1+\alpha)^n\geq 1+n\alpha+\frac{n(n-1)}{2}\alpha^2$, where $\alpha\geq 0$, $n\in\mathbb{N}$.

7.18. $k!\geq\left(\frac{k+1}{e}\right)^k$, where $k\in\mathbb{N}$.

7.19. $\sum_{i=0}^{n}\left|\sin(2^i x)\right|\leq 1+\frac{\sqrt{3}}{2}n$, where $n=0,1,2,\dots$.

7.20. $\cos\alpha+\dots+\frac{\cos n\alpha}{n}\geq-\frac{1}{2}$, where $n\in\mathbb{N}, 0\leq\alpha\leq\frac{\pi}{2}$.

7.21. Let the numbers $a_1,\dots,a_n$ $(n\geq 2)$ be greater than 1 and $|a_{k+1}-a_k|<1$ for $k=1,\dots,n-1$. Prove that the sum $\frac{a_1}{a_2}+\frac{a_2}{a_3}+\dots+\frac{a_{n-1}}{a_n}+\frac{a_n}{a_1}$ is less than $2n-1$.

7.22. Let $1=x_1\leq x_2\leq\dots\leq x_{n+1}$. Prove that
(a) $\frac{\sqrt{x_2-x_1}}{x_2}+\dots+\frac{\sqrt{x_{n+1}-x_n}}{x_{n+1}}\leq\frac{\sqrt{4n-3}}{2}$, where $n\in\mathbb{N}$,
(b) $\frac{\sqrt{x_2-x_1}}{x_2}+\dots+\frac{\sqrt{x_{n+1}-x_n}}{x_{n+1}}<1+\frac{1}{2}+\dots+\frac{1}{n^2}$, where $x_2,\dots,x_{n+1}\in\mathbb{N}$.

7.23. Prove that, if $\alpha_1>0,\dots,\alpha_n>0$, $\beta_1>0,\dots,\beta_n>0$, and $\alpha_1+\dots+\alpha_n\leq\beta_1+\dots+\beta_n\leq\pi$, then $\frac{\cos\beta_1}{\sin\alpha_1}+\dots+\frac{\cos\beta_n}{\sin\alpha_n}\leq\frac{\cos\alpha_1}{\sin\alpha_1}+\dots+\frac{\cos\alpha_n}{\sin\alpha_n}$.

7.24. $2(a^{2012}+1)(b^{2012}+1)(c^{2012}+1)\geq(1+abc)(a^{2011}+1)(b^{2011}+1)(c^{2011}+1)$, where $a>0, b>0, c>0$.

7.25. (a) $b_1\geq\sqrt{b_2}$,
(b) *Newton's inequality*: $b_k^2\geq b_{k-1}b_{k+1}$, for $k=2,\dots,n-1$, where $n\geq 2, n\in\mathbb{N}, a_1>0,\dots,a_n>0$, and

$$b_k=\frac{1}{C_n^k}(a_1\cdots a_{k-1}a_k+a_1\cdots a_{k-1}a_{k+1}+\dots+a_1\cdots a_{k-1}a_n+\dots+a_{n-k+1}\cdots a_{n-1}a_n).$$

Proofs

7.1. (a) We proceed by mathematical induction on n.
The given inequality holds for $n=1$, since $\frac{1}{2}\leq\frac{1}{\sqrt{4}}$.
Assume that the given inequality holds for $n=k$, and let us prove that it holds for $n=k+1$.
For $n=k$, we have $\frac{1}{2}\cdots\frac{2k-1}{2k}\leq\frac{1}{\sqrt{3k+1}}$.
Multiplying both sides of this inequality by $\frac{2(k+1)-1}{2(k+1)}$, we deduce that

$$\frac{1}{2}\cdots\frac{2k-1}{2k}\cdot\frac{2k+1}{2k+2}\leq\frac{1}{\sqrt{3k+1}}\cdot\frac{2k+1}{2k+2}.$$

Taking into consideration that $\frac{1}{\sqrt{3k+1}} \cdot \frac{2k+1}{2k+2} \le \frac{1}{\sqrt{3k+4}}$ (Problem 1.21), we obtain $\frac{1}{2} \cdots \frac{2k-1}{2k} \cdot \frac{2(k+1)-1}{2(k+1)} \le \frac{1}{\sqrt{3(k+1)+1}}$.
Therefore, the given inequality holds for every positive integer n.
(b) In order to prove the given inequality, let us prove the following inequality:

$$\frac{1}{n+1} + \frac{1}{n+2} + \cdots + \frac{1}{2n} + \frac{1}{4n+1} \le \frac{25}{36}, \text{ where } \ n \ge 2, \quad n \in \mathbb{N}. \qquad (1)$$

We proceed by mathematical induction on n.
Note that inequality (1) holds for $n = 2$, since $\frac{1}{3} + \frac{1}{4} + \cdots + \frac{1}{9} = \frac{25}{36}$.
Assume that inequality (1) holds for $n = k$, where $k \ge 2, \quad k \in \mathbb{N}$. Let us prove that inequality (1) holds for $n = k + 1$.
We have

$$\frac{1}{k+2} + \frac{1}{k+3} + \cdots + \frac{1}{2(k+1)} + \frac{1}{4(k+1)+1} = \frac{1}{k+1} + \frac{1}{k+2}$$
$$+ \cdots + \frac{1}{2k} + \frac{1}{4k+1} + \left(\frac{1}{2k+1} + \frac{1}{2k+2} - \frac{1}{k+1} + \frac{1}{4k+5} - \frac{1}{4k+1}\right) \le \frac{25}{36}$$
$$+ \left(\frac{1}{2k+1} + \frac{1}{2k+2} - \frac{1}{k+1} + \frac{1}{4k+5} - \frac{1}{4k+1}\right) \le \frac{25}{36} + \frac{1}{(2k+1)(2k+2)}$$
$$- \frac{4}{(4k+1)(4k+5)} \le \frac{25}{36}.$$

Then inequality (1) holds for $n = k + 1$.
Therefore, the given inequality holds for every positive integer $n \ge 2$.

7.2. (a) We proceed by induction on n.
The given inequality holds for $n = 1$, since $\sqrt{a} \le \frac{1+\sqrt{4a+1}}{2}$.

Assume that the inequality $\underbrace{\sqrt{a + \sqrt{a + \cdots + \sqrt{a}}}}_{k} \le \frac{1+\sqrt{4a+1}}{2}$ holds. We have

$$\underbrace{\sqrt{a + \sqrt{a + \cdots + \sqrt{a}}}}_{k+1} = \sqrt{a + \underbrace{\sqrt{a + \cdots + \sqrt{a}}}_{k}} \le \sqrt{a + \frac{1+\sqrt{4a+1}}{2}} = \frac{1+\sqrt{4a+1}}{2},$$

and thus it follows that the given inequality holds for $n = k + 1$.
Therefore, the inequality holds for every positive integer n.

(b) We prove the inequality $\sqrt{k\sqrt{(k+1)\sqrt{\ldots\sqrt{n}}}} < k + 1$, where $1 \le k \le n$ and $k \in N$, by induction on $p = n - k$ (n is assumed to be a constant).
For $p = 0$, we have $\sqrt{n} < n + 1$, whose proof is obvious.
Assume that the given inequality holds for $p = n - m$, and let us prove that it holds for $p = n - m + 1 = n - (m - 1)$.

For $p = n - m$, we have $\sqrt{m\sqrt{(m+1)\sqrt{\ldots\sqrt{n}}}} < m + 1$, and hence we obtain

$$\sqrt{(m-1)\sqrt{m\sqrt{(m+1)\ldots\sqrt{n}}}} < \sqrt{(m-1)(m+1)} < m.$$

Hence, the inequality $\sqrt{k\sqrt{(k+1)\sqrt{\ldots\sqrt{n}}}} < k+1$ holds for $1 \leq k \leq n, \quad k \in \mathbb{N}$.

For $k = 2$, we have $\sqrt{2\sqrt{3\sqrt{4\sqrt{\ldots\sqrt{n}}}}} < 3$, which ends the proof.

7.3. We proceed by induction on n.
For $n = 1$, we have $x_1^2 \leq x_1^2$.
Assume that the inequality $x_1^2 + \cdots + (2k-1)x_k^2 \leq (x_1 + \cdots + x_k)^2$ holds. Adding $(2k+1)x_{k+1}^2$ to both sides of this inequality, we obtain the following inequality:

$$x_1^2 + \cdots + (2k-1)x_k^2 + (2k+1)x_{k+1}^2 \leq (x_1 + \cdots + x_k)^2 + (2k+1)x_{k+1}^2. \tag{7.4}$$

Let us prove that

$$(x_1 + \cdots + x_k)^2 + (2k+1)x_{k+1}^2 \leq (x_1 + \cdots + x_{k+1})^2, \tag{7.5}$$

which is equivalent to $(2k+1)x_{k+1}^2 \leq (x_1 + \cdots + x_{k+1})^2 - (x_1 + \cdots + x_k)^2 = (2(x_1 + \cdots + x_k) + x_{k+1})x_{k+1}$, or to the following inequality:

$$0 \leq x_{k+1}(2(x_1 + \cdots + x_k) - 2kx_{k+1}) = 2x_{k+1}((x_1 - x_{k+1}) + \cdots + (x_k - x_{k+1})).$$

This inequality holds because according to the assumption of the problem, we have $0 \leq x_{k+1} \leq x_i, \quad i = 1, \ldots, k$.
Hence, from (7.4) and (7.5), it follows that $x_1^2 + \cdots + (2k-1)x_k^2 + (2k+1)x_{k+1}^2 \leq (x_1 + \cdots + x_{k+1})^2$, which is the given inequality for $n = k+1$.
Therefore, the given inequality holds for every positive integer n.

7.4. (a) We proceed by induction on n.
For $n = 2$, we have $|\sin(x_1 + x_2)| \leq |\sin x_1| + |\sin x_2|$.
Indeed, since $\sin(x_1 + x_2) = \sin x_1 \cos x_2 + \sin x_2 \cos x_1$ and $|\cos\alpha| \leq 1$, it follows that $|\sin(x_1 + x_2)| \leq |\sin x_1 \cos x_2| + |\sin x_2 \cos x_1| \leq |\sin x_1| + |\sin x_2|$.
Assume that the given inequality holds for $n = k$, and let us prove that it holds for $n = k+1$.
Indeed, we have

$$|\sin((x_1 + \cdots + x_k) + x_{k+1})| \leq |\sin(x_1 + \cdots + x_k)| + |\sin x_{k+1}| \leq (|\sin x_1| + \cdots + |\sin x_k|) + |\sin x_{k+1}|,$$

which is the given inequality for $n = k+1$.
Therefore, the given inequality holds for every positive integer n.
(b) As for $x \in [0, \pi]$, we have that $\sin x \geq 0$, and then from part (a), it follows that $\sin x_1 + \cdots + \sin x_n \geq |\sin(x_1 + \cdots + x_n)| \geq \sin(x_1 + \cdots + x_n)$.
(c) We have

$$|\cos x_1| + |\cos x_2| + |\cos x_3| + |\cos x_4| + |\cos x_5| = \left|\sin\left(\frac{\pi}{2} - x_1\right)\right| + \cdots + \left|\sin\left(\frac{\pi}{2} - x_5\right)\right|,$$

and then from part (a) it follows that

$$\left|\sin\left(\frac{\pi}{2}-x_1\right)\right|+\cdots+\left|\sin\left(\frac{\pi}{2}-x_5\right)\right|\geq\left|\sin\left(\left(\frac{\pi}{2}-x_1\right)+\cdots+\left(\frac{\pi}{2}-x_5\right)\right)\right|=\left|\sin\frac{5\pi}{2}\right|=1.$$

Therefore, $|\cos x_1|+|\cos x_2|+|\cos x_3|+|\cos x_4|+|\cos x_5|\geq 1$.

7.5. (a) Let us first prove the given inequality for $n=1,3,5,7,\ldots,2k+1,\ldots$.
For $n=1$, we have $f(x_1)\geq f(x_1)$.
Assume that the given inequality holds for $n=2k-1$ $(k\in\mathbb{N})$, and let us prove that it holds for $n=2k+1$.
We have $f(x_1)-f(x_2)+\cdots+f(x_{2k-1})\geq f(x_1-x_2+\cdots+x_{2k-1})$, and therefore,

$$f(x_1)-f(x_2)+\cdots+f(x_{2k-1})-f(x_{2k})+f(x_{2k+1})\geq f(x_1-x_2+x_3-\cdots+x_{2k-1})-f(x_{2k})+f(x_{2k+1}).$$

Note that $x_1\geq x=x_1-x_2+x_3-\cdots+x_{2k-1}\geq y=x_{2k}\geq z=x_{2k+1}\geq 0$, since

$$x_1-x=(x_2-x_3)+\cdots+(x_{2k-2}-x_{2k-1})\geq 0 \text{ and } x-y=(x_1-x_2)+\cdots+(x_{2k-1}-x_{2k})\geq 0.$$

Therefore, $f(x)-f(y)+f(x)\geq f(x-y+z)$.
Thus

$$\begin{aligned}&f(x_1)-f(x_2)+\cdots+f(x_{2k-1})-f(x_{2k})+f(x_{2k+1})\\&\geq f(x_1-x_2+\cdots+x_{2k-1})-f(x_{2k})+f(x_{2k+1})\\&\geq f(x_1-x_2+x_3-\cdots+x_{2k-1}-x_{2k}+x_{2k+1}),\end{aligned}$$

from which it follows that the given inequality holds for $n=2k+1$.
Hence, the given inequality holds for an arbitrary odd number n.
Now let us prove that the given inequality holds for $n=2k$. Let $x_1\geq x_2\geq\cdots\geq x_{2k}$.
Let us take $x_{2k+1}=0$, from which we obtain

$$f(x_1)-f(x_2)+\cdots-f(x_{2k})+f(0)\geq f(x_1-x_2+\cdots-x_{2k}+0), \text{ or}$$
$$f(x_1)-f(x_2)+\cdots-f(x_{2k})\geq f(x_1-x_2+\cdots-x_{2k}), \text{ as } f(0)\leq 0.$$

(b) Let us prove that $\tan x-\tan y+\tan z\geq\tan(x-y+z)$, where $\frac{\pi}{2}>x\geq y\geq z\geq 0$.
The last inequality can be rewritten as $\tan x-\tan y\geq\tan(x-y+z)-\tan z$, or $\frac{\sin(x-y)}{\cos x\cos y}\geq\frac{\sin(x-y)}{\cos z\cos(x-y+z)}$.
If $x=y$, then the proof is obvious.
If $x\neq y$, then the given inequality is equivalent to the following inequality: $\cos z\cos(x-y+z)\geq\cos x\cos y$, which holds because $0\leq z\leq y<\frac{\pi}{2}$ and $0\leq x-y+z\leq x<\frac{\pi}{2}$.
Let us consider the function $f(x)=\tan x$. Since $f(0)=0$ and the assumption of Problem 7.5(a) holds, it follows that the given inequality holds.

(c) The proof follows from Problem 7.5(a) and Problem 9.34.

7.6. (a) Let us begin by proving the more general inequality

$$(x_1 + \cdots + x_n)^2 \geq 4(x_1x_2 + x_2x_3 + \cdots + x_{n-1}x_n + x_nx_1), \qquad (7.6)$$

where $n \geq 4, \quad x_1 > 0, \ldots, x_n > 0$.

We proceed by induction on n.

If $n = 4$, then the inequality $(x_1 + x_2 + x_3 + x_4)^2 \geq 4(x_1x_2 + x_2x_3 + x_3x_4 + x_4x_1)$ holds, since

$$(x_1 + x_2 + x_3 + x_4)^2 - 4(x_1x_2 + x_2x_3 + x_3x_4 + x_4x_1) = (x_1 - x_2 + x_3 - x_4)^2 \geq 0.$$

Assume that (7.6) holds for $n = k$, and let us prove that (7.6) holds for $n = k + 1$.

Let $\max(x_1, \ldots, x_{k+1}) = x_i$. It follows that

$$(x_1 + \cdots + (x_{i-2} + x_{i-1}) + x_i + \cdots + x_{k+1})^2$$
$$\geq 4(x_1x_2 + x_2x_3 + \cdots + x_{i-3}(x_{i-2} + x_{i-1}) + (x_{i-2} + x_{i-1})x_i + x_ix_{i+1}$$
$$+ \cdots + x_kx_{k+1} + x_{k+1}x_1) > 4(x_1x_2 + x_2x_3 + \cdots + x_{i-3}x_{i-2} + x_{i-2}x_i + x_{i-1}x_i$$
$$+ x_ix_{i+1} + \cdots + x_kx_{k+1} + x_{k+1}x_1) \geq 4(x_1x_2 + x_2x_3 + \cdots + x_{i-3}x_{i-2} + x_{i-2}x_{i-1}$$
$$+ x_{i-1}x_i + x_ix_{i+1} + \cdots + x_kx_{k+1} + x_{k+1}x_1)$$

(here we assume that $x_0 = x_{k+1}, \quad x_{-1} = x_k$).

(b) If $n \geq 4$, then we have

$$x_1\sqrt{x_n^2 + x_2^2} + x_2\sqrt{x_1^2 + x_3^2} + \cdots + x_{n-1}\sqrt{x_{n-2}^2 + x_n^2} + x_n\sqrt{x_{n-1}^2 + x_1^2}$$
$$\leq x_1(x_n + x_2) + x_2(x_1 + x_3) + \cdots + x_{n-1}(x_{n-2} + x_n) + x_n(x_{n-1} + x_1)$$
$$= 2(x_1x_2 + x_2x_3 + x_{n-1}x_n + x_nx_1) \leq \frac{1}{2}(x_1 + \cdots + x_n)^2.$$

The last inequality holds by to Problem 7.6(a).

If $n = 3$, then without loss of generality one can assume that $\max(x_1, x_2, x_3) = x_3$.

Thus, it follows that

$$x_1\sqrt{x_3^2 + x_2^2} + x_2\sqrt{x_1^2 + x_3^2} + x_3\sqrt{x_1^2 + x_2^2} \leq x_1\left(x_3 + \frac{x_2}{2}\right) + x_2\left(x_3 + \frac{x_1}{2}\right)$$
$$+ \frac{x_3^2 + x_1^2 + x_2^2}{2} = \frac{1}{2}(x_1 + x_2 + x_3)^2.$$

7.7. We proceed by induction on n.

If $n = 1$, then $\frac{1}{2}x_1^2 \leq x_1^2$. Obviously, the last inequality holds.

Assume that the given inequality holds for $n = k$, and let us prove that it holds for $n = k + 1$.

If $n = k$, we have $\frac{1}{2}(x_1 + \cdots + x_k)^2 \leq (x_1 + \cdots + kx_k)\max(x_1, \ldots, x_k)$. Adding to both sides of this inequality the expression

$x_{k+1}(x_1 + \cdots + x_k) + \frac{x_{k+1}^2}{2}$, we obtain on the left-hand side the expression $\frac{1}{2}(x_1 + \cdots + x_{k+1})^2$, and on the right-hand side the expression $(x_1 + \cdots + kx_k)\max(x_1, \ldots, x_k) + x_{k+1}(x_1 + \cdots + x_k) + \frac{x_{k+1}^2}{2}$, which is not greater than $(x_1 + \cdots + (k+1)x_{k+1})\max(x_1, \ldots, x_{k+1})$.

Indeed, since $\max(x_1, \ldots, x_k) \leq \max(x_1, \ldots, x_{k+1}) = c$, $\frac{x_{k+1}^2}{2} \leq x_{k+1}^2 \leq cx_{k+1}$, we have

$$(x_1 + \cdots + kx_k)\max(x_1, \ldots, x_k) + x_{k+1}(x_1 + \cdots + x_k) + \frac{x_{k+1}^2}{2} \leq (x_1 + \cdots + kx_k)c$$
$$+ kcx_{k+1} + cx_{k+1} = (x_1 + \cdots + (k+1)x_{k+1})c.$$

Therefore, the given inequality holds for every positive integer n.

7.8. Let us first prove that for arbitrary numbers $x_1, \ldots, x_n$ belonging to the domain I, we have

$$\frac{f(x_1) + \cdots + f(x_n)}{n} \geq f\Big(\frac{x_1 + \cdots + x_n}{n}\Big). \tag{7.7}$$

We proceed by induction on n.

If $n = 2$, then inequality (7.7) coincides with the condition of Theorem 11.1.

Assume that for $n = k$, inequality (7.7) holds, and let us prove that it holds for $n = k + 1$.

Let us consider the following expression:

$$f(x_1) + \cdots + f(x_k) + f(x_{k+1}) + (k-1)f\Big(\frac{x_1 + \cdots + x_{k+1}}{k+1}\Big) = A.$$

Note that

$$A = (f(x_1) + \cdots + f(x_k)) + f(x_{k+1}) + \underbrace{f\Big(\frac{x_1 + \cdots + x_{k+1}}{k+1}\Big) + \cdots + f\Big(\frac{x_1 + \cdots + x_{k+1}}{k+1}\Big)}_{k-1}$$

$$\geq kf\Big(\frac{x_1 + \cdots + x_k}{k}\Big) + kf\left(\frac{x_{k+1} + (k-1)\frac{x_1 + \cdots + x_{k+1}}{k+1}}{k}\right)$$

$$\geq 2kf\left(\frac{\frac{x_1 + \cdots + x_k}{k} + \frac{x_{k+1} + (k-1)\frac{x_1 + \cdots + x_{k+1}}{k+1}}{k}}{2}\right) = 2kf\Big(\frac{x_1 + \cdots + x_{k+1}}{k+1}\Big).$$

Hence, we deduce that $f(x_1) + \cdots + f(x_{k+1}) + (k-1)f\big(\frac{x_1 + \cdots + x_{k+1}}{k+1}\big) \geq 2kf\big(\frac{x_1 + \cdots + x_{k+1}}{k+1}\big)$,
and therefore,

$$\frac{f(x_1) + \cdots + f(x_{k+1})}{k+1} \geq f\Big(\frac{x_1 + \cdots + x_{k+1}}{k+1}\Big).$$

Thus, it follows that inequality (7.7) holds.

Now let us prove the inequality (11.3). Let $\alpha_i = \frac{p_i}{q_i}$, $i = 1, \ldots, n$, where $p_i, q_i \in \mathbb{N}$, $i = 1, \ldots, n$.

Let us denote the least common multiple of the numbers $q_1, \ldots, q_n$ by q Therefore, $\frac{p_i}{q_i} = \frac{l_i}{q}$, where $l_i \in \mathbb{N}$, $i = 1, \ldots, n$.
Let us estimate the left-hand side of the given inequality, $\alpha_1 f(x_1) + \cdots + \alpha_n f(x_n) = \frac{l_1 f(x_1) + \cdots + l_n f(x_n)}{q} = B$.
Applying inequality (7.7) to the numbers $\underbrace{x_1, \ldots, x_1}_{l_1}, \ldots, \underbrace{x_n, \ldots, x_n}_{l_n}$, note that their total number is equal to q and since $\alpha_1 + \cdots + \alpha_n = 1$, we obtain $B \geq f\left(\frac{l_1 x_1 + l_2 x_2 + \cdots + l_n x_n}{q}\right) = f(\alpha_1 x_1 + \cdots + \alpha_n x_n)$.

7.9. We proceed by induction on n.
If $n = 2$, then inequality (11.7) coincides with the condition (11.5) of Theorem 11.2.
Assuming that (11.7) holds for $n = k$, let us prove that it holds for $n = k+1$, that is, that $\beta_1 f(a_1) + \cdots + \beta_{k+1} f(a_{k+1}) \geq f(\beta_1 \alpha_1 + \cdots + \beta_{k+1} \alpha_{k+1})$, where

$$\beta_1 \geq 0, \ldots, \beta_{k+1} \geq 0 \text{ and } \beta_1 + \cdots + \beta_{k+1} = 1, \; a_1, \ldots, a_n \in I.$$

We have

$$\begin{aligned}
&\beta_1 f(a_1) + \cdots + \beta_{k+1} f(a_{k+1}) \\
&\quad = (\beta_1 + \cdots + \beta_k)\left(\frac{\beta_1}{\beta_1 + \cdots + \beta_k} f(a_1) + \ldots + \frac{\beta_k}{\beta_1 + \cdots + \beta_k} f(a_k)\right) + \beta_{k+1} f(a_{k+1}) \\
&\quad \geq (\beta_1 + \cdots + \beta_k) f\left(\frac{\beta_1 a_1 + \cdots + \beta_k a_k}{\beta_1 + \cdots + \beta_k}\right) + \beta_{k+1} f(a_{k+1}) \\
&\quad \geq f\left((\beta_1 + \cdots + \beta_k)\frac{\beta_1 a_1 + \cdots + \beta_k a_k}{\beta_1 + \cdots + \beta_k} + \beta_{k+1} a_{k+1}\right) = f(\beta_1 a_1 + \cdots + \beta_k a_k + \beta_{k+1} a_{k+1}),
\end{aligned}$$

therefore $\beta_1 f(a_1) + \cdots + \beta_{k+1} f(a_{k+1}) \geq f(\beta_1 \alpha_1 + \cdots + \beta_{k+1} \alpha_{k+1})$.
We have used (11.7) for $n = k$, substituting x_i by a_i, and α_i by $\frac{\beta_i}{\beta_1 + \cdots + \beta_k}$, $i = 1, \ldots, k$.
Hence inequality (11.7) holds for every positive integer n.

7.10. (a) We proceed by induction on n.
For $n = 1$, the given inequality holds because $a_1 \leq a_1$.
Assume that the given inequality holds for $n = k$, and let us prove that it holds for $n = k + 1$.
For $n = k$, we have $a_1 + \cdots + a_k^k \leq k a_1 \cdots a_k$. On the other hand, by to the assumption of the problem, we have $1 \leq a_{k+1}$. Multiplying these inequalities, we obtain $a_1 + \cdots + a_k^k \leq k a_1 \cdots a_k a_{k+1}$. By to the assumption of the problem, we have $a_{k+1}^{k+1} \leq a_1 \cdots a_{k+1}$. Summing these inequalities, we deduce that

$$a_1 + \cdots + a_k^k + a_{k+1}^{k+1} \leq k a_1 \cdots a_{k+1} + a_1 \cdots a_{k+1} = (k+1) a_1 \cdots a_{k+1}.$$

Therefore, the given inequality holds for every positive integer n.
(b) We proceed by induction on n.
For $n = 1$, the given inequality holds because $a_1 \geq a_1$.

Assume that the given inequality holds for $n = k$, and let us prove that it holds for $n = k + 1$.
For $n = k$, we have $a_1 + \cdots + a_k^k \geq ka_1 \cdots a_k$. On the other hand, by assumption, we have $1 \geq a_{k+1}$. Multiplying these two inequalities, we deduce that $a_1 + \cdots + a_k^k \geq ka_1 \cdots a_k a_{k+1}$.
By to the assumption of the problem, we have $a_{k+1}^{k+1} \geq a_1 \cdots a_{k+1}$.
Summing these inequalities, we obtain $a_1 + \cdots + a_k^k + a_{k+1}^{k+1} \geq (k+1)a_1 \cdots a_{k+1}$.
Therefore, the given inequality holds for every positive integer n.

7.11. (a) We proceed by induction on n.
For $n = 2$, we have $\frac{a_2^2}{a_1} \geq 4(a_2 - a_1)$, $\left(a_2^2 - 4a_1a_2 + 4a_1^2 = (a_2 - 2a_1)^2 \geq 0\right)$.
Assume that the given inequality holds for $n = k$, and let us prove that it holds for $n = k + 1$.
For $n = k$, we have $\frac{a_2^2}{a_1} + \cdots + \frac{a_k^2}{a_{k-1}} \geq 4(a_k - a_1)$.
Let us add to both sides of this inequality the expression $4(a_{k+1} - a_k)$. Then we obtain $\frac{a_2^2}{a_1} + \cdots + \frac{a_k^2}{a_{k-1}} + 4(a_{k+1} - a_k) \geq 4(a_{k+1} - a_1)$.
It is left to prove that

$$\frac{a_2^2}{a_1} + \cdots + \frac{a_k^2}{a_{k-1}} + \frac{a_{k+1}^2}{a_k} \geq \frac{a_2^2}{a_1} + \cdots + \frac{a_k^2}{a_{k-1}} + 4(a_{k+1} - a_k).$$

Indeed, this inequality is equivalent to the following inequality:

$$\frac{a_{k+1}^2}{a_k} \geq 4(a_{k+1} - a_k), \text{ or } (a_{k+1} - 2a_k)^2 \geq 0.$$

Therefore, the given inequality holds for every positive integer n.
(b) We proceed by induction on n.
For $n = 2$, we have

$$\left(\frac{a_1^3}{b_1c_1} + \frac{a_2^3}{b_2c_2}\right)(b_1 + b_2)(c_1 + c_2) = a_1^3 + a_2^3$$
$$+ \left(a_1^3\frac{c_2}{c_1} + a_1^3\frac{b_2}{b_1} + a_2^3\frac{b_1c_1}{b_2c_2}\right) + \left(a_2^3\frac{c_1}{c_2} + a_2^3\frac{b_1}{b_2} + a_1^3\frac{b_2c_2}{b_1c_1}\right)$$
$$\geq a_1^3 + a_2^3 + 3\sqrt[3]{a_1^3\frac{c_2}{c_1} \cdot a_1^3\frac{b_2}{b_1} \cdot a_2^3\frac{b_1c_1}{b_2c_2}} + 3\sqrt[3]{a_2^3\frac{c_1}{c_2} \cdot a_2^3\frac{b_1}{b_2} \cdot a_1^3\frac{b_2c_2}{b_1c_1}} = (a_1 + a_2)^3,$$

and therefore,

$$\frac{a_1^3}{b_1c_1} + \frac{a_2^3}{b_2c_2} \geq \frac{(a_1 + a_2)^3}{(b_1 + b_2)(c_1 + c_2)}.$$

Assume that the given inequality holds for $n = k$, and let us prove that it holds for $n = k + 1$.
For $n = k$, we have $\frac{a_1^3}{b_1c_1} + \cdots + \frac{a_k^3}{b_kc_k} \geq \frac{(a_1 + \cdots + a_k)^3}{(b_1 + \cdots + b_k)(c_1 + \cdots + c_k)}$.

Let us add to both sides of this inequality the expression $\frac{a_{k+1}^3}{b_{k+1}c_{k+1}}$. Then we obtain

$$\frac{a_1^3}{b_1c_1}+\cdots+\frac{a_k^3}{b_kc_k}+\frac{a_{k+1}^3}{b_{k+1}c_{k+1}} \geq \frac{(a_1+\cdots+a_k)^3}{(b_1+\cdots+b_k)(c_1+\cdots+c_k)}+\frac{a_{k+1}^3}{b_{k+1}c_{k+1}}$$
$$\geq \frac{((a_1+\cdots+a_k)+a_{k+1})^3}{((b_1+\cdots+b_k)+b_{k+1})((c_1+\cdots+c_k)+c_{k+1})},$$

and hence the given inequality holds for $n=k+1$.
Therefore, the given inequality holds for every positive integer n.

7.12. (a) We proceed by induction on n.
For $n=2$, we have

$$(x_2+x_1)(f(x_2)-f(x_1)) \geq (x_2+x_1)(f(x_2)-f(x_1)).$$

For $n=k$, we have

$$(x_2+x_1)(f(x_2)-f(x_1))+\cdots+(x_k+x_{k-1})(f(x_k)-f(x_{k-1})) \geq (x_k+x_1)(f(x_k)-f(x_1)).$$

Let us add to both sides of this inequality the expression $(x_{k+1}+x_k)(f(x_{k+1})-f(x_k))$, from which we obtain

$$(x_2+x_1)(f(x_2)-f(x_1))+\cdots+(x_k+x_{k-1})(f(x_k)-f(x_{k-1}))$$
$$+(x_{k+1}+x_k)(f(x_{k+1})-f(x_k)) \geq (x_k+x_1)(f(x_k)-f(x_1))$$
$$+(x_{k+1}+x_k)(f(x_{k+1})-f(x_k)).$$

In order to prove that the given inequality holds for $n=k+1$, it is sufficient to prove
the following inequality:

$$(x_k+x_1)(f(x_k)-f(x_1))+(x_{k+1}+x_k)(f(x_{k+1})-f(x_k)) \geq (x_{k+1}+x_1)(f(x_{k+1})-f(x_1)), \text{ or}$$
$$(x_{k+1}-x_k)f(x_1)+(x_k-x_1)f(x_{k+1}) \geq (x_{k+1}-x_1)f(x_k). \tag{7.8}$$

Since $f(x)$ is a convex function, it follows that

$$\frac{x_{k+1}-x_k}{x_{k+1}-x_1}f(x_1)+\frac{x_k-x_1}{x_{k+1}-x_1}f(x_{k+1}) \geq f\left(\frac{x_{k+1}-x_k}{x_{k+1}-x_1}x_1+\frac{x_k-x_1}{x_{k+1}-x_1}x_{k+1}\right)=f(x_k),$$

which implies inequality (7.8).
This ends the proof of the given inequality for $n=k+1$.
Therefore, the given inequality holds for every positive integer n $(n>1)$.
(b) Note that one can rewrite inequality (a) of this problem in the following way:

$$x_nf(x_1)+x_1f(x_2)+x_2f(x_3)+\cdots+x_{n-1}f(x_n) \geq x_2f(x_1)+x_3f(x_2)+\cdots+x_nf(x_{n-1})+x_1f(x_n),$$

where on taking $f(x)=-\sqrt{x}$, we obtain

$$a(-\sqrt{c})+c\left(-\sqrt{b}\right)+b(-\sqrt{a})\geq b(-\sqrt{c})+a\left(-\sqrt{b}\right)+c(-\sqrt{a}).$$

Thus, it follows that $a\sqrt{c}+c\sqrt{b}+b\sqrt{a}\leq b\sqrt{c}+a\sqrt{b}+c\sqrt{a}$.
(c) Consider the function $f(x)=-\ln x$. By to inequality (a), we have
$x_n(-\ln x_1)+x_1(-\ln x_2)+\cdots+x_{n-1}(-\ln x_n)\geq x_2(-\ln x_1)+\cdots+x_n(-\ln x_{n-1})+x_1(-\ln x_n)$, or
$x_1^{x_n}\cdot x_2^{x_1}\cdots x_n^{x_{n-1}}\leq x_1^{x_2}\cdot x_2^{x_3}\cdots x_{n-1}^{x_n}\cdot x_n^{x_1}$.

(d) Consider the function

$$f(x)=\frac{x}{(a+b+c-x)(a+b+c+x)}=\frac{1}{2}\left(\frac{1}{a+b+c-x}-\frac{1}{a+b+c+x}\right).$$

We have $f'(x)=\frac{1}{2}\left(\frac{1}{(a+b+c-x)^2}+\frac{1}{(a+b+c+x)^2}\right)$.
If $0\leq x\leq a$, then $f''(x)=\frac{1}{(a+b+c-x)^3}-\frac{1}{(a+b+c+x)^3}\geq 0$, and thus it follows that $af(c)+cf(b)+bf(a)\geq bf(c)+af(b)+cf(a)$, or

$$\begin{aligned}&a\cdot\frac{c}{(a+b)(a+b+2c)}+c\cdot\frac{b}{(a+c)(a+c+2b)}+b\cdot\frac{a}{(b+c)(b+c+2a)}\\&\geq b\cdot\frac{c}{(a+b)(a+b+2c)}+a\cdot\frac{b}{(a+c)(a+c+2b)}+c\cdot\frac{a}{(b+c)(b+c+2a)}.\end{aligned}$$

7.13. Note that the given inequality is equivalent to the following inequality:

$$\begin{aligned}&(a_1+\cdots+a_{n-1})n(n+1)+(a_1+\cdots+a_{n-1})n(n-1)+(a_n+a_{n+1})n(n-1)\\&\geq 2\left(n^2-1\right)(a_1+\cdots+a_{n-1})+2\left(n^2-1\right)a_n,\end{aligned}$$

or

$$2(a_1+\cdots+a_{n-1})+n(n-1)a_{n+1}\geq\left(n^2+n-2\right)a_n. \quad (7.9)$$

We prove this inequality by induction.
For $n=2$, we obtain the following obvious inequality: $2a_1+2a_3\geq 4a_2$.
Assume that the given inequality holds for $n=k$, that is,

$$2(a_1+\cdots+a_{k-1})+k(k-1)a_{k+1}\geq\left(k^2+k-2\right)a_k, \quad (7.10)$$

and let us prove that it holds for $n=k+1$.
Let us add to both sides of inequality (7.10) the expression $2a_k+(k+1)ka_{k+2}-k(k-1)a_{k+1}$, from which we deduce that

$$\begin{aligned}&2(a_1+\cdots+a_{k-1}+a_k)+(k+1)ka_{k+2}\geq\left(k^2+k\right)a_k+\left(k^2+k\right)a_{k+2}-k(k-1)a_{k+1}\\&=\left(k^2+k\right)(a_k+a_{k+2})-k(k-1)a_{k+1}\geq\left(k^2+k\right)\cdot 2a_{k+1}-k(k-1)a_{k+1}=\left(k^2+3k\right)a_{k+1}.\end{aligned}$$

Hence, we obtain $2(a_1+\cdots+a_k)+(k+1)ka_{k+2}\geq\left(k^2+3k\right)a_{k+1}$, and thus it follows that the given inequality holds for $n=k+1$.
Therefore, inequality (7.9) holds for every positive integer n.

7.14. (a) Let us rewrite the given inequality as $a_1 + a_3 + \cdots + a_{2n-1} \geq \frac{n}{n+1}(a_0 + a_2 + \cdots + a_{2n})$ and prove it by induction.
For $n = 1$, we obtain the following obvious inequality: $a_1 \geq \frac{a_0+a_2}{2}$.
Assume that the given inequality holds for $n = k$, and let us prove that it holds for $n = k + 1$, that is,

$$a_1 + a_3 + \cdots + a_{2k-1} \geq \frac{k}{k+1}(a_0 + a_2 + \cdots + a_{2k}), \text{ or}$$

$$a_1 + a_2 + \cdots + a_{2k-1} + a_{2k+1} \geq \frac{k}{k+1}(a_0 + a_2 + \cdots + a_{2k}) + a_{2k+1}.$$

Let us prove that
$\frac{k}{k+1}(a_0 + a_2 + \cdots + a_{2k}) + a_{2k+1} \geq \frac{k+1}{k+2}(a_0 + a_2 + \cdots + a_{2k} + a_{2k+2})$,
or $a_{2k+1} \geq \frac{1}{(k+1)(k+2)}(a_0 + a_2 + \cdots + a_{2k}) + \frac{k+1}{k+2}a_{2k+2}$.
Since $a_{2k+1} \geq \frac{a_{2k}+a_{2k+2}}{2}$, it follows that in order to complete the proof of the last inequality, it is sufficient to prove that $\frac{a_{2k}+a_{2k+2}}{2} \geq \frac{1}{(k+1)(k+2)}(a_0 + \cdots + a_{2k}) + \frac{k+1}{k+2}a_{2k+2}$, or

$$\frac{k(k+3)}{2}a_{2k} - \frac{k(k+1)}{2}a_{2k+2} \geq a_0 + a_2 + \cdots + a_{2k-2}.$$

We prove this inequality by induction.
For $k = 1$, we obtain the obvious inequality $2a_2 - a_4 \geq a_0$, since $a_2 \geq \frac{a_1+a_3}{2} \geq \frac{1}{2}\left(\frac{a_0+a_2}{2} + \frac{a_2+a_4}{2}\right)$.
Let $k = m$, then the following inequality holds:

$$\frac{m(m+3)}{2}a_{2m} - \frac{m(m+1)}{2}a_{2m+2} \geq a_0 + \cdots + a_{2m-2}.$$

Therefore,

$$\frac{m(m+3)}{2}a_{2m} - \frac{m(m+1)}{2}a_{2m+2} + a_{2m} \geq a_0 + a_2 + \cdots + a_{2m-2} + a_{2m}.$$

It is sufficient to prove that
$$\frac{m(m+3)}{2}a_{2m} - \frac{m(m+1)}{2}a_{2m+2} + a_{2m} \leq \frac{(m+1)(m+4)}{2}a_{2m+2} - \frac{(m+1)(m+2)}{2}a_{2m+4}, \text{ or}$$
$$(m+1)(m+2)a_{2m+2} \geq \frac{(m+1)(m+2)}{2}a_{2m} + \frac{(m+1)(m+2)}{2}a_{2m+4}.$$

Note that this inequality holds.
(b) Let $a_k = -a^k$. As $-a^k \geq \frac{-a^{k-1}-a^{k+1}}{2}$. Then by to inequality (a) of this problem, we have $\frac{-a-a^3-\cdots-a^{2n-1}}{n} \geq \frac{-1-a^2-\cdots-a^{2n}}{n+1}$, and therefore $\frac{1+a^2+\cdots+a^{2n}}{a+a^3+\cdots+a^{2n-1}} \geq \frac{n+1}{n}$.

7.15. We proceed by induction on n.
For $n = 1$, we have $1 > \ln 2$.
Assume that the given inequality holds for $n = k$, that is, $1 + \frac{1}{2} + \cdots + \frac{1}{k} > \ln(k+1)$.

Let us add to both sides of this inequality the expression $\frac{1}{k+1}$. Then $1 + \frac{1}{2} + \cdots + \frac{1}{k} + \frac{1}{k+1} > \ln(k+1) + \frac{1}{k+1}$.
Now let us prove that $\ln(k+1) + \frac{1}{k+1} > \ln(k+2)$, which is equivalent to

$$\frac{1}{k+1} > \ln(k+2) - \ln(k+1) = \ln\frac{k+2}{k+1}, \text{ or } \ 1 > \ln\left(\frac{k+2}{k+1}\right)^{k+1}.$$

Since $\lim_{n\to\infty}\left(1+\frac{1}{n}\right)^n = e$ and the sequence $\left(1+\frac{1}{n}\right)^n$ is monotonically increasing (Problem 3.16(a)), it follows that

$$e > \left(1 + \frac{1}{k+1}\right)^{k+1}, \text{ or } \ 1 > \ln\left(\frac{k+2}{k+1}\right)^{k+1}.$$

7.16. We proceed by induction on n.
For $n = 1$, we have $1 \leq 1$.
Assume that the given inequality holds for $n = k$, that is, $1+\frac{1}{2\sqrt{2}}+\cdots+\frac{1}{k\sqrt{k}} \leq 3 - \frac{2}{\sqrt{k}}$, and let us add to both sides of the last inequality the expression $\frac{1}{(k+1)\sqrt{k+1}}$. Then we obtain

$$1 + \frac{1}{2\sqrt{2}} + \cdots + \frac{1}{k\sqrt{k}} + \frac{1}{(k+1)\sqrt{k+1}} \leq 3 - \frac{2}{\sqrt{k}} + \frac{1}{(k+1)\sqrt{k+1}}.$$

In order to complete the proof, it is sufficient to prove that $3 - \frac{2}{\sqrt{k}} + \frac{1}{(k+1)\sqrt{k+1}} \leq 3 - \frac{2}{\sqrt{k+1}}$.
Indeed, since $\frac{1}{k+1} \leq \frac{2}{\sqrt{k(k+1)}+k}$, we have

$$\frac{1}{(k+1)\sqrt{k+1}} \leq \frac{2}{\sqrt{k(k+1)}\left(\sqrt{k+1}+\sqrt{k}\right)} = \frac{2\left(\sqrt{k+1}-\sqrt{k}\right)}{\sqrt{k(k+1)}} = \frac{2}{\sqrt{k}} - \frac{2}{\sqrt{k+1}}.$$

Therefore, the given inequality holds for every positive integer n.

7.17. We proceed by induction on n.
For $n = 1$, we have $1 + \alpha \geq 1 + \alpha$.
Assume that the given inequality holds for $n = k$, and let us prove that it holds for $n = k + 1$.
For $n = k$, we have $(1+\alpha)^k \geq 1 + k\alpha + \frac{k(k-1)}{2}\alpha^2$.
Hence, we deduce that

$$(1+\alpha)^{k+1} \geq (1+\alpha)\left(1 + k\alpha + \frac{k(k-1)}{2}\alpha^2\right) = 1 + k\alpha + \frac{k(k-1)}{2}\alpha^2 + \alpha + k\alpha^2$$
$$+ \frac{k(k-1)}{2}\alpha^3 = 1 + (k+1)\alpha + \frac{k(k+1)}{2}\alpha^2 + \frac{k(k-1)}{2}\alpha^3 \geq 1 + (k+1)\alpha + \frac{k(k+1)}{2}\alpha^2.$$

It follows that $(1+\alpha)^{k+1} \geq 1 + (k+1)\alpha + \frac{k(k+1)}{2}\alpha^2$, and therefore, the given inequality holds for every positive integer n.

7.18. We proceed by induction on k.
For $k = 1$, we have $1 > \frac{2}{e}$.

Assume that the given inequality holds for $k = m$, that is, $m! > \left(\frac{m+1}{e}\right)^m$, and let us prove that it holds true for $k = m + 1$, that is, $(m+1)! > \left(\frac{m+2}{e}\right)^{m+1}$. We have $m! > \left(\frac{m+1}{e}\right)^m$, and therefore

$$(m+1)! > e\left(\frac{m+1}{e}\right)^{m+1}. \qquad (7.11)$$

Since $\lim_{n\to\infty}\left(1+\frac{1}{n}\right)^n = e$ and the sequence $\left(1+\frac{1}{n}\right)^n$ is monotonically increasing (Problem 3.16(a)), it follows that

$$e > \left(1+\frac{1}{m+1}\right)^{m+1}. \qquad (7.12)$$

Multiplying inequalities (7.11) and (7.12), we obtain $(m+1)! > \left(\frac{m+2}{e}\right)^{m+1}$. Therefore, the given inequality holds for every positive integer k.

7.19. Let us first prove that

$$2|\sin x| + |\sin 2x| \le \frac{3\sqrt{3}}{2}. \qquad (7.13)$$

We have $2|\sin x|+|\sin 2x| \le \frac{2}{\sqrt{3}}\sqrt{(3-3|\cos x|)(1+|\cos x|)^3} \le \frac{2}{\sqrt{3}}\sqrt{\left(\frac{6}{4}\right)^4} = \frac{3\sqrt{3}}{2}$.

We proceed by induction on n.
For $n = 0$, we obtain the following obvious inequality: $|\sin x| \le 1$.
Assume that the given inequality holds for $n \le k$, that is, $|\sin x| + |\sin 2x| + \cdots + |\sin 2^n x| \le 1 + \frac{\sqrt{3}}{2}n$, for all x Let us prove that it holds for $n = k + 1$.
Let us consider the following two cases.
(a) If $|\sin x| \le \frac{\sqrt{3}}{2}$, then $|\sin x| + \left(|\sin 2x| + \cdots + \left|\sin(2^k \cdot 2x)\right|\right) \le \frac{\sqrt{3}}{2} + 1 + \frac{\sqrt{3}}{2}k = 1 + \frac{\sqrt{3}}{2}(k+1)$.
(b) If $|\sin x| > \frac{\sqrt{3}}{2}$, then from (7.13), it follows that $|\sin x| + |\sin 2x| < \sqrt{3}$, whence

$$|\sin x| + |\sin 2x| + \left(|\sin 4x| + \cdots + \left|\sin(2^{k-1}\cdot 4x)\right|\right) < \sqrt{3} + \left(1 + \frac{\sqrt{3}}{2}(k-1)\right) = 1 + \frac{\sqrt{3}}{2}(k+1).$$

Therefore, the given inequality holds for every positive integer n.

7.20. Consider the function $f_n(\alpha) = \cos\alpha + \frac{\cos 2\alpha}{2} + \cdots + \frac{\cos n\alpha}{n}$ in $\left[0, \frac{\pi}{2}\right]$ and suppose it attains its minimum value in this interval at the point α_n.
Let us consider the following three cases.
(a) If $\alpha_n = 0$, then $f_n(\alpha) \ge 1 + \frac{1}{2} + \cdots + \frac{1}{n} > -\frac{1}{2}$.
(b) If $\alpha_n = \frac{\pi}{2}$, then $f_n(\alpha) \ge f_n\left(\frac{\pi}{2}\right) \ge -\frac{1}{2}$, as $f_1\left(\frac{\pi}{2}\right) = 0$, $f_2\left(\frac{\pi}{2}\right) = f_3\left(\frac{\pi}{2}\right) = -\frac{1}{2}$, $f_4\left(\frac{\pi}{2}\right) = f_5\left(\frac{\pi}{2}\right) = -\frac{1}{2} + \frac{1}{4}$, $f_6\left(\frac{\pi}{2}\right) = f_7\left(\frac{\pi}{2}\right) = -\frac{1}{2} + \left(\frac{1}{4} - \frac{1}{6}\right)$, and so on.

(c) If $0 < \alpha_n < \frac{\pi}{2}$, then $f_n'(\alpha_n) = 0$, that is, $-\sin\alpha_n - \sin 2\alpha_n - \cdots - \sin n\alpha_n = 0$.
We obtain that $2\sin\frac{\alpha_n}{2}\sin\alpha_n + \cdots + 2\sin\frac{\alpha_n}{2}\sin n\alpha_n = 0$, whence $\cos\frac{\alpha_n}{2} = \cos\left(n\alpha_n + \frac{\alpha_n}{2}\right)$.
Thus, $\sin\frac{\alpha_n}{2} = \pm\sin\left(n\alpha_n + \frac{\alpha_n}{2}\right)$, and it follows that

$$\cos n\alpha_n = \cos\left(n\alpha_n + \frac{\alpha_n}{2}\right)\cos\frac{\alpha_n}{2} + \sin\left(n\alpha_n + \frac{\alpha_n}{2}\right)\sin\frac{\alpha_n}{2} = \cos^2\frac{\alpha_n}{2} \pm \sin^2\frac{\alpha_n}{2} > 0.$$

We deduce that $f_n(\alpha) \geq f_{n-1}(\alpha_n) = \cos\alpha_n + \cdots + \frac{\cos(n-1)\alpha_n}{n-1} \geq -\frac{1}{2}$, since $\alpha_n \in \left(0, \frac{\pi}{2}\right)$, and for $(n-1)$, the statement holds (for $n = 1$, we have $f_1(\alpha) = \cos\alpha \geq 0 > -\frac{1}{2}$).

7.21. We proceed by induction on n.
For $n = 2$, we have $\frac{a_1}{a_2} + \frac{a_2}{a_1} = 2 + \frac{(a_1-a_2)^2}{a_1a_2} < 2 + \frac{1}{a_1a_2} < 3$.
Let $n \geq 3$. Consider the following two cases.
(a) There exists $i \in \{2, \ldots, n-1\}$ such that

$$(a_i - a_{i-1})(a_i - a_{i+1}) \geq 0. \tag{7.14}$$

Then $\frac{a_{i-1}}{a_i} + \frac{a_i}{a_{i+1}} \leq \frac{a_{i-1}}{a_{i+1}} + 2$, since $(a_i - a_{i-1})(a_i - a_{i+1}) \leq |a_i - a_{i-1}||a_i - a_{i+1}| < 1 < a_ia_{i+1}$.
Taking this into consideration and the induction hypothesis, we obtain

$$\frac{a_1}{a_2} + \frac{a_2}{a_3} + \cdots + \frac{a_{i-1}}{a_i} + \frac{a_i}{a_{i+1}} + \cdots + \frac{a_{n-1}}{a_n} + \frac{a_n}{a_1} \leq \frac{a_1}{a_2} + \cdots + \frac{a_{i-2}}{a_{i-1}} + \frac{a_{i-1}}{a_{i+1}} + \frac{a_{i+1}}{a_{i+2}} + \cdots + \frac{a_n}{a_1} + 2 < 2(n-1) - 1 + 2 = 2n - 1,$$

since from (7.14) it follows that $|a_{i-1} - a_{i+1}| \leq \max(|a_i - a_{i-1}|, |a_i - a_{i+1}|) < 1$.
(b) If $(a_i - a_{i-1})(a_i - a_{i+1}) < 0, \quad i = 2, \ldots, n-1$, then either $a_1 < a_2 < \cdots < a_n$, or $a_1 > a_2 > \cdots > a_n$. In the first case, we have $\frac{a_1}{a_2} + \frac{a_2}{a_3} + \cdots + \frac{a_{n-1}}{a_n} + \frac{a_n}{a_1} \leq \underbrace{1 + \cdots + 1}_{n-1} + \frac{a_n}{a_1} < 2n - 1$,
since $a_n = (a_n - a_{n-1}) + (a_{n-1} - a_{n-2}) + \cdots + (a_2 - a_1) + a_1 < na_1$.
In the second case, we have $\frac{a_1}{a_2} + \frac{a_2}{a_3} + \cdots + \frac{a_{n-1}}{a_n} + \frac{a_n}{a_1} \leq \underbrace{2 + \cdots + 2}_{n-1} + 1 = 2n - 1$,
since $a_i = a_{i+1} + |a_{i+1} - a_i| < a_{i+1} + 1 < 2a_{i+1}, \quad i = 1, \ldots, n-1$.

7.22. (a) We proceed by induction on n.
For $n = 1$, we need to prove that $\frac{\sqrt{x_2 - 1}}{x_2} \leq \frac{1}{2}$, or $(x_2 - 2)^2 \geq 0$.
Assume that the inequality holds for $n = k$, and let us prove that it holds for $n = k + 1$, where $k \in \mathbb{N}$. We need to prove that if $1 = x_1 \leq x_2 \leq \cdots \leq x_{k+1} \leq x_{k+2}$, then $\frac{\sqrt{x_2 - x_1}}{x_2} + \frac{\sqrt{x_3 - x_2}}{x_3} + \cdots + \frac{\sqrt{x_{k+2} - x_{k+1}}}{x_{k+2}} \leq \frac{\sqrt{4k+1}}{2}$.
For the $k+1$ numbers $1 \leq \frac{x_3}{x_2} \leq \cdots \leq \frac{x_{k+2}}{x_2}$, we have $\frac{\sqrt{\frac{x_3}{x_2} - 1}}{\frac{x_3}{x_2}} + \cdots + \frac{\sqrt{\frac{x_{k+2}}{x_2} - \frac{x_{k+1}}{x_2}}}{\frac{x_{k+1}}{x_2}} \leq \frac{\sqrt{4k-3}}{2}$, and therefore,

$$\frac{\sqrt{x_2-x_1}}{x_2}+\frac{\sqrt{x_3-x_2}}{x_3}+\cdots+\frac{\sqrt{x_{k+2}-x_{k+1}}}{x_{k+2}} \le \frac{\sqrt{x_2-x_1}}{x_2}+\frac{1}{\sqrt{x_2}}\left(\frac{\sqrt{\frac{x_3}{x_2}-1}}{\frac{x_3}{x_2}}+\cdots+\frac{\sqrt{\frac{x_{k+2}}{x_2}-\frac{x_{k+1}}{x_2}}}{\frac{x_{k+2}}{x_2}}\right)$$

$$\le \frac{\sqrt{x_2-x_1}}{x_2}\cdot 1+\frac{1}{\sqrt{x_2}}\frac{\sqrt{4k-3}}{2} \le \sqrt{\left(\left(\frac{\sqrt{x_2-x_1}}{x_2}\right)^2+\left(\frac{1}{\sqrt{x_2}}\right)^2\right)\left(1^2+\left(\frac{\sqrt{4k-3}}{2}\right)^2\right)}$$

$$=\sqrt{\frac{2x_2-1}{x_2^2}\cdot\frac{4k+1}{4}} \le \frac{\sqrt{4k+1}}{2}$$

(see Chapter 4, inequality (4.1)). Hence, the given inequality holds for $n = k + 1$.
Therefore the given inequality holds for every positive integer n.
(b) We proceed by induction on n.
For $n = 1$, we have $\frac{\sqrt{x_2-x_1}}{x_2} \le \frac{1}{2} < 1$.
Assume that the given inequality holds for $n = k$, and let us prove that it holds for $n = k + 1$, where $k \in \mathbb{N}$.
Let $1 = x_1 \le x_2 \le \cdots \le x_{k+1} \le x_{k+2}$ be positive integers. Then we have

$$\frac{\sqrt{x_2-x_1}}{x_2}+\frac{\sqrt{x_3-x_2}}{x_3}+\cdots+\frac{\sqrt{x_{k+2}-x_{k+1}}}{x_{k+2}} \le \frac{x_2-x_1}{x_2}+\frac{x_3-x_2}{x_3}+\cdots+\frac{x_{k+2}-x_{k+1}}{x_{k+2}}$$

$$\le\left(\frac{1}{x_1+1}+\cdots+\frac{1}{x_2}\right)+\left(\frac{1}{x_2+1}+\cdots+\frac{1}{x_3}\right)+\cdots+\left(\frac{1}{x_{k+1}+1}+\cdots+\frac{1}{x_{k+2}}\right)$$

$$\le \frac{1}{2}+\frac{1}{3}+\cdots+\frac{1}{x_{k+2}}.$$

If $x_{k+2} \le (k+1)^2$, then $\frac{1}{2}+\frac{1}{3}+\cdots+\frac{1}{x_{k+2}} \le \frac{1}{2}+\cdots+\frac{1}{(k+1)^2} < 1+\frac{1}{2}+\cdots+\frac{1}{(k+1)^2}$, and therefore, the given inequality holds for $n = k + 1$.
If $x_{k+2} > (k + 1)^2$, then

$$\frac{\sqrt{x_{k+2}-x_{k+1}}}{x_{k+2}} \le \frac{\sqrt{x_{k+2}-1}}{x_{k+2}} = \sqrt{\frac{1}{x_{k+2}}-\left(\frac{1}{x_{k+2}}\right)^2} < \sqrt{\frac{1}{(k+1)^2}-\frac{1}{(k+1)^4}} = \frac{\sqrt{k^2+2k}}{(k+1)^2}.$$

Hence, we obtain

$$\frac{\sqrt{x_2-x_1}}{x_2}+\cdots+\frac{\sqrt{x_{k+1}-x_k}}{x_{k+1}}+\frac{\sqrt{x_{k+2}-x_{k+1}}}{x_{k+2}} \le 1+\frac{1}{2}+\cdots+\frac{1}{k^2}$$

$$+\frac{\sqrt{x_{k+2}-x_{k+1}}}{x_{k+2}} \le 1+\frac{1}{2}+\cdots+\frac{1}{k^2}+\frac{\sqrt{k^2+2k}}{(k+1)^2} < 1+\frac{1}{2}+\cdots+\frac{1}{k^2}+\frac{2k+1}{(k+1)^2}$$

$$=1+\frac{1}{2}+\cdots+\frac{1}{k^2}+\underbrace{\frac{1}{(k+1)^2}+\cdots+\frac{1}{(k+1)^2}}_{2k+1} < 1+\frac{1}{2}+\frac{1}{3}+\cdots+\frac{1}{(k+1)^2}.$$

Hence, the given inequality holds for $n = k + 1$.
Therefore, the given inequality holds for every positive integer n.

Remark If $1 = x_1 \leq x_2 \leq \cdots \leq x_n \leq x_{n+1}$ are positive integers, then

$$\frac{\sqrt{x_2 - x_1}}{x_2} + \cdots + \frac{\sqrt{x_{n+1} - x_n}}{x_{n+1}} \leq \left(\sum_{i=1}^{n^2} \frac{1}{i}\right) - \frac{1}{2}.$$

7.23. We proceed by induction on n.
For $n = 1$, we have $0 < \alpha_1 \leq \beta_1 \leq \pi$, and therefore $\frac{\cos \beta_1}{\sin \alpha_1} \leq \frac{\cos \alpha_1}{\sin \alpha_1}$.
For $n = 2$, we have $\alpha_1 > 0,\ \alpha_2 > 0,\ \beta_1 \geq 0,\ \beta_2 \geq 0$ and $\alpha_1 + \alpha_2 \leq \beta_1 + \beta_2 \leq \pi$, and we need to prove that $\frac{\cos \beta_1}{\sin \alpha_1} + \frac{\cos \beta_2}{\sin \alpha_2} \leq \frac{\cos \alpha_1}{\sin \alpha_1} + \frac{\cos \alpha_2}{\sin \alpha_2}$.
Let α_1 and α_2 be constants. Consider the expression $\frac{\cos x_1}{\sin \alpha_1} + \frac{\cos x_2}{\sin \alpha_2}$, where $\alpha_1 + \alpha_2 \leq x_1 + x_2 \leq \pi,\ x_1 \geq 0,\ x_2 \geq 0$. Assume that this expression attains its maximum value at $x_1 = \beta_1,\ x_2 = \beta_2$.
Let $\alpha_1 \leq \alpha_2$. Then one can assume that $\beta_1 \leq \beta_2$. Otherwise, we have $\cos \beta_1 < \cos \beta_2$, and therefore $(\cos \beta_1 - \cos \beta_2)(\sin \alpha_2 - \sin \alpha_1) \leq 0$, or

$$\frac{\cos \beta_1}{\sin \alpha_1} + \frac{\cos \beta_2}{\sin \alpha_2} \leq \frac{\cos \beta_2}{\sin \alpha_1} + \frac{\cos \beta_1}{\sin \alpha_2}.$$

If $\beta_1 = 0$, then

$$\frac{\cos \beta_1}{\sin \alpha_1} + \frac{\cos \beta_2}{\sin \alpha_2} - \frac{\cos \alpha_1}{\sin \alpha_1} - \frac{\cos \alpha_2}{\sin \alpha_2} \leq \frac{1}{\sin \alpha_1} + \frac{\cos(\alpha_1 + \alpha_2)}{\sin \alpha_2} - \frac{\cos \alpha_1}{\sin \alpha_1} - \frac{\cos \alpha_2}{\sin \alpha_2}$$

$$= \tan \frac{\alpha_1}{2} - \frac{2 \sin \frac{\alpha_1}{2} \sin(\frac{\alpha_1}{2} + \alpha_2)}{\sin \alpha_2} = -\frac{\sin^2 \frac{\alpha_1}{2} \sin(\alpha_1 + \alpha_2)}{\cos \frac{\alpha_1}{2} \sin \alpha_2} \leq 0.$$

If $0 < \beta_1 \leq \beta_2$, then on decreasing the value of β_1, the value of the expression $\frac{\cos \beta_1}{\sin \alpha_1} + \frac{\cos \beta_2}{\sin \alpha_2}$ increases, and therefore $\beta_1 + \beta_2 = \alpha_1 + \alpha_2$. Let us prove that $\frac{\sin \beta_1}{\sin \alpha_1} = \frac{\sin \beta_2}{\sin \alpha_2}$.
Note that the function $f(x) = \frac{\cos(\beta_1 + x)}{\sin \alpha_1} + \frac{\cos(\beta_2 - x)}{\sin \alpha_2}$ in $[-\beta_1, \beta_2]$ attains its maximum value at the point $x = 0$, and hence by Fermat's theorem on stationary points, $f'(0) = -\frac{\sin \beta_1}{\sin \alpha_1} + \frac{\sin \beta_2}{\sin \alpha_2} = 0$.
Lemma. *If* $\alpha_i > 0,\quad \beta_i > 0,\quad i = 1, \ldots, n,\quad \alpha_1 + \cdots + \alpha_n = \beta_1 + \cdots + \beta_n \leq \pi$, *and*

$$\frac{\sin \beta_1}{\sin \alpha_1} = \cdots = \frac{\sin \beta_n}{\sin \alpha_n} = \lambda, \tag{7.15}$$

then $\lambda = 1$.
Proof. Let $\lambda \neq 1$. Then without loss of generality one can assume that $\lambda < 1$. Let $\alpha_1 \leq \cdots \leq \alpha_n$. Then from (7.15) it follows that $\beta_1 \leq \cdots \leq \beta_n$ and $\beta_1 < \alpha_1, \ldots, \beta_{n-1} < \alpha_{n-1} \leq \frac{\pi}{2}$.
Note that

$$\sin(\beta_1 + \beta_2) = \lambda(\sin \alpha_1 \cos \beta_2 + \sin \alpha_2 \cos \beta_1) > \lambda(\sin \alpha_1 \cos \alpha_2 + \sin \alpha_2 \cos \alpha_1) = \lambda \sin(\alpha_1 + \alpha_2).$$

In a similar way, we obtain

$$\begin{aligned}\sin((\beta_1+\beta_2)+\beta_3) &= \lambda(\sin(\beta_1+\beta_2)\cos\beta_3+\sin\beta_3\cos(\beta_1+\beta_2))\\ &> \lambda(\sin(\alpha_1+\alpha_2)\cos\beta_3+\sin\alpha_3\cos(\beta_1+\beta_2))\\ &> \lambda(\sin(\alpha_1+\alpha_2)\cos\alpha_3+\sin\alpha_3\cos(\alpha_1+\alpha_2))\\ &= \lambda\ \sin(\alpha_1+\alpha_2+\alpha_3),\end{aligned}$$

and so on.
In general, we have $\sin(\beta_1+\cdots+\beta_{n-1}) > \lambda\ \sin(\alpha_1+\cdots+\alpha_{n-1})$.
Therefore, $\sin(\varphi-\beta_n) > \frac{\sin\beta_n}{\sin\alpha_n}\sin(\varphi-\alpha_n)$, where $\varphi=\alpha_1+\cdots+\alpha_n=\beta_1+\cdots+\beta_n$.
Hence, $\sin\alpha_n\sin(\varphi-\beta_n) > \sin\beta_n\sin(\varphi-\alpha_n)$, or

$$\sin\alpha_n\sin\varphi\ \cos\beta_n-\sin\alpha_n\sin\beta_n\cos\varphi > \\ \sin\beta_n\sin\varphi\ \cos\alpha_n-\sin\beta_n\sin\alpha_n\cos\varphi,\ \sin\varphi\ \sin(\alpha_n-\beta_n)>0.$$

Since $0<\varphi\le\pi$, we must have $\varphi\ne\pi$ and $\sin\varphi>0$, whence $\alpha_n-\beta_n>0$.
It follows that $\alpha_n>\beta_n$.
We obtain $\beta_1<\alpha_1,\ldots,\beta_{n-1}<\alpha_{n-1},\ \beta_n<\alpha_n$, and therefore $\beta_1+\cdots+\beta_n<\alpha_1+\cdots+\alpha_n$, which leads to a contradiction. Thus $\lambda=1$. This ends the proof of the lemma.
Since $\sin\alpha_1=\sin\beta_1$ and $\sin\alpha_2=\sin\beta_2$, it follows that $\alpha_1=\beta_1$ and $\alpha_2=\beta_2$, or $\alpha_1=\beta_1$ and $\pi-\alpha_2=\beta_2(\alpha_1+\alpha_2=\alpha_1+\pi-\alpha_2,\ \ \alpha_2=\frac{\pi}{2}$, whence $\alpha_2=\beta_2)$.
Therefore, $\frac{\cos\beta_1}{\sin\alpha_1}+\frac{\cos\beta_2}{\sin\alpha_2}=\frac{\cos\alpha_1}{\sin\alpha_1}+\frac{\cos\alpha_2}{\sin\alpha_2}$.
Let $n\ge3$ and suppose that the given inequality holds for $n-1$. Let us prove that it holds for n.
Let $0<\alpha_1\le\cdots\le\alpha_n$ be constants. Consider the expression $\frac{\cos x_1}{\sin\alpha_1}+\cdots+\frac{\cos x_n}{\sin\alpha_n}$, where $\alpha_1+\cdots+\alpha_n\le x_1+\cdots+x_n\le\pi,\ \ x_1\ge0,\ldots,x_n\ge0$, and suppose that this expression attains its maximum value at $x_1=\beta_1,\ldots,x_n=\beta_n$. Then $\beta_1\le\beta_2\le\cdots\le\beta_n$ (see the case $n=2$).
Without loss of generality one can assume that $\beta_1>0$, for otherwise, for $n-1$ we would have

$$\frac{\cos\beta_2}{\sin(\alpha_1+\alpha_2)}+\frac{\cos\beta_3}{\sin\alpha_3}+\cdots+\frac{\cos\beta_n}{\sin\alpha_n}\le\frac{\cos(\alpha_1+\alpha_2)}{\sin(\alpha_1+\alpha_2)}+\frac{\cos\alpha_3}{\sin\alpha_3}+\cdots+\frac{\cos\alpha_n}{\sin\alpha_n}. \tag{7.16}$$

Note that

$$\frac{1}{\sin\alpha_1}+\frac{\cos\beta_2}{\sin\alpha_2}-\frac{\cos\alpha_1}{\sin\alpha_1}-\frac{\cos\alpha_2}{\sin\alpha_2}\le\frac{\cos\beta_2}{\sin(\alpha_1+\alpha_2)}-\frac{\cos(\alpha_1+\alpha_2)}{\sin(\alpha_1+\alpha_2)}. \tag{7.17}$$

Indeed, since

$$\cos\beta_2\left(\frac{1}{\sin\alpha_2}-\frac{1}{\sin(\alpha_1+\alpha_2)}\right)\le\frac{1}{\sin\alpha_1}-\frac{1}{\sin(\alpha_1+\alpha_2)}$$

$(\alpha_2+(\alpha_1+\alpha_2)\le\alpha_1+\alpha_2+\alpha_3\le\pi,\quad \sin(\alpha_1+\alpha_2)\ge\sin\alpha_2)$, it is sufficient to prove that $\frac{1-\cos\alpha_1}{\sin\alpha_1}+\frac{1-\cos\alpha_2}{\sin\alpha_2}-\frac{1-\cos(\alpha_1+\alpha_2)}{\sin(\alpha_1+\alpha_2)}\le 0$, or

$$\tan\frac{\alpha_1}{2}+\tan\frac{\alpha_2}{2}-\tan\left(\frac{\alpha_1}{2}+\frac{\alpha_2}{2}\right)\le 0,$$

$$\frac{-\left(\tan\frac{\alpha_1}{2}+\tan\frac{\alpha_2}{2}\right)\tan\frac{\alpha_1}{2}\tan\frac{\alpha_2}{2}}{1-\tan\frac{\alpha_1}{2}\tan\frac{\alpha_2}{2}}\le 0.$$

The last inequality holds because $0<\frac{\alpha_1}{2}\le\frac{\alpha_2}{2}<\frac{\pi}{4}$. Summing (7.16) and (7.17), we obtain the inequality for n. Then we have $0<\beta_1\le\beta_2\le\cdots\le\beta_n$, and if the value of β_1 decreases, then the value of the expression $\frac{\cos\beta_1}{\sin\alpha_1}+\cdots+\frac{\cos\beta_n}{\sin\alpha_n}$ increases, and therefore $\alpha_1+\cdots+\alpha_n=\beta_1+\cdots+\beta_n$. Then $\frac{\sin\beta_1}{\sin\alpha_1}=\cdots=\frac{\sin\beta_n}{\sin\alpha_n}$ (see the case $n=2$), and therefore by the lemma, we have $\alpha_1=\beta_1,\ldots,\alpha_n=\beta_n$.

7.24. We first prove by induction that if $a>0$ and $n\in\mathbb{N}$, then

$$2(1+a^{n+1})^3\ge(1+a^3)(1+a^n)^3. \qquad (1)$$

For $n=1$, we need to prove that $2(1+a^2)^3\ge(1+a^3)(1+a)^3$, which is equivalent to the following obvious inequality: $(a-1)^4(a^2+a+1)\ge 0$.
Assume that the inequality holds for $n=k$, where $k\in\mathbb{N}$, and let us prove that it holds for $n=k+1$.
We have $2(1+a^{k+1})^3\ge(1+a^3)(1+a^k)^3$, as $(1+a^{k+2})(1+a^k)\ge(1+a^{k+1})^2$, whence

$$\begin{aligned}2(1+a^{k+2})^3&=2(1+a^{k+1})^3\cdot\left(\frac{1+a^{k+2}}{1+a^{k+1}}\right)^3\ge(1+a^k)^3(1+a^3)\cdot\left(\frac{1+a^{k+1}}{1+a^k}\right)^3\\&=(1+a^3)(1+a^{k+1})^3,\end{aligned}$$

and therefore $2(1+a^{k+2})^3\ge(1+a^3)(1+a^{k+1})^3$.
Thus, inequality (1) holds.
Note that $(1^3+a^3)(1^3+b^3)(1^3+c^3)\ge(1+abc)^3$ (Problem 6.2) and

$$\begin{aligned}&(2(a^{2012}+1)(b^{2012}+1)(c^{2012}+1))^3=2(a^{2012}+1)^3\cdot 2(b^{2012}+1)^3\cdot 2(c^{2012}+1)^3\\&\ge(1+a^3)(1+a^{2011})^3\cdot(1+b^3)(1+b^{2011})^3\cdot(1+c^3)(1+c^{2011})^3\\&\ge((1+abc)(a^{2011}+1)(b^{2011}+1)(c^{2011}+1))^3,\end{aligned}$$

and therefore $2(a^{2012}+1)(b^{2012}+1)(c^{2012}+1)\ge(1+abc)(a^{2011}+1)(b^{2011}+1)(c^{2011}+1)$.

7.25. (a) The inequality $b_1^2\ge b_2$ follows from the inequality of Problem 2.2.
(b) Let $b_0=1$, $b_{n+1}=0$. We then need to prove that $b_k^2\ge b_{k-1}b_{k+1}$, for $k=1,2,\ldots,n$.

For the numbers $a_1, \ldots, a_{n-1}$, the numbers $\overline{b}_1, \ldots, \overline{b}_{n-1}$ are defined in a similar way as we defined the numbers $b_1, \ldots, b_n$ for the numbers $a_1, \ldots, a_n$, and $\overline{b}_0 = 1, \overline{b}_n = 0$.
For $n = 2$, we have $b_0 = 1, b_1 = \frac{a_1+a_2}{2}, b_2 = a_1a_2, b_3 = 0$, and hence $b_1^2 \geq b_0b_2, b_2^2 \geq b_1b_3$.
Assume that this statement holds for $n-1$ numbers, and let us prove that it holds for n $(n \geq 3)$ numbers.
Note that $b_k = \frac{\overline{b}_k C_{n-1}^k + \overline{b}_{k-1} C_{n-1}^{k-1} a_n}{C_n^k} = \frac{n-k}{n}\overline{b}_k + \frac{k}{n}\overline{b}_{k-1} \cdot a_n$, for $k = 1, 2, \ldots, n$.
For $k = 2, \ldots, n-1$, we have

$$
\begin{aligned}
n^2(b_k^2 - b_{k-1}b_{k+1}) &= \left((n-k)\overline{b}_k + k\overline{b}_{k-1} \cdot a_n\right)^2 \\
&- \left((n-k+1)\overline{b}_{k-1} + (k-1)\overline{b}_{k-2} \cdot a_n\right)\left((n-k-1)\overline{b}_{k+1} + (k+1)\overline{b}_k \cdot a_n\right) \\
&= \left((n-k)^2 - 1\right)(\overline{b}_k^2 - \overline{b}_{k-1}\overline{b}_{k+1}) + (k^2-1)a_n^2(\overline{b}_{k-1}^2 - \overline{b}_{k-2}\overline{b}_k) \\
&+ (k-1)(n-k-1)(\overline{b}_k\overline{b}_{k-1} - \overline{b}_{k-2}\overline{b}_{k+1})a_n \geq 0,
\end{aligned}
$$

as $\overline{b}_k^2 \cdot \overline{b}_{k-1}^2 \geq \overline{b}_{k-1}\overline{b}_{k+1} \cdot \overline{b}_{k-2}\overline{b}_k$, and therefore $b_k^2 - b_{k-1}b_{k+1} \geq 0$.
This ends the proof.

Problems for Independent Study

Prove the following inequalities (1–20).

1. $\left(\frac{a_2}{a_1} - \frac{a_1}{a_2}\right) + \cdots + \left(\frac{a_n}{a_{n-1}} - \frac{a_{n-1}}{a_n}\right) \leq \frac{a_n}{a_1} - \frac{a_1}{a_n}$, where $n \geq 2,\quad 0 < a_1 \leq \cdots \leq a_n$.
2. $(1+a_1)\cdots(1+a_n) \geq 1 + a_1 + \cdots + a_n$, where $a_1 > -1, \ldots, a_n > -1$ and the numbers $a_1, \ldots, a_n$ have the same sign.
3. $C \leq D \leq 2C$, where $C = (a_1 - b_1)^2 + \cdots + (a_n - b_n)^2,\quad D = (a_1 - b_n)^2 + (a_2 - b_n)^2 + \cdots + (a_n - b_n)^2$,

$$b_k = \frac{a_1 + \cdots + a_k}{k}, k = 1, \ldots, n.$$

4. $\sqrt[n]{n} > \sqrt[n+1]{n+1}$, where $n \geq 3,\quad n \in \mathbb{N}$.
5. (a) $\left(1 + \frac{1}{n}\right)^k < 1 + \frac{k}{n} + \frac{k^2}{n^2}$, where $k \leq n,\quad n, k \in \mathbb{N}$.
   (b) $\left(1 + \frac{m}{n}\right)^{\frac{n}{m}} < 3$, where $m \in \mathbb{N},\ n \in \mathbb{N}$.
   Hint. See Problems 5(a) and 3.6(c).
6. $1 - x + \frac{x^2}{2!} - \frac{x^3}{3!} + \cdots + \frac{x^{2k}}{(2k)!} > 0$, where $k \in \mathbb{N}$.
7. $\sum_{i=1}^{n} \frac{1}{a_i} \leq \frac{1}{a_1a_n}\left(n(a_1 + a_n) - \sum_{i=1}^{n} a_i\right)$, where $0 < a_1 \leq \cdots \leq a_n$.
8. $\frac{a_1+\cdots+a_k}{k} < \frac{a_1+\cdots+a_n}{n} < \frac{a_{k+1}+\cdots+a_n}{n-k}$, where $a_1 < \cdots < a_n,\quad n > k,\quad n, k \in \mathbb{N}$.

9. $a_1^2 - a_2^2 + \cdots - a_{2n}^2 + a_{2n+1}^2 \geq (a_1 - a_2 + \cdots - a_{2n} + a_{2n+1})^2$, where $a_1 \geq a_2 \geq \cdots \geq a_{2n+1} \geq 0$.
10. $x_1 + x_1(x_2 - x_1) + \cdots + x_{n-1}(x_n - x_{n-1}) \leq \frac{(n-1)x_n^2+2x_n+n-1}{2n}$, where $n \geq 2$.
11. $1^1 \cdot \ldots \cdot n^n > (2n)!$, where $n \geq 5$, $n \in \mathbb{N}$.
12. $\frac{(2m_1)!}{m_1!} \cdots \frac{(2m_n)!}{m_n!} \geq 2^S$, where $m_1, \ldots, m_n \in \mathbb{Z}_0$ and $m_1 + \cdots + m_n = S$.
13. $\frac{1}{a+b} + \cdots + \frac{1}{a+nb} < \frac{n}{\sqrt{a(a+nb)}}$, where $a > 0$, $b > 0$ and $n \in \mathbb{N}$.
14. $\sum\limits_{k=1}^{n-1} \frac{n}{n-k} \cdot \frac{1}{2^{k-1}} < 4$, where $n \geq 2$, $n \in \mathbb{N}$.
15. (a) $\left(1 + \frac{1}{n}\right)^k < 1 + \frac{k}{n} + \frac{k^2}{2n^2}$, where $k, n \in \mathbb{N}$ and $(k-1)^2 < n$;
(b) $\left(1 + \frac{1}{n}\right)^n \left(1 + \frac{1}{4n}\right) < \left(1 + \frac{1}{n+1}\right)^{n+1} \left(1 + \frac{1}{4(n+1)}\right)$, where $n \in \mathbb{N}$.
16. $\sum\limits_{i=0}^{n} \left|\cos 2^i x\right| \geq \frac{n}{2}$, where $n \in \mathbb{N}$.
17. (a) $\sin \alpha + \frac{\sin 2\alpha}{2} + \cdots + \frac{\sin n\alpha}{n} \geq 0$, where $n \in \mathbb{N}$ and $0 \leq \alpha \leq \pi$,
(b) $\cos \alpha + \frac{\cos 2\alpha}{2} + \cdots + \frac{\cos n\alpha}{n} \geq -1$, where $n \in \mathbb{N}$.
18. (a) $\frac{x_1}{x_2} + \frac{x_2}{x_3} + \cdots + \frac{x_n}{x_1} \geq \frac{x_2}{x_1} + \frac{x_3}{x_2} + \cdots + \frac{x_1}{x_n}$, where $0 < x_1 \leq \cdots \leq x_n$, $n \geq 2$;
(b) $x_1^3 + \cdots + x_n^3 \geq x_1 + \cdots + x_n$, where $x_1 > 0, \ldots, x_n > 0$ and $x_1 \cdot \ldots \cdot x_n = 1$.
Hint. Note that $\frac{x_1^3}{1 \cdot (x_1 + \ldots + x_n)} + \cdots + \frac{x_n^3}{1 \cdot (x_1 + \cdots + x_n)} \geq \frac{(x_1 + \cdots + x_n)^3}{n \cdot n(x_1 + \cdots + x_n)}$.
19. $(1 + a_1)(2 + a_2) \cdot \ldots \cdot (n + a_n) \leq 2 \cdot n!$, where $n \geq 2$, $a_1 > 0, \ldots, a_n > 0$ and $a_1 + \cdots + a_n = 1$.
Hint. Let $a_i = i\alpha_i, i = 1, \ldots, n$, and we need to prove that
$(1 + \alpha_1)(1 + \alpha_2) \cdots (1 + \alpha_n) \leq 2$, having that $\alpha_1 + 2\alpha_2 + \cdots + n\alpha_n = 1$.
Note that if $(n-1)\alpha_{n-1} + n\alpha_n = (n-1)\beta_{n-1}$, then $(1+\alpha_{n-1})(1+\alpha_n) \leq 1+\beta_{n-1}$.
20. $(1 - a)(a^{k_1} + a^{k_2} + \cdots + a^{k_n})^2 < (1 + a)(a^{2k_1} + a^{2k_2} + \cdots + a^{2k_n})$, where $n \geq 2$, $n \in \mathbb{N}$,
$0 < a < 1$, $0 \leq k_1 \leq \cdots \leq k_n$ and $k_1, k_2, \ldots, k_n \in \mathbb{N}$.
Hint. Note that $a^{k_1}(a^{k_2} + \cdots + a^{k_n}) < a^{2k_1+1} + a^{2k_1+2} + \cdots = \frac{a^{2k_1+1}}{1-a}$.
21. For arbitrary positive integer $n > 1$ find the smallest possible value of C if the inequality $\frac{a_1-b_1}{a_1+b_1} + \cdots + \frac{a_n-b_n}{a_n+b_n} < C$ holds for all positive numbers $a_1, \ldots, a_n, b_1, \ldots, b_n$ satisfying the equality $a_1 + \cdots + a_n = b_1 + \cdots + b_n$.
22. Prove that for an arbitrary positive integer $n > 1$, one has $\frac{1-x_1 \cdots x_n}{1-y_1 \cdots y_n} < \frac{1-x_1}{1-y_1} + \cdots + \frac{1-x_n}{1-y_n}$, where $0 < y_i \leq x_i < 1$, $i = 1, \ldots, n$.
Hint. Note that $\frac{1-ab}{1-cd} < \frac{1-a}{1-cd} + \frac{1-b}{1-cd} < \frac{1-a}{1-c} + \frac{1-b}{1-d}$, where $0 < c \leq a < 1$, $0 < d \leq b < 1$.

Chapter 8
A Useful Inequality

In 1997 *Nairi Sedrakyan* has published an article (in Russian) [17] called "On the applications of a useful inequality," in which the author proves a very useful inequality and provides some applications.

In 1998, this inequality was re-published by the author, this time in Armenian, in the book [16], in which the author devotes an entire chapter to its applications. In 2002 this book was published in Moscow in Russian [15].

Russian-speaking reader sometimes calls this inequality *Sedrakyan's inequality* in reference to Sedrakyan's 1997 article [17] on its numerous applications, while the English-speaking reader sometimes calls this inequality *Engel's form* or *Titu's lemma,* in reference to the book [4] published in 1998, and the book [1], published in 2003. Nevertheless, even though Sedrakyan in his article [17] stated and proved this inequality without using the Cauchy–Bunyakovsky–Schwarz inequality, it turns out that this inequality is nothing but another form of that inequality. Probably this form was known even before this article, with the difference that in his article [17], Sedrakyan noticed that written in this form, the inequality has very useful *new* applications, and he provided numerous ways in which this inequality can be used as a mathematical proof technique to prove inequalities of various types. In this chapter we consider that inequality, along with its generalizations and applications.

Lemma 1 *(A useful inequality) Let* $a_1, a_2, \ldots, a_n$ *be real numbers and* $b_1, b_2, \ldots, b_n$ *positive real numbers. Then*

$$\frac{a_1^2}{b_1} + \cdots + \frac{a_n^2}{b_n} \geq \frac{(a_1 + \cdots + a_n)^2}{b_1 + \cdots + b_n}. \tag{8.1}$$

Moreover, equality holds if and only if $\frac{a_1}{b_1} = \cdots = \frac{a_n}{b_n}$

H. Sedrakyan and N. Sedrakyan, *Algebraic Inequalities*, Problem Books in Mathematics, https://doi.org/10.1007/978-3-319-77836-5_8

Generalizations of inequality (8.1).

Lemma 2 *(Generalization 1)*

$$\frac{a_1^3}{b_1c_1}+\frac{a_2^3}{b_2c_2}+\cdots+\frac{a_n^3}{b_nc_n}\geq\frac{(a_1+a_2+\cdots+a_n)^3}{(b_1+b_2+\cdots+b_n)(c_1+c_2+\cdots+c_n)}, \tag{8.2}$$

where $a_i>0,\ b_i>0,\ >c_i>0,\ i=1,2,\ldots,n.$

Lemma 3 *(Generalization 2)*

$$\frac{a_1^n}{b_1^{n-1}}+\cdots+\frac{a_n^n}{b_n^{n-1}}\geq\frac{(a_1+\cdots+a_n)^n}{(b_1+\cdots+b_n)^{n-1}}, \tag{8.3}$$

where $a_i>0, b_i>0, i=1,2,\ldots,n.$

Lemma 4 *(Generalization 3)*

$$\frac{a_{1,1}^n}{a_{1,2}\cdot a_{1,3}\cdots a_{1,n}}+\frac{a_{2,1}^n}{a_{2,2}\cdot a_{2,3}\cdots a_{2,n}}+\cdots+\frac{a_{m,1}^n}{a_{m,2}\cdot a_{m,3}\cdots a_{m,n}}\geq$$
$$\geq\frac{\left(a_{1,1}+a_{2,1}+\cdots+a_{m,1}\right)^n}{\left(a_{1,2}+a_{2,2}+\cdots+a_{m,2}\right)\cdot\left(a_{1,3}+a_{2,3}+\cdots+a_{m,3}\right)\cdots\left(a_{1,n}+a_{2,n}+\cdots+a_{m,n}\right)}, \tag{8.4}$$

where $a_{i,j}>0,\ i=1,2,\ldots,m,\quad j=1,2,\ldots,n.$

Proof of Lemma 1 Let us at first prove that for all real numbers $a,\ b$ and positive real numbers $x,\ y$, one has

$$\frac{a^2}{x}+\frac{b^2}{y}\geq\frac{(a+b)^2}{x+y}. \tag{8.5}$$

Note that this inequality is equivalent to the inequality $(ay-bx)^2\geq 0$.
Obviously, here the equality holds here if and only if $ay-bx=0$, that is $\frac{a}{x}=\frac{b}{y}$.
Now, let us proceed to the proof of this *useful inequality*.
Applying inequality (8.5) several times, we obtain

$$\begin{aligned}\frac{a_1^2}{b_1}+\frac{a_2^2}{b_2}+\frac{a_3^2}{b_3}+\cdots+\frac{a_n^2}{b_n}&\geq\frac{(a_1+a_2)^2}{b_1+b_2}+\frac{a_3^2}{b_3}+\cdots+\frac{a_n^2}{b_n}\\&\geq\frac{(a_1+a_2+a_3)^2}{b_1+b_2+b_3}+\cdots+\frac{a_n^2}{b_n}\\&\geq\cdots\\&\geq\frac{(a_1+\cdots+a_n)^2}{b_1+\cdots+b_n}.\end{aligned}$$

Obviously, equality holds if and only if $\frac{a_1}{b_1}=\frac{a_2}{b_2}=\cdots=\frac{a_n}{b_n}$.

This ends the proof.

We provide proofs of Lemmas 2–4 after the remark following Lemma 6.

Let us consider the following examples and applications of this *useful inequality* (Lemma 1).

Cauchy–Bunyakovsky–Schwarz inequality. Let $x_1, x_2, \dots, x_n$ and $y_1, y_2, \dots, y_n$ be real numbers, then $\left(x_1^2 + \dots + x_n^2\right)\left(y_1^2 + \dots + y_n^2\right) \geq (x_1 y_1 + \dots + x_n y_n)^2$.

Proof For real numbers $a_1, a_2, \dots, a_n$ and positive real numbers $b_1, b_2, \dots, b_n$, we have $\frac{a_1^2}{b_1} + \dots + \frac{a_n^2}{b_n} \geq \frac{(a_1+\dots+a_n)^2}{b_1+\dots+b_n}$.

Let us take $a_i = x_i y_i$ and $b_i = y_i^2$. Then for all $1 \leq i \leq n$, we obtain $\left(x_1^2 + \dots + x_n^2\right)\left(y_1^2 + \dots + y_n^2\right) \geq (x_1 y_1 + \dots + x_n y_n)^2$.

This ends the proof.

Remark We have obtained a very beautiful inequality (8.1). However, it turns out that this is one of the forms of the Cauchy–Bunyakovsky–Schwarz inequality. Indeed, making a change of variables $x_i \frac{a_i}{\sqrt{b_i}}$ and $y_i = \sqrt{b_i}$ we obtain Cauchy–Bunyakovsky–Schwarz inequality $\left(x_1^2 + \dots + x_n^2\right)\left(y_1^2 + \dots + y_n^2\right) \geq (x_1 y_1 + \dots + x_n y_n)^2$.

Note that in order to prove certain inequalities the form (8.1) is more convenient than the classical form of Cauchy–Bunyakovsky–Schwarz inequality.

Example 8.1 Prove that $\frac{1}{a+b} + \frac{1}{b+c} + \frac{1}{a+c} \geq \frac{9}{2(a+b+c)}$, where $a > 0,\ b > 0,\ c > 0$.

Proof Note that according to Lemma 1 (for $n = 3$), we have

$$\frac{1}{a+b} + \frac{1}{b+c} + \frac{1}{a+c} = \frac{1^2}{a+b} + \frac{1^2}{b+c} + \frac{1^2}{a+c} \geq \frac{(1+1+1)^2}{(a+b+b+c+a+c)} = \frac{9}{2(a+b+c)}.$$

This ends the proof.

Example 8.2 Prove that $\frac{12}{a+b+c+d} \leq \frac{1}{a+b} + \frac{1}{a+c} + \frac{1}{a+d} + \frac{1}{b+c} + \frac{1}{b+d} + \frac{1}{c+d} \leq \frac{3}{4}\left(\frac{1}{a} + \frac{1}{b} + \frac{1}{c} + \frac{1}{d}\right)$, where $a > 0,\ b > 0,\ c > 0,\ d > 0$.

Proof Note that according to Lemma 1 (for $n = 6$), we have

$$\frac{1}{a+b} + \frac{1}{a+c} + \frac{1}{a+d} + \frac{1}{b+c} + \frac{1}{b+d} + \frac{1}{c+d} \geq \frac{(1+1+1+1+1+1)^2}{3(a+b+c+d)} = \frac{12}{a+b+c+d}.$$

On the other hand, on applying Lemma 1 (for $n = 2$) multiple times for each of the following sets of parentheses, we obtain

$$3\left(\frac{1}{a}+\frac{1}{b}+\frac{1}{c}+\frac{1}{d}\right)=\left(\frac{1}{a}+\frac{1}{b}\right)+\left(\frac{1}{a}+\frac{1}{c}\right)+\left(\frac{1}{a}+\frac{1}{d}\right)+\left(\frac{1}{b}+\frac{1}{c}\right)+\left(\frac{1}{b}+\frac{1}{d}\right)+\left(\frac{1}{c}+\frac{1}{d}\right)\geq$$

$$\geq\frac{(1+1)^2}{a+b}+\frac{(1+1)^2}{a+c}+\frac{(1+1)^2}{a+d}+\frac{(1+1)^2}{b+c}+\frac{(1+1)^2}{b+d}+\frac{(1+1)^2}{c+d}.$$

Thus, it follows that

$$\frac{12}{a+b+c+d}\leq\frac{1}{a+b}+\frac{1}{a+c}+\frac{1}{a+d}+\frac{1}{b+c}+\frac{1}{b+d}+\frac{1}{c+d}$$
$$\leq\frac{3}{4}\left(\frac{1}{a}+\frac{1}{b}+\frac{1}{c}+\frac{1}{d}\right),$$

which completes the proof.

A more general inequality. Let us try to obtain a more general inequality.

To this end, we write a_i^2 as the product of two factors.

On proving this inequality for $n=2$, we obtain new conditions. It turns out that the following lemma holds.

Lemma 5 *(Generalization 4) If* $\frac{a_1}{c_1}\geq\cdots\geq\frac{a_n}{c_n}$, $\frac{b_1}{c_1}\geq\cdots\geq\frac{b_n}{b_n}$ *(that is* $\frac{a_i}{c_i}$ *and* $\frac{b_i}{c_i}$ *have the same order) and* $c_i>0$, $i=1,\ldots,n$, *then the following inequality holds:*

$$\sum_{i=1}^{n}\frac{a_ib_i}{c_i}\geq\frac{\sum\limits_{i=1}^{n}a_i\cdot\sum\limits_{i=1}^{n}b_i}{\sum\limits_{i=1}^{n}c_i}. \tag{8.6.1}$$

If $a_i=b_i$, *then inequality (8.6.1) becomes inequality (8.1).*

Proof In order to prove inequality (8.6.1), we use the following fact: if $\frac{a_1}{c_1}\geq\cdots\geq\frac{a_n}{c_n}$ and $c_i>0$, $i=1,\ldots,n$, then $\frac{a_1+\cdots+a_k}{c_1+\cdots+c_k}\geq\frac{a_n}{c_n}$ for all $k=1,\ldots,n$ (see Problem 1. 11).

Let us prove inequality (8.6.1) by induction on n

For $n=2$, we have that $\frac{a_1b_1}{c_1}+\frac{a_2b_2}{c_2}\geq\frac{(a_1+a_2)(b_1+b_2)}{c_1+c_2}$, or $(a_1c_2-a_2c_1)(b_1c_2-b_2c_1)\geq 0$, which follows from $\frac{a_1}{c_1}\geq\frac{a_2}{c_2}$ and $\frac{b_1}{c_1}\geq\frac{b_1}{c_2}$.

Assume that inequality (8.6.1) holds for $n=k$ Then we have $\frac{a_1b_1}{c_i}+\cdots+\frac{a_kb_k}{c_k}\geq\frac{(a_1+\cdots+a_k)(b_1+\cdots+b_k)}{c_1+\cdots+c_k}$.

From the above-mentioned fact and the inequality for $n=2$ it follows that $\frac{a_1b_1}{c_1}+\cdots+\frac{a_kb_k}{c_k}+\frac{a_{k+1}b_{k+1}}{c_{k+1}}\geq\frac{(a_1+\cdots+a_k)(b_1+\cdots+b_k)}{c_1+\cdots+c_k}+\frac{a_{k+1}\cdot b_{k+1}}{c_{k+1}}\geq\frac{(a_1+\cdots+a_{k+1})(b_1+\cdots+b_{k+1})}{c_1+\cdots+c_{k+1}}$ or $\frac{a_1b_1}{c_1}+\cdots+\frac{a_{k+1}b_{k+1}}{c_{k+1}}\geq\frac{(a_1+\cdots+a_{k+1})(b_1+\cdots+b_{k+1})}{c_1+\cdots+c_{k1}}$.

In a similar way, one can prove the following lemma.

Lemma 6 *(Generalization 5) If $\frac{a_i}{c_i}$ and $\frac{b_i}{c_i}$ have opposite orders and $c_i > 0,\ i = 1, \ldots, n$, then the following inequality holds:*

$$\sum_{i=1}^{n} \frac{a_i b_i}{c_i} \leq \frac{\sum\limits_{i=1}^{n} a_i \cdot \sum\limits_{i=1}^{n} b_i}{\sum\limits_{i=1}^{n} c_i}. \tag{8.6.2}$$

Remark **(Chebyshev's inequality)** If the conditions $a_1 \geq \cdots \geq a_n,\quad b_1 \geq \cdots \geq b_n$, and $0 < c_1 \geq \cdots \geq c_n$ hold, then the conditions of inequality (8.6.1) hold, and therefore this inequality holds as well.

From inequalities (8.1) and (8.6.1) we obtain the following well-known inequalities:

$$\sum_{i=1}^{n} \frac{1}{c_i} \geq \frac{n^2}{\sum\limits_{i=1}^{n} c_i}, \quad \text{where } c_i > 0, \quad i = 1, \ldots, n, \tag{8.7}$$

$$\frac{1}{n} \sum_{i=1}^{n} a_i^2 \geq \left(\frac{\sum\limits_{i=1}^{n} a_i^2}{n} \right)^2, \tag{8.8}$$

$$\sum_{i=1}^{n} a_i b_i \geq \frac{1}{n} \sum_{i=1}^{n} a_i \cdot \sum_{i=1}^{n} b_i, \tag{8.9}$$

where $a_1 \geq \cdots \geq a_n$ and $b_1 \geq \cdots \geq b_n$. Inequality (8.9) is called *Chebyshev's inequality*.

Note that inequality (8.7) follows from Chebyshev's inequality (taking $a_i = c_i, b_i = -\frac{1}{c_i}$).

Note also that inequality (8.8) follows from Chebyshev's inequality (taking $a_i = b_i$).

Now, taking $a_i = c_i x_i, \quad b_i = c_i y_i, \quad P_i = \frac{c_i}{\sum\limits_{i=1}^{n} c_i}$, we obtain the following form of Chebyshev's inequality:

Chebyshev's inequality (alternative form). If x_i and y_i have the *same order*, $\sum\limits_{i=1}^{n} P_i = 1, \quad P_i > 0, i = 1, \ldots, n$, then for the means $Mz = \sum\limits_{i=1}^{n} z_i P_i$, one has $Mx \cdot My \leq M(xy)$.

Remark Therefore, the Cauchy–Bunyakovsky–Schwarz inequality is a particular case of Chebyshev's inequality.

Proof of Lemma 2 Indeed, inequality (8.2) is one of the forms of the Cauchy–Bunyakovsky–Schwarz inequality

$$\left(a_1^3 + a_2^3 + \cdots + a_n^3\right) \cdot \left(b_1^3 + b_2^3 + \cdots + b_n^3\right) \cdot \left(c_1^3 + c_2^3 + \cdots + c_n^3\right) \geq (a_1b_1c_1 + a_2b_2c_2 + \cdots + a_nb_nc_n)^3, \quad (8.10)$$

where $a_i > 0,\ b_i > 0,\ c_i > 0,\ i = 1, 2, \ldots, n$.

Note that in inequality (8.10), by making the change of variables $a_i = \frac{x_i}{\sqrt[3]{y_i z_i}}$, $b_i = \sqrt[3]{y_i}$, $c_i = \sqrt[3]{z_i}$, we obtain for the numbers $x_i,\ y_i,\ z_i$ inequality (8.2).

In order to prove inequality (8.10), note that if the variables $a_1, a_2, \ldots, a_n$ are replaced by variables $\lambda a_1, \lambda a_2, \ldots, \lambda a_n$, where λ is any positive number, then we obtain an equivalent inequality.

Choose a number λ such that $(\lambda a_1)^3 + (\lambda a_2)^3 + \cdots + (\lambda a_n)^3 = 1$.

Therefore, without loss of generality one can assume that $a_1^3 + a_2^3 + \cdots + a_n^3 = 1,\ \ b_1^3 + b_2^3 + \cdots + b_n^3 = 1,\ \ c_1^3 + c_2^3 + \cdots + c_n^3 = 1$, and hence it is sufficient to prove that $1 \geq a_1b_1c_1 + a_2b_2c_2 + \cdots + a_nb_nc_n$ Indeed, if $x > 0,\ y > 0,\ z > 0$, then $x^3 + y^3 + z^3 - 3xyz = \frac{1}{2}(x + y + z)\left((x - y)^2 + (y - z)^2 + (z - x)^2\right) \geq 0$, and therefore $a_ib_ic_i \leq \frac{a_i^3+b_i^3+c_i^3}{3},\ i = 1, 2, \ldots, n$.

Summing all these inequalities, we deduce that $1 \geq a_1b_1c_1+a_2b_2c_2+\cdots+a_nb_nc_n$.

This ends the proof.

Proof of Lemma 3 Lemma 3 is a direct consequence of Lemma 4.

Proof of Lemma 4 See Problem 45 in the end of this chapter.

Let us consider the following examples (applications of Lemma 2).

Example 8.3 Prove that $\frac{a^3}{bc} + \frac{b^3}{ca} + \frac{c^3}{ab} \geq a + b + c$, where $a > 0,\ b > 0,\ c > 0$.

Proof According to inequality (8.2) we have that, $\frac{a^3}{bc} + \frac{b^3}{ca} + \frac{c^3}{ab} \geq \frac{(a+b+c)^3}{(b+c+a)+(c+a+b)} = a + b + c$.

This ends the proof.

Example 8.4 Prove that $\frac{a^3}{x} + \frac{b^3}{y} + \frac{c^3}{z} \geq \frac{(a+b+c)^3}{3(x+y+z)}$ where $a > 0,\ b > 0,\ c > 0, x > 0, y > 0, z > 0$.

Proof According to inequality (8.2), we have $\frac{a^3}{x} + \frac{b^3}{y} + \frac{c^3}{z} = \frac{a^3}{1 \cdot x} + \frac{b^3}{1.y} + \frac{c^3}{1.z} \geq \frac{(a+b+c)^3}{(1+1+1)(x+y+z)} = \frac{(a+b+c)^3}{3(x+y+z)}$.

This ends the proof.

Example 8.5 Prove that $\frac{a}{b^2c^2} + \frac{b}{c^2a^2} + \frac{c}{a^2b^2} \geq \frac{9}{a+b+c}$, where $a^2 + b^2 + c^2 = 3abc$ and $a > 0,\ b > 0,\ c > 0$.

Proof Since

$$\frac{a}{b^2c^2} + \frac{b}{c^2a^2} + \frac{c}{a^2b^2} = \frac{\left(a^2\right)^3}{a^2 \cdot a^3b^2c^2} + \frac{\left(b^2\right)^3}{b^2 \cdot a^3b^2c^2} + \frac{\left(c^2\right)^3}{c^2 \cdot a^3b^2c^2},$$

then according to inequality (8.2), we have $\frac{a}{b^2c^2} + \frac{b}{c^2a^2} + \frac{c}{a^2b^2} \geq \frac{\left(a^2+b^2+c^2\right)^3}{\left(a^2+b^2+c^2\right)\left(a^3b^2c^2+a^2b^2c^3+a^2b^2c^3\right)} = \frac{9}{a+b+c}$.

This ends the proof.

Problems

Prove the following inequalities (8.1)–(8.8).

8.1. *Nesbitt's inequality*: $\frac{x_1}{x_2+x_3}+\frac{x_2}{x_3+x_1}+\frac{x_3}{x_1+x_2}\geq\frac{3}{2}$, where $x_1>0, x_2>0, x_3>0$.

8.2. $\frac{a}{b+c}+\frac{b}{c+d}+\frac{c}{d+a}+\frac{d}{a+b}\geq 2$, where $a>0,\ b>0,\ c>0,\ d>0$.

8.3. $\frac{x_1}{x_2+x_n}+\frac{x_2}{x_3+x_1}+\cdots+\frac{x_{n-1}}{x_n+x_{n-2}}+\frac{x_n}{x_1+x_{n-1}}\geq 2$, where $n\geq 4$ and $x_1>0,\ldots,x_n>0$.

8.4. (a) $\frac{a^3}{a^2+ab+b^2}+\frac{b^3}{b^2+bc+c^2}+\frac{c^3}{c^2+ca+a^2}\geq\frac{a+b+c}{3}$, where $a>0,\ b>0,\ c>0$.
(b) $\frac{a}{b}+\frac{b}{c}+\frac{c}{a}\geq\frac{a+b}{b+c}+\frac{b+c}{a+b}+1$, where $a>0,\ b>0,\ c>0$.
(c) $\frac{a}{b}+\frac{b}{c}+\frac{c}{a}\geq\frac{a+c}{b+c}+\frac{b+a}{c+a}+\frac{c+b}{a+b}$, where $a>0,\ b>0,\ c>0$.
(d) $2\left(\frac{a}{b}+\frac{b}{c}+\frac{c}{a}\right)\geq\frac{a+b}{b+c}+\frac{b+c}{a+b}+\frac{a+b}{a+c}+\frac{a+c}{a+b}+\frac{a+c}{b+c}+\frac{b+c}{a+c}$ where $a>0,\ b>0,\ c>0$.

8.5. $\frac{a_1}{p-2a_1}+\cdots+\frac{a_n}{p-2a_n}\geq\frac{n}{n-2}$, where p is the perimeter of a polygon with side lengths $a_1,\ldots,a_n$.

8.6. (a) $\frac{1}{a^3(b+c)}+\frac{1}{b^3(a+c)}+\frac{1}{c^3(a+b)}\geq\frac{3}{2}$, where $abc=1$ and $a,b,c>0$.
(b) $\frac{1}{1+x_1}+\cdots+\frac{1}{1+x_1+\cdots+x_n}<\sqrt{\frac{1}{x_1}+\cdots+\frac{1}{x_n}}$, where $x_i>0,\quad i=1,\ldots,n$
(c) $\frac{x}{y^2-z}+\frac{y}{z^2-x}+\frac{z}{x^2-y}>1$, where $2<x<4,\quad 2<y<4,\quad 2<z<4$.

8.7. $8\left(x^3+y^3+z^3\right)^2\geq 9\left(x^2+yz\right)\left(y^2+xz\right)\left(z^2+xy\right)$, where $x>0,\ y>0\ z>0$.

8.8. $\left(\frac{1}{n}\sum_{i=1}^{n}a_i\right)\left(\frac{1}{n}\sum_{i=1}^{n}b_i\right)\left(\frac{1}{n}\sum_{i=1}^{n}c_i\right)\cdots\left(\frac{1}{n}\sum_{i=1}^{n}d_i\right)\leq\frac{1}{n}\sum_{i=1}^{n}a_i\cdot b_i\cdot c_i\cdots d_i$, where $0\leq a_1\leq\cdots\leq a_n,\quad 0\leq b_1\leq\cdots\leq b_n,\quad 0\leq c_1\leq\cdots\leq c_n,\cdots,0\leq d_1\leq\cdots\leq d_n$.

8.9. Consider a sequence with positive terms (x_k) where $1=x_0\geq x_1\geq\cdots\geq x_n\geq\cdots$. Prove that there exists a positive integer n such that for every such sequence (x_k), one has $\frac{x_0^2}{x_1}+\frac{x_1^2}{x_2}+\cdots+\frac{x_{n-1}^2}{x_n}\geq 3.999$.

8.10. Let M be an arbitrary point in the interior of a given triangle ABC. Consider perpendiculars MA_1, MB_1, MC_1 drawn from the point M to lines BC, CA, AB, respectively. For which point M in the interior of triangle ABC does the quantity $\frac{BC}{MA_1}+\frac{CA}{MB_1}+\frac{AB}{MC_1}$ attain the smallest possible value?

8.11. Let G be the intersection point of the medians of triangle $A_1A_2A_3$ and let C be the circumcircle of triangle $A_1A_2A_3$ Let the lines GA_1, GA_2, GA_3 intersect the circle C a second time at points B_1, B_2, B_3, respectively.
Prove that $GA_1+GA_2+GA_3\leq GB_1+GB_2+GB_3$.

8.12. Prove that $a_1+a_2+\cdots+a_{2k}-2k\sqrt[2k]{a_1\cdot a_2\cdots a_{2k}}\geq\frac{(a_1-a_2+\cdots+a_{2k-1}-a_{2k})^2}{2(a_1+a_2+\cdots+a_{2k})}$, where $k\in\mathbb{N}$ and $a_1,a_2,\ldots,a_{2k}>0$.

8.13. Prove that $\sqrt{a_1a_2}+\sqrt{a_1a_3}+\cdots+\sqrt{a_{n-1}a_n}\leq\frac{n}{2}\leq\frac{n-1}{2}(a_1+a_2+\cdots+a_n)$, where $n\in\mathbb{N}, n\geq 2, a_1,\ldots,a_n\geq 0$ and $\frac{1}{1+a_1}+\cdots+\frac{1}{1+a_n}=n-1$.

8.14. Prove that $\frac{1}{1+a_1+\cdots+a_{n-1}}+\frac{1}{1+a_1+\cdots+a_{n-2}+a_n}+\cdots+\frac{1}{1+a_2+\cdots+a_n}\leq 1$, where $n\in\mathbb{N}, n\geq 2, a_1,\ldots,a_n\geq 0$, and $a_1\ldots a_n=1$.

8.15. Prove that $\left(a^2+bc\right)\left(b^2+cd\right)\left(c^2+da\right)\left(d^2+ab\right)\geq(a+1)(b+1)(c+1)(d+1)$, where $abcd=1$ and $a>0,\ b>0,\ c>0,\ d>0$.

8.16. Prove that $\frac{1}{a^5(b+2c)^2}+\frac{1}{b^5(c+2a)^2}+\frac{1}{c^5(a+2b)^2}\geq\frac{1}{3}$, where $abd=1$ and $a>0,\ b>0,\ c>0$.

8.17. Prove that $\frac{a}{a^3+b^3+c^2}+\frac{b}{a^2+b^3+c^3}+\frac{c}{a^3+b^2+c^3}\le 1$, where $ab+bc+ac\ge 3$ and $a>0,\ b>0,\ c>0$.

8.18. Prove that

(a) $\frac{x^3}{(1+y)(1+z)}+\frac{y^3}{(1+x)(1+z)}+\frac{z^3}{(1+x)(1+y)}\ge\frac{3}{4}$, where $x>0,\ y>0,\ z>0$ and $xyz=1$.

(b) $\frac{x}{1-x}+\frac{y}{1-y}+\frac{z}{1-z}\ge\frac{3\sqrt[3]{xyz}}{1-\sqrt[3]{xyz}}$, where $0<x<1,\ 0<y<1,\ 0<z<1$.

(c) $\frac{x}{(y+z)(y+z-x)}+\frac{y}{(x+z)(x+z-y)}+\frac{z}{(x+y)(x+y-z)}\ge\frac{9}{2(x+y+z)}$, where x,y,z are the side lengths of some triangle.

(d) $\frac{a^3}{b+c+d}+\frac{b^3}{a+c+d}+\frac{c^3}{a+b+d}+\frac{d^3}{a+b+c}\ge\frac{1}{3}$, where $a>0,b>0,c>0,d>0$ and $ab+bc+cd+da=1$.

(e) $\left(1+\frac{a_1^3}{a_2^2}\right)\left(1+\frac{a_2^3}{a_3^2}\right)\cdots\left(1+\frac{a_n^3}{a_1^2}\right)\ge(1+a_1)(1+a_2)\cdots(1+a_n)$, where $a_1,\ldots,a_n>0$.

(f) $\left(x_1^2+\cdots+x_n^2\right)\left(\frac{1}{x_1^2+x_1x_2}+\cdots+\frac{1}{x_n^2+x_nx_1}\right)\ge\frac{n^2}{2}$, where $n\in\mathbb{N}, n\ge 2, x_1,\ldots,x_n>0$.

(g) $(a_1a_2+a_2a_3+\cdots+a_na_1)\left(\frac{a_1}{a_2^2+a_2}+\frac{a_2}{a_3^2+a_3}+\cdots+\frac{a_n}{a_1^2+a_1}\right)\ge\frac{n}{n+1}$, where $n\in\mathbb{N}, n\ge 3, a_1,\ldots,a_n>0$ and $a_1+\cdots+a_n=1$.

8.19. Find the smallest value of the expression $\frac{x_1^5}{x_2+x_3+\cdots+x_n}+\frac{x_2^5}{x_1+x_3+\cdots+x_n}+\cdots+\frac{x_n^5}{x_1+x_2+\cdots+x_{n-1}}$, where $n\ge 3,\ x_1>0,\ x_2>0,\ldots,x_n>0$, and $x_1^2+x_2^2+\cdots+x_n^2=1$.

8.20. Prove that for all positive numbers x,y,z such that $xyz=1$, the following inequality holds: $\frac{x^3}{x^2+y}+\frac{y^3}{y^2+z}+\frac{z^3}{z^2+x}\ge\frac{3}{2}$.

Proofs

8.1. In similar problems one often needs to write $\frac{a^2}{ab}$ instead of $\frac{a}{b}$. It follows that $\frac{x_1}{x_2+x_3}+\frac{x_2}{x_3+x_1}+\frac{x_3}{x_1+x_2}=\frac{x_1^2}{x_1(x_2+x_3)}+\frac{x_2^2}{x_2(x_3+x_1)}+\frac{x_3^2}{x_3(x_1+x_2)}\ge\frac{(x_1+x_2+x_3)^2}{2(x_1x_2+x_2x_3+x_3x_1)}\ge\frac{3}{2}$ (here we have used that $x_1x_2+x_2x_3+x_3x_1\le x_1^2+x_2^2+x_3^2$).

8.2. We have that, $\frac{a^2}{a(b+c)}+\frac{b^2}{b(c+d)}+\frac{c^2}{c(d+a)}+\frac{d^2}{d(a+b)}\ge\frac{(a+b+c+d)^2}{ab+2ac+ad+bc+2bd+cd}\ge 2$. In order to prove the last inequality, it is sufficient to expand $(a+b+c+d)^2$ and use that $a^2+c^2\ge 2ac,\quad b^2+d^2\ge 2bd$.

8.3. We have $\frac{x_1^2}{x_1(x_2+x_n)}+\frac{x_2^2}{x_2(x_3+x_1)}+\cdots+\frac{x_n^2}{x_n(x_1+x_{n-1})}\ge\frac{(x_1+x_2+\cdots+x_n)^2}{2(x_1x_2+x_2x_3+\cdots+x_{n-1}x_n+x_nx_1)}=A_n$. If $n\ge 4$, then $A_n\ge 2$ (see the proof Problem 7.6).

8.4. (a) Using inequality (8.1), we obtain $\frac{a^4}{a^3+a^2b+ab^2}+\frac{b^4}{b^3+b^2c+bc^2}+\frac{c^4}{c^3+c^2a+ca^2}\ge\frac{(a^2+b^2+c^2)^2}{(a+b+c)(a^2+b^2+c^2)}\ge\frac{a+b+c}{3}$.

Here we have used that $a^2+b^2+c^2\ge\frac{1}{3}(a+b+c)^2$, or inequality (8.6.1) for $n=3$.

(b) By inequality (8.1), it follows that $\frac{a}{b}+\frac{b}{c}+\frac{c}{a}=\frac{a}{b}+\frac{b^2}{bc}+\frac{c^2}{ac}\geq\frac{a}{b}+\frac{(b+c)^2}{bc+ac}=\frac{a}{b}+\frac{b}{c}\cdot\frac{b+c}{a+b}+\frac{b+c}{a+b}$.
Hence, in order to complete the proof, one needs to prove that $\frac{a}{b}+\frac{b}{c}\cdot\frac{b+c}{a+b}\geq\frac{a+b}{b+c}+1$, which is true because it is equivalent to the inequality $\left(ac-b^2\right)^2\geq 0$.
(c) Without loss of generality one can assume that $(b-a)(b-c)\leq 0$.
On the other hand, by Problem 8.4 (b), we have $\frac{a}{b}+\frac{b}{c}+\frac{c}{a}\geq\frac{a+b}{b+c}+\frac{b+c}{a+b}+1$.
Therefore, it is sufficient to prove that $\frac{a+b}{b+c}+1\geq\frac{a+c}{b+c}+\frac{a+b}{a+c}$, which holds because it is equivalent to the inequality $(b-a)(b-c)\leq 0$.
(d) In a similar way, one can prove that $\frac{a}{b}+\frac{b}{c}+\frac{c}{a}\geq\frac{a+b}{b+c}+\frac{a+c}{a+b}+\frac{b+c}{a+c}$.

8.5. Without loss of generality one can assume that $a_1\geq\cdots\geq a_n$, whence $0<p-2a_1\leq\cdots\leq p-2a_n$.
According to the remark of inequality (8.6.1), we have that $\frac{a_1\cdot 1}{p-2a_1}+\cdots+\frac{a_n\cdot 1}{p-2a_n}\geq\frac{(a_1+\cdots+a_n)n}{np-2(a_1+\cdots+a_n)}=\frac{n}{n-2}$.

8.6. (a) We have that, $\frac{\left(\frac{1}{a}\right)^2}{a(b+c)}+\frac{\left(\frac{1}{b}\right)^2}{b(a+c)}+\frac{\left(\frac{1}{c}\right)^2}{c(a+b)}\geq\frac{\left(\frac{1}{a}+\frac{1}{b}+\frac{1}{c}\right)^2}{2(ab+bc+ac)}=\frac{ab+bc+ac}{2}\geq\frac{3\sqrt[3]{a^2b^2c^2}}{2}=\frac{3}{2}$.
(b) Set $\frac{1}{1+x_1+\cdots+x_i}=y_i,\quad i=1,\ldots,n$, in which case $1>y_1>\cdots>y_n>0$ and $\frac{1}{x_i}=\frac{y_{i-1}y_i}{y_{i-1}-y_i},\; i=1,\ldots,n,\quad y_0=1$.
From inequality (8.1), we deduce that

$$\sqrt{\frac{1}{x_1}+\cdots+\frac{1}{x_n}}=\sqrt{\sum_{i=1}^{n}\frac{y_{i-1}y_i}{y_{i-1}-y_i}}=\sqrt{\sum_{i=1}^{n}\left(y_i+\frac{y_i^2}{y_{i-1}-y_i}\right)}$$
$$>\sqrt{\sum_{i=1}^{n}\frac{y_i^2}{y_{i-1}-y_i}}\geq\frac{y_1+\cdots+y_n}{\sqrt{1-y_n}}>y_1+\cdots+y_n,$$

and therefore $\sqrt{\frac{1}{x_1}+\cdots+\frac{1}{x_n}}>\frac{1}{1+x_1}+\cdots+\frac{1}{1+x_1+\cdots+x_n}$.
(c) We have that, $\frac{x}{y^2-z}+\frac{y}{z^2-x}+\frac{z}{x^2-y}=\frac{x^2}{x(y^2-z)}+\frac{y^2}{y(z^2-x)}+\frac{z^2}{z(x^2-y)}\geq$
$\geq\frac{(x+y+z)^2}{xy^2+yz^2+zx^2-xz-yx-zy}>\frac{(x+y+z)^2}{4xy+4yz+4zx-xz-yx-zy}\geq 1$, and therefore $\frac{x}{y^2-z}+\frac{y}{z^2-x}+\frac{z}{x^2-y}>1$.

8.7 We have

$$\left(x^2+yz\right)\left(y^2+xy\right)\left(z^2+xy\right)\leq\left(\frac{x^2+y^2+z^2+xy+yz+xz}{3}\right)^3$$
$$=\left(\frac{2}{3}\left(x^2+y^2+z^2\right)\right)^3. \tag{8.11}$$

On the other hand, from inequality (8.1) we obtain
$\frac{x^3+y^3+z^3}{x^2+y^2+z^2}=\frac{x^4}{x\left(x^2+y^2+z^2\right)}+\frac{y^4}{x\left(x^2+y^2+z^2\right)}+\frac{z^4}{x\left(x^2+y^2+z^2\right)}\geq\frac{x^2+y^2+z^2}{x+y+z}\geq\sqrt{\frac{x^2+y^2+z^2}{3}}$,

whence

$$3\left(x^3+y^3+z^3\right)^2 \geq \left(x^2+y^2+z^2\right)^3. \tag{8.12}$$

From (8.11) and (8.12) it follows that $(x^2+yz)(y^2+xz)(z^2+xy) \leq \frac{8}{9}(x^3+y^3+z^3)^2$.

8.8. Using Chebyshev's inequality, we obtain

$$\left(\frac{1}{n}\sum_{i=1}^{n}a_i\right)\left(\frac{1}{n}\sum_{i=1}^{n}b_i\right)\left(\frac{1}{n}\sum_{i=1}^{n}c_i\right)\cdots\left(\frac{1}{n}\sum_{i=1}^{n}d_i\right) \leq \left(\frac{1}{n}\sum_{i=1}^{n}a_ib_i\right)\left(\frac{1}{n}\sum_{i=1}^{n}c_i\right)\cdots\left(\frac{1}{n}\sum_{i=1}^{n}d_i\right) \leq$$
$$\leq \left(\frac{1}{n}\sum_{i=1}^{n}a_ib_ic_i\right)\cdots\left(\frac{1}{n}\sum_{i=1}^{n}d_i\right) \leq \cdots \leq \frac{1}{n}\sum_{i=1}^{n}a_ib_ic_i\cdots d_i.$$

8.9. From inequality (8.1), we obtain $\frac{x_0^2}{x_1}+\frac{x_1^2}{x_2}+\cdots+\frac{x_{n-1}^2}{x_n} \geq \frac{(x_0+x_1+\cdots+x_{n-1})^2}{x_1+x_2+\cdots+x_n} = K_n$.
Let us prove that there exists a positive integer n_0 such that for $n > n_0$, $K_n \geq 3.999$.
We have

$$\left(x_0+x_1+\cdots+x_{n-1}\right)^2 - 3.999(x_1+x_2+\cdots+x_n) = \left(1-\left(x_1+x_2+\cdots+x_{n-1}\right)\right)^2 +$$
$$+0.001\left(x_1+x_2+\cdots+x_{n-1}\right) - 3.999x_n \geq 0.001(n-1)x_n - 3.999x_n \geq 0, \text{ if } n \geq \frac{3.999}{0.001}+1.$$

Therefore, one can take n_0 equal to 4000.

8.10. By transforming the given expression and using inequality (8.1), we obtain that

$$\frac{BC^2}{BC\cdot MA_1}+\frac{CA^2}{CA\cdot MB_1}+\frac{AB^2}{AB\cdot MC_1}$$
$$\geq \frac{(BC+CA+AB)^2}{BC\cdot MA_1+CA\cdot MB_1+AB\cdot MC_1} = \frac{4p^2}{2S} = 2\frac{p}{r}.$$

Therefore, the smallest possible value of the expression $\frac{BC}{MA_1}+\frac{CA}{MB_1}+\frac{AB}{MC_1}$ is equal to $\frac{2p}{r}$, when $\frac{BC}{BC\cdot MA_1} = \frac{CA}{CA\cdot MB_1} = \frac{AB}{AB\cdot MC_1}$ that is, when $MA_1 = MB_1 = MC_1$
Therefore, the point M is the incenter of triangle ABC.

8.11. Let us denote by A_1', A_2', A_3' and a_1, a_2, a_3 the midpoints and the lenghts of sides A_2A_3, A_1A_3, A_1A_2, respectively.
Letting $a_1 \leq a_2 \leq a_3$, one can easily prove that $GA_3 \leq GA_2 \leq GA_1$.
We have $\frac{3}{2}GA_1 \cdot B_1A_1' = \frac{1}{4}a_1^2$, and hence $GB_1 = \frac{GA_1}{2}+\frac{a_1^2}{6GA_1}$.
Hence, it is sufficient to prove that $\frac{a_1^2}{3GA_1}+\frac{a_2^2}{3GA_2}+\frac{a_3^2}{3GA_3} \geq GA_1+GA_2+GA_3$.

From inequality (8.6.1), it follows that $\frac{a_1^2}{3GA_1}+\frac{a_2^2}{3GA_2}+\frac{a_3^2}{3GA_3}\geq\frac{a_1^2+a_2^2+a_3^2}{GA_1+GA_2+GA_3}=\frac{3(GA_1^2+GA_2^2+GA_3^2)}{GA_1+GA_2+GA_3}\geq GA_1+GA_2+GA_3$,

and therefore, $\frac{a_1^2}{3GA_1}+\frac{a_2^2}{3GA_2}+\frac{a_3^2}{3GA_3}\geq GA_1+GA_2+GA_3$.

8.12. From Problem 2.1, it follows that

$$\sqrt{a_1a_2}+\sqrt{a_1a_2}+\sqrt{a_3a_4}+\sqrt{a_3a_4}+\cdots+\sqrt{a_{2k-1}a_{2k}}+\sqrt{a_{2k-1}a_{2k}}$$
$$\geq 2k\sqrt[2k]{a_1\cdot a_2\cdots a_{2k}},$$

and hence we obtain that

$$a_1+a_2+\cdots+a_{2k}-2k\sqrt[2k]{a_1\cdot a_2\cdots a_{2k}}\geq\left(\sqrt{a_1}-\sqrt{a_2}\right)^2+\cdots+\left(\sqrt{a_{2k-1}}-\sqrt{a_{2k}}\right)^2=$$

$$\geq\frac{(a_1-a_2)^2}{\left(\sqrt{a_1}+\sqrt{a_2}\right)^2}+\cdots+\frac{(a_{2k-1}-a_{2k})^2}{\left(\sqrt{a_{2k-1}}+\sqrt{a_{2k}}\right)^2}\geq\frac{(a_1-a_2)^2}{2(a_1+a_2)}+\cdots+\frac{(a_{2k-1}-a_{2k})^2}{2(a_{2k-1}+a_{2k})}.$$

From inequality (8.1), it follows that $\frac{(a_1-a_2)^2}{2(a_1+a_2)}+\cdots+\frac{(a_{2k-1}-a_{2k})^2}{2(a_{2k-1}+a_{2k})}\geq\frac{(a_1-a_2+\cdots+a_{2k-1}-a_{2k})^2}{2(a_1+a_2+\cdots+a_{2k})}$,

It follows that $a_1+a_2+\cdots+a_{2k}-2k\sqrt[2k]{a_1\cdot a_2\cdots a_{2k}}\geq\frac{(a_1-a_2+\cdots+a_{2k-1}-a_{2k})^2}{2(a_1+a_2+\cdots+a_{2k})}$,

8.13. We have $n-1=\frac{1}{1+a_1}+\cdots+\frac{1}{1+a_n}\geq\frac{n^2}{(1+a_1)+\cdots+(1+a_n)}$, and hence $\frac{n-1}{2}(a_1+a_2+\cdots+a_n)\geq\frac{n}{2}$.

Note that $1=\frac{a_1}{1+a_1}+\cdots+\frac{a_n}{1+a_n}\geq\frac{\left(\sqrt{a_1}+\cdots+\sqrt{a_n}\right)^2}{(1+a_1)+\cdots+(1+a_n)}$, and therefore $\frac{n}{2}\geq\sqrt{a_1a_2}+\sqrt{a_1a_3}+\cdots+\sqrt{a_{n-1}a_n}$.

8.14. Letting $S=a_1+\cdots+a_n$ we need to prove that $\frac{S-a_1}{1+S-a_1}+\cdots+\frac{S-a_n}{1+S-a_n}\geq n-1$. From inequalities (8.1) and (3.5), it follows that

$$\frac{S-a_1}{1+S-a_1}+\cdots+\frac{S-a_n}{1+S-a_n}=\frac{\left(\sqrt{S-a_1}\right)^2}{1+S-a_1}+\cdots+\frac{\left(\sqrt{S-a_n}\right)^2}{1+S-a_n}\geq\frac{\left(\sqrt{S-a_1}+\cdots+\sqrt{S-a_n}\right)^2}{n+(n-1)S}=$$
$$=\frac{(n-1)S+2\sum\limits_{i\neq j}\sqrt{S-a_i}\sqrt{S-a_j}}{n+(n-1)S}\geq\frac{(n-1)S+2\sum\limits_{i\neq j}\left(S-a_i-a_j+\sqrt{a_ia_j}\right)}{n+(n-1)S}=$$
$$=\frac{(n-1)S+n(n-1)S-2\sum\limits_{i\neq j}(a_i+a_j)+2\sum\limits_{i\neq j}\sqrt{a_ia_j}}{n+(n-1)S}=\frac{(n-1)^2S+2\sum\limits_{i\neq j}\sqrt{a_ia_j}}{n+(n-1)S}\geq$$
$$\geq\frac{(n-1)^2S+2C_n^2\sqrt[C_n^2]{\prod\limits_{i\neq j}\sqrt{a_ia_j}}}{n+(n-1)S}=\frac{(n-1)^2S+2C_n^2}{n+(n-1)S}=n-1,$$

whence we obtain $\frac{S-a_1}{1+S-a_1}+\cdots+\frac{S-a_n}{1+S-a_n}\geq n-1$.

8.15. Since $a^2+bc=\frac{a^3}{a\cdot 1}+\frac{1}{a\cdot d}$, it follows from inequality (8.2) that $a^2+bc\geq\frac{(a+1)^3}{2a(d+1)}$.

In a similar way, we obtain $b^2+cd\geq\frac{(b+1)^3}{2b(a+1)}$, $c^2+da\geq\frac{(c+1)^3}{2c(b+1)}$, $d^2+ab\geq\frac{(d+1)^3}{2d(c+1)}$.

Multiplying these inequalities, we obtain that

$\left(a^2+bc\right)\left(b^2+cd\right)\left(c^2+da\right)\left(d^2+ab\right)\geq\frac{(a+1)^2(b+1)^2(c+1)^2(d+1)^2}{8}$.

Since $(a+1)(b+1)(c+1)(d+1) \geq 2\sqrt{a}\cdot 2\sqrt{b}\cdot 2\sqrt{c}\cdot 2\sqrt{d} = 8$, we have $(a^2+bc)(b^2+cd)(c^2+da)(d^2+ab) \geq \frac{(a+1)^2(b+1)^2(c+1)^2(d+1)^2}{8} \geq (a+1)(b+1)(c+1)(d+1)$, and therefore, $(a^2+bc)(b^2+cd)(c^2+da)(d^2+ab) \geq (a+1)(b+1)(c+1)(d+1)$.

8.16. Since

$$\frac{1}{a^5(b+2c)^2}+\frac{1}{b^5(c+2a)^2}+\frac{1}{c^5(a+2b)^2}=$$
$$=\frac{\left(\frac{1}{a}\right)^3}{(ab+2ac)(ab+2ac)}+\frac{\left(\frac{1}{b}\right)^3}{(bc+2ab)(bc+2ab)}+\frac{\left(\frac{1}{c}\right)^3}{(ac+2bc)(ac+2bc)},$$

then from inequality (8.2), it follows that $\frac{1}{a^5(b+2c)^2}+\frac{1}{b^5(c+2a)^2}+\frac{1}{c^5(a+2b)^2} \geq \frac{ab+bc+ac}{9}$.

Note that, $\frac{ab+bc+ac}{9} \geq \frac{3\sqrt[3]{ab\cdot bc\cdot ac}}{9} = \frac{1}{3}$, and therefore, $\frac{1}{a^5(b+2c)^2}+\frac{1}{b^5(c+2a)^2}+\frac{1}{c^5(a+2b)^2} \geq \frac{1}{3}$.

8.17. From inequality (8.2), it follows that, $a^3+b^3+c^2 = \frac{a^3}{1\cdot 1}+\frac{b^3}{1\cdot 1}+\frac{c^3}{1\cdot c} \geq \frac{(a+b+c)^3}{3(2+c)}$, and therefore, $\frac{a}{a^3+b^3+c^2} \leq \frac{6a+3ac}{(a+b+c)^3}$.

In a similar way, we obtain $\frac{b}{a^2+b^3+c^3} \leq \frac{6b+3ab}{(a+b+c)^3}$, $\frac{c}{a^3+b^2+c^3} \leq \frac{6c+3bc}{(a+b+c)^3}$.

Summing these inequalities, we deduce that $\frac{a}{a^3+b^3+c^2}+\frac{b}{a^2+b^3+c^3}+\frac{c}{a^3+b^2+c^3} \leq \frac{6a+3ac}{(a+b+c)^3}+\frac{6b+3ab}{(a+b+c)^3}+\frac{6c+3bc}{(a+b+c)^3}$.

Note that $\frac{6a+3ac}{(a+b+c)^3}+\frac{6b+3ab}{(a+b+c)^3}+\frac{6c+3bc}{(a+b+c)^3} = \frac{6}{(a+b+c)^2}+\frac{3(ab+bc+ac)}{(a+b+c)^2(a+b+c)} \leq$
$\leq \frac{6}{3(ab+bc+ac)}+\frac{1}{a+b+c} \leq \frac{2}{3}+\frac{1}{3} = 1$, and therefore, $\frac{a}{a^3+b^3+c^2}+\frac{b}{a^2+b^3+c^3}+\frac{c}{a^3+b^2+c^3} \leq 1$.

8.18 (a) From inequality (8.2), it follows that,

$$\frac{x^3}{(1+y)\cdot(1+z)}+\frac{y^3}{(1+z)\cdot(1+x)}+\frac{z^3}{(1+x)\cdot(1+y)} \geq \frac{(x+y+z)^3}{(1+y+1+z+1+x)(1+z+1+x+1+y)} =$$
$$=\left(\frac{x+y+z}{3+x+y+z}\right)^2(x+y+z) \geq \frac{3}{4}, \text{ as } x+y+z \geq 3\sqrt[3]{xyz}=3 \text{ and } \frac{x+y+z}{3+x+y+z} \geq \frac{1}{2}.$$

Therefore, $\frac{x^3}{(1+y)(1+z)}+\frac{y^3}{(1+x)(1+z)}+\frac{z^3}{(1+x)(1+y)} \geq \frac{3}{4}$.

(b) From inequality (8.2), it follows that

$$\frac{x}{1-x}+\frac{y}{1-y}+\frac{z}{1-z} = \frac{\left(\sqrt[3]{x}\right)^3}{1\cdot(1-x)}+\frac{\left(\sqrt[3]{y}\right)^3}{1\cdot(1-y)}+\frac{\left(\sqrt[3]{z}\right)^3}{1\cdot(1-z)} \geq \frac{\left(\sqrt[3]{x}+\sqrt[3]{y}+\sqrt[3]{z}\right)^3}{3(3-x-y-z)} \geq \frac{27\sqrt[3]{xyz}}{3(3-3\sqrt[3]{xyz})} =$$
$$=\frac{3\sqrt[3]{xyz}}{(1-\sqrt[3]{xyz})}, \text{ as } \sqrt[3]{x}+\sqrt[3]{y}+\sqrt[3]{z} \geq 3\sqrt[3]{\sqrt[3]{x}\cdot\sqrt[3]{y}\cdot\sqrt[3]{z}} \text{ and } x+y+z \geq 3\sqrt[3]{xyz}.$$

Therefore, $\frac{x}{1-x}+\frac{y}{1-y}+\frac{z}{1-z} \geq \frac{3\sqrt[3]{xyz}}{1-\sqrt[3]{xyz}}$.

(c) From inequality (8.2), it follows that

$$\frac{x}{(y+z)(y+z-x)}+\frac{y}{(x+z)(x+z-y)}+\frac{z}{(x+y)(x+y-z)} =$$
$$=\frac{x^3}{(xy+xz)(xy+xz-x^2)}+\frac{y^3}{(xy+yz)(xy+yz-y^2)}+\frac{z^3}{(xz+yz)(xz+yz-z^2)} \geq$$
$$\geq \frac{(x+y+z)^3}{2(xy+yz+xz)(2xy+2yz+2xz-x^2-y^2-z^2)} \geq \frac{9}{2(x+y+z)},$$

as $x^2+y^2+z^2-xy-yz-xz=0$, $5\left((x-y)^2+(y-z)^2+(z-x)^2\right)\geq 0$ and

$$\frac{(x+y+z)^2}{xy+yz+xz}\geq 3,\quad \frac{(x+y+z)^2}{2xy+2yz+2xz-x^2-y^2-z^2}\geq 3.$$

Therefore, $\frac{x}{(y+z)(y+z-x)}+\frac{y}{(x+z)(x+z-y)}+\frac{z}{(x+y)(x+y-z)}\geq\frac{9}{2(x+y+z)}$.

(d) From inequality (8.2), it follows that

$$\frac{a^3}{1\cdot(b+c+d)}+\frac{b^3}{1\cdot(a+c+d)}+\frac{c^3}{1\cdot(a+b+d)}+\frac{d^3}{1\cdot(a+b+c)}\geq$$
$$\geq\frac{(a+b+c+d)^3}{4(3a+3b+3c+3d)}=\frac{(a-b+c-d)^2+4(ab+bc+cd+da)}{12}\geq\frac{1}{3},$$

and therefore, $\frac{a^3}{b+c+d}+\frac{b^3}{a+c+d}+\frac{c^3}{a+b+d}+\frac{d^3}{a+b+c}\geq\frac{1}{3}$.

(e) From inequality (8.2), it follows that

$1+\frac{x^3}{y^2}=\frac{1^3}{1\cdot 1}+\frac{x^3}{y\cdot y}\geq\frac{(1+x)^3}{(1+y)(1+y)}$, whence we obtain

$$\left(1+\frac{a_1^3}{a_2^2}\right)\left(1+\frac{a_2^3}{a_3^2}\right)\cdots\left(1+\frac{a_n^3}{a_1^2}\right)\geq\frac{(1+a_1)^3}{(1+a_2)^2}\cdot\frac{(1+a_2)^3}{(1+a_3)^2}\cdots\frac{(1+a_n)^3}{(1+a_1)^2}$$
$$=(1+a_1)(1+a_2)\cdots(1+a_n).$$

(f) From inequality (8.2), it follows that,

$\frac{1^3}{x_1\cdot(x_1+x_2)}+\cdots+\frac{1^3}{x_n\cdot(x_n+x_1)}\geq\frac{n^3}{2(x_1+\cdots+x_n)^2}$, and therefore $(x_1^2+\cdots+x_n^2)\left(\frac{1}{x_1^2+x_1x_2}+\cdots+\frac{1}{x_n^2+x_nx_1}\right)\geq\frac{n^3(x_1^2+\cdots+x_n^2)}{2(x_1+\cdots+x_n)^2}\geq\frac{n^2}{2}$, since $\frac{x_1^2+\cdots+x_n^2}{n}\geq\left(\frac{x_1+\cdots+x_n}{n}\right)^2$.

Therefore, $\left(x_1^2+\cdots+x_n^2\right)\left(\frac{1}{x_1^2+x_1x_2}+\cdots+\frac{1}{x_n^2+x_nx_1}\right)\geq\frac{n^2}{2}$.

(g) For $A=a_1a_2+a_2a_3+\cdots+a_{n-1}a_n+a_na_1<\frac{1}{n}$, from inequality (8.2), it follows that

$$\frac{a_1}{a_2^2+a_2}+\frac{a_2}{a_3^2+a_3}+\cdots+\frac{a_{n-1}}{a_n^2+a_n}+\frac{a_n}{a_1^2+a_1}=\frac{a_1^3}{a_1a_2(a_1a_2+a_1)}+\frac{a_2^3}{a_2a_3(a_2a_3+a_2)}+\cdots+\frac{a_n^3}{a_na_1(a_na_1+a_n)}\geq$$
$$\geq\frac{(a_1+a_2+\cdots+a_n)^3}{A(A+a_1+a_2+\cdots+a_n)}=\frac{1}{A(A+1)}>\frac{1}{A}\cdot\frac{n}{n+1},\text{ and therefore,}$$
$$(a_1a_2+a_2a_3+\cdots+a_{n-1}a_n+a_na_1)\left(\frac{a_1}{a_2^2+a_2}+\frac{a_2}{a_3^2+a_3}+\cdots+\frac{a_{n-1}}{a_n^2+a_n}+\frac{a_n}{a_1^2+a_1}\right)\geq\frac{n}{n+1}.$$

For $A\geq\frac{1}{n}$, from inequality (3), we have

$$\frac{a_1}{a_2^2+a_2}+\frac{a_2}{a_3^2+a_3}+\cdots+\frac{a_{n-1}}{a_n^2+a_n}+\frac{a_n}{a_1^2+a_1}=\frac{\left(\sqrt{\frac{a_1}{a_2}}\right)^2}{a_2+1}+\frac{\left(\sqrt{\frac{a_2}{a_3}}\right)^2}{a_3+1}+\cdots+\frac{\left(\sqrt{\frac{a_n}{a_1}}\right)^2}{a_1+1}\geq$$
$$\geq\frac{\left(\sqrt{\frac{a_1}{a_2}}+\sqrt{\frac{a_2}{a_3}}+\sqrt{\frac{a_n}{a_1}}\right)^2}{n+1}\geq\frac{n^2}{n+1},\text{ as}\sqrt{\frac{a_1}{a_2}}+\sqrt{\frac{a_2}{a_3}}+\cdots+\sqrt{\frac{a_n}{a_1}}\geq n\sqrt[n]{\sqrt{\frac{a_1}{a_2}}\cdot\sqrt{\frac{a_2}{a_3}}\cdots\sqrt{\frac{a_n}{a_1}}}=n.$$

Hence, we obtain that,

$$(a_1a_2+a_2a_3+\cdots+a_{n-1}a_n+a_na_1)\left(\frac{a_1}{a_2^2+a_2}+\frac{a_2}{a_3^2+a_3}+\cdots+\frac{a_{n-1}}{a_n^2+a_n}+\frac{a_n}{a_1^2+a_1}\right)\geq\frac{1}{n}\cdot\frac{n^2}{n+1}=\frac{n}{n+1}.$$

8.19. Letting $x_1+x_2+\cdots+x_n=S$, $\frac{x_1^5}{x_2+x_3+\cdots+x_n}+\frac{x_2^5}{x_1+x_3+\cdots+x_n}+\cdots+\frac{x_n^5}{x_1+x_2+\cdots+x_{n-1}}=A$, from inequality (8.2), it follows that $A=\frac{(x_1^2)^3}{x_1(S-x_1)}+\frac{(x_2^2)^3}{x_2(S-x_2)}+\cdots+\frac{(x_n^2)^3}{x_n(S-x_n)}\geq \frac{(x_1^2+x_2^2+\cdots+x_n^2)^3}{S\cdot(S-x_1+S-x_2+\cdots+S-x_n)}=\frac{1}{(n-1)S^2}\geq\frac{1}{n(n-1)}$, as $1=\frac{x_1^3}{1\cdot x_1}+\frac{x_2^3}{1\cdot x_2}+\cdots+\frac{x_n^3}{1\cdot x_n}\geq\frac{S^3}{nS}=\frac{S^2}{n}$.

We have obtained $A\geq\frac{1}{n(n-1)}$ and for $x_1=x_2=\cdots=x_n=\frac{1}{\sqrt{n}}$, we have $A=\frac{1}{n(n-1)}$, and therefore, the smallest possible value of the given expression is equal to $\frac{1}{n(n-1)}$.

8.20. From inequality (8.2), it follows that $\frac{x^3}{x^2+y}+\frac{y^3}{y^2+z}+\frac{z^3}{z^2+x}\geq\frac{(x+y+z)^3}{3(x^2+y+y^2+z+z^2+x)}$.

It is sufficient to prove that, $2(x+y+z)^3\geq 9\big(x^2+y^2+z^2+x+y+z\big)$.

Letting $\sqrt{x+y+z}=d$, we have $xy+yz+zx=\sqrt{(xy+yz+zx)^2}\geq\sqrt{3xy\cdot yz+3xy\cdot zx+3yz\cdot zx}=\sqrt{3}d.$

One needs to prove that, $2(x+y+z)^3+18(xy+yz+zx)\geq 9\big((x+y+z)^2+x+y+z\big)$.

Let us prove that, $2d^5+18\sqrt{3}\geq 9d^3+9d$.

Note that, $2d^5-9d^3-9d+18\sqrt{3}=(d-\sqrt{3})^2(2d^3+4\sqrt{3}d^2+9d+6\sqrt{3})\geq 0.$

Problems for Independent Study

Prove the following inequalities (1–22, 24–45, 47, 48).

1. $2\sum\limits_{i=1}^{n}\frac{a_i^2}{a_i+a_{i+1}}\geq\sum\limits_{i=1}^{n}a_i$, where $a_{n+1}=a_1,\ a_i>0\quad i=1,\ldots,n.$
2. $\sum\limits_{i=1}^{n}\left(x_i+\frac{1}{x_i}\right)^2\geq\frac{(n^2+1)^2}{n}$, where $x_1+\cdots+x_n=1,\quad x_i>0,\quad i=1,\ldots,n.$
3. $a+b+c\leq\frac{a^2+b^2}{2c}+\frac{b^2+c^2}{2a}+\frac{a^2+c^2}{2b}\leq\frac{a^3}{bc}+\frac{b^3}{ac}+\frac{c^3}{ab}$, where $a>0,\ b>0,\ c>0.$
4. (a) $\frac{a_1b_1}{a_1+b_1}+\cdots+\frac{a_nb_n}{a_n+b_n}\leq\frac{(a_1+\cdots+a_n)(b_1+\cdots+b_n)}{(a_1+\cdots+a_n)+(b_1+\cdots+b_n)}$, where $a_i>0,\ b_i>0,\ i=1,\ldots,n,$

   (b) $\frac{a_1+b_1}{a_1+b_1+aa_1b_1}+\cdots+\frac{a_n+b_n}{a_n+b_n+aa_nb_n}\geq\frac{n^2}{a(a_1+\cdots+a_n)+2n}+\frac{n^2}{a(b_1+\cdots+b_n)+2n}$, where $a>0,\quad a_i>0,\quad b_i>0,\quad i=1,\ldots,n,$

   (c) $\sum\limits_{i=1}^{n}\frac{a_ib_i}{\sqrt{a_i^2+b_i^2}}\leq\frac{\sum\limits_{i=1}^{n}a_i\sum\limits_{i=1}^{n}b_i}{\sqrt{\left(\sum\limits_{i=1}^{n}a_i\right)^2+\left(\sum\limits_{i=1}^{n}b_i\right)^2}}$, where $a_i>0,\ b_i>0,\ i=1,\ldots,n,$

   (d) $\frac{a_1+b_1}{a_1+b_1+na_1b_1}+\cdots+\frac{a_n+b_n}{a_n+b_n+na_nb_n}+\frac{n}{2}\left(\frac{a_1+\cdots+a_n}{a_1+\cdots+a_n+2}+\frac{b_1+\cdots+b_n}{b_1+\cdots+b_n+2}\right)\geq n$, where $a_i>0,\ b_i>0,\quad i=1,\ldots,n,$

   (e) $\frac{c_1}{a_1\cdot b_1}+\cdots+\frac{c_n}{a_n\cdot b_n}\geq\frac{n^2(c_1+\cdots+c_n)}{(a_1+\cdots+a_n)(b_1+\cdots+b_n)}$, where $a_i>0,\ b_i>0,\ c_i>0,\quad i=1,\ldots,n,$
   $\left(\frac{a_i}{c_i}-\frac{a_j}{c_j}\right)\left(\frac{b_i}{c_i}-\frac{b_j}{c_j}\right)\leq 0,\quad i,j\in\{1,\ldots,n\}.$

(f) $\frac{1}{\sqrt{a_1b_1-c_1d_1}}+\cdots+\frac{1}{\sqrt{a_nb_n-c_nd_n}} \geq \frac{n^2}{\sqrt{(a_1+\cdots+a_n)(b_1+\cdots+b_n)-\left(\sqrt{c_1d_1}+\cdots+\sqrt{c_nd_n}\right)^2}}$,
where $a_i > 0, b_i > 0, c_i > 0, d_i > 0, a_ib_i \geq c_id_i, \quad i = 1, \ldots, n.$

Remark Letting $\sqrt{a_ib_i - c_id_i} = x_i$, we have

$$\left(\frac{1}{\sqrt{a_1b_1-c_1d_1}}+\cdots+\frac{1}{\sqrt{a_nb_n-c_nd_n}}\right)\sqrt{(a_1+\cdots+a_n)(b_1+\cdots+b_n)-\left(\sqrt{c_1d_1}+\cdots+\sqrt{c_nd_n}\right)^2}$$
$$=\left(\frac{1}{x_1}+\cdots+\frac{1}{x_n}\right)\sqrt{(a_1+\cdots+a_n)\left(\frac{x_1^2}{a_1}+\cdots+\frac{x_n^2}{a_n}\right)+(a_1+\cdots+a_n)\left(\frac{c_1d_1}{a_1}+\cdots+\frac{c_nd_n}{a_n}\right)-\left(\sqrt{c_1d_1}+\cdots+\sqrt{c_nd_n}\right)^2}.$$

5. $\frac{x_1}{x_2+x_3}+\frac{x_2}{x_3+x_4}+\cdots+\frac{x_{n-1}}{x_n+x_1}+\frac{x_n}{x_1+x_2} \geq \frac{n}{2}$, where $n = 5$ or $n = 6, x_i > 0, \quad i = 1, \ldots, n.$
6. $\left(1+\frac{a_1^2}{a_2}\right)\left(1+\frac{a_2^2}{a_3}\right)\cdots\left(1+\frac{a_n^2}{a_1}\right) \geq (1+a_1)\cdots(1+a_n)$, where $n \geq 2, \quad a_1 > 0, \ldots, a_n > 0.$

Remark Note that $1+\frac{a^2}{b} = \frac{1^2}{1}+\frac{a^2}{b} \geq \frac{(1+a)^2}{1+b}$, where $b > 0$.

7. $\frac{a}{b+c}+\frac{b}{c+a}+\frac{c}{a+b}+\frac{ab+bc+ca}{a^2+b^2+c^2} \leq \frac{5}{2}$, where a, b, c are the side lengths of some triangle.

Remark Note that

$$\frac{b+c-a}{b+c}+\frac{c+a-b}{c+a}+\frac{a+b-c}{a+b}=$$
$$=\frac{(b+c-a)^2}{(b+c)(b+c-a)}+\frac{(c+a-b)^2}{(c+a)(c+a-b)}+\frac{(a+b-c)^2}{(a+b)(a+b-c)} \geq \frac{(a+b+c)^2}{2\left(a^2+b^2+c^2\right)}.$$

8. $\frac{a}{b+2c+3d}+\frac{b}{c+2d+3a}+\frac{c}{d+2a+3b}+\frac{d}{a+2b+3c} \geq \frac{2}{3}$, where $a > 0, b > 0, c > 0 d > 0.$
9. $\sum_{i=1}^{n} \frac{x_i}{\sqrt{1+x_i}} \geq \frac{\sum_{i=1}^{n}\sqrt{x_i}}{\sqrt{n-1}}$, where $n \geq 2, \quad x_i > 0, \quad i = 1, \ldots, n$, and $\sum_{i=1}^{n} x_i = 1.$
10. $\frac{x_1^2}{1+x_2+x_3+\cdots+x_n}+\frac{x_2^2}{1+x_1+x_3+\cdots+x_n}+\cdots+\frac{x_n^2}{1+x_1+x_2+\cdots+x_{n-1}} \geq \frac{4}{3n-2}$, where $n \geq 2$, $x_i > 0, \ i = 1, \ldots, n$ and $x_1 + \cdots + x_n = 2.$
11. $x_1^k + \cdots + x_n^k \geq x_1 + \cdots + x_n$, where $k \in \mathbb{N}$ and $n \geq 2, \quad x_i > 0, i = 1, \ldots, n, \quad x_1 \cdots x_n = 1.$
12. $\sqrt{a_1} + \cdots + \sqrt{a_n} \leq a_1 + \cdots + a_n$, where $n \geq 2, \quad a_i > 0, \quad i = 1, \ldots, n$ and $a_1 \cdots a_n = 1.$
13. (a) $\frac{(x+y)z}{(x+y)^2+z^2} \leq \frac{4z}{4z+3x+3y}$, where $x > 0, \ y > 0, \ z > 0$;
    (b) $\frac{(b+c-a)^2}{(b+c)^2+a^2}+\frac{(c+a-b)^2}{(c+a)^2+b^2}+\frac{(a+b-c)^2}{(a+b)^2+c^2} \geq \frac{3}{5}$, where $a > 0, \ b > 0, \ c > 0.$
14. (a) $\frac{x_1}{\sqrt{x_1}+\sqrt{x_2}}+\frac{x_2}{\sqrt{x_2}+\sqrt{x_3}}+\cdots+\frac{x_n}{\sqrt{x_n}+\sqrt{x_1}} \geq \frac{1}{2}\left(\sqrt{x_1}+\cdots+\sqrt{x_n}\right)$, where $n \geq 3, \quad x_1 > 0, \ldots, x_n > 0.$
    (b) $\frac{b\sqrt{a}}{4b\sqrt{c}-c\sqrt{a}}+\frac{c\sqrt{b}}{4c\sqrt{a}-a\sqrt{b}}+\frac{a\sqrt{c}}{4a\sqrt{b}-b\sqrt{c}} \geq 1$, where $a, b, c \in (1, 2).$

Remark Note that $\frac{b\sqrt{a}}{4b\sqrt{c}-c\sqrt{a}} = \frac{\left(\sqrt{ab}\right)^2}{4b\sqrt{ac}-ac}$.

15. $\frac{(a_1+\cdots+a_n)^2}{2\left(a_1^2+\cdots+a_n^2\right)} \leq \frac{a_1}{a_2+a_3}+\frac{a_2}{a_3+a_4}+\cdots+\frac{a_n}{a_1+a_2}$, where $n \geq 3, \quad a_1 > 0, \ldots, a_n > 0.$
16. $\frac{x^9+y^9}{x^6+x^3y^3+y^6}+\frac{y^9+z^9}{y^6+y^3z^3+z^6}+\frac{z^9+x^9}{z^6+z^3x^3+x^6} \geq 2$, where $x > 0, \ y > 0, \ z > 0$ and $xyz = 1.$

17. $\frac{x}{1+x^2}+\frac{y}{1+y^2}+\frac{z}{1+z^2}\leq\frac{3\sqrt{3}}{4}$, where $x^2+y^2+z^2=1$.
18. $\frac{a^3}{b+2c}+\frac{b^3}{c+2a}+\frac{c^3}{a+2b}\geq\frac{a^2+b^2+c^2}{3}$, where $a>0,\ b>0,\ c>0$.
19. $\frac{2n}{3n+1}\leq\frac{1}{n+1}+\frac{1}{n+2}+\cdots+\frac{1}{2n}\leq\frac{3n+1}{4(n+1)}$, where $n\in\mathbb{N}$
20. $\frac{a_1}{a_2^2+1}+\frac{a_2}{a_3^2+1}+\cdots+\frac{a_n}{a_1^2+1}\geq\frac{4}{5}\left(a_1\sqrt{a_1}+a_2\sqrt{a_2}+\cdots+a_n\sqrt{a_n}\right)^2$, where $n\geq 4$, $a_1>0,\ldots,a_n>0$ and $a_1^2+\cdots+a_n^2=1$.
21. $\frac{a}{2a+b}+\frac{b}{2b+c}+\frac{c}{2c+a}\leq 1$, where $a>0,\ b>0,\ c>0$.

Remark We have $\frac{1}{2}-\frac{a}{2a+b}+\frac{1}{2}+\frac{b}{2b+c}=\frac{b^2}{2b(2a+b)}+\frac{c^2}{2c(2b+c)}\geq\frac{(b+c)^2}{2(2ab+(b+c)^2)}\geq\frac{c}{2c+a}$.

22. $\frac{a}{b^2c^2}+\frac{b}{c^2a^2}+\frac{c}{a^2b^2}\geq\frac{9}{a+b+c}$, where $a>0,\ b>0,\ c>0$ and $a^2+b^2+c^2=3abc$.
23. Solve the following system of equations:

$$\begin{cases} x_1+x_2+\cdots+x_n=n, \\ x_1^4+x_2^4+\cdots+x_n^4=x_1^3+x_2^3+\cdots+x_n^3 \end{cases}.$$

24. $\frac{1}{a(b+a)}+\frac{1}{b(b+c)}+\frac{1}{c(c+a)}\geq\frac{27}{2(a+b+c)^2}$, where $a>0,\ b>0,\ c>0$.
25. $a^4b+b^4c+c^4d+d^4a\geq abcd(a+b+c+d)$, where $a>0,\ b>0,\ c>0,\ d>0$
26. $\frac{a}{b(b+c)^2}+\frac{b}{c(a+c)^2}+\frac{c}{a(a+b)^2}\geq\frac{9}{4(ab+bc+ac)}$ where $a>0,\ b>0,\ c>0$.

Remark We have $\frac{a}{b(b+c)^2}=\frac{a^3}{ab\cdot a(b+c)^2}$.

27. $3\left(x^2-x+1\right)\left(y^2-y+1\right)\left(z^2-z+1\right)\geq(xyz)^2-xyz+1$.

Remark Prove that $3\left(t^2-t+1\right)^3\geq t^6+t^3+1$.

28. $2\left(a^{2012}+1\right)\left(b^{2012}+1\right)\left(c^{2012}+1\right)\geq(1+abc)\left(a^{2011}+1\right)\left(b^{2011}+1\right)\left(c^{2011}+1\right)$, where $a>0,\ b>0,\ c>0$.
29. $\frac{a}{(b+c)^2}+\frac{b}{(a+c)^2}+\frac{c}{(a+b)^2}\geq\frac{9}{4(a+b+c)}$, where $a>0, b>0, c>0$.

Remark Note that $\frac{a}{(b+c)^2}=\frac{a^3}{(ab+ac)\cdot(ab+ac)}$.

30. $x^8+y^8+z^8\geq x^2y^2z^2(xy+yz+zx)$.

Remark Note that $xyz\neq 0$, whence $\frac{x^8+y^8+z^8}{x^2y^2z^2}=\frac{\left(x^2\right)^3}{y^2\cdot z^2}+\frac{\left(y^2\right)^3}{z^2\cdot x^2}+\frac{\left(z^2\right)^3}{x^2\cdot y^2}$.

31. $\frac{x^3}{x+yz}+\frac{y^3}{y+zx}+\frac{z^3}{z+xy}\geq\frac{1}{4}$, where $x+y+z=1$ and $x>0,\ y>0,>z>0$.
32. $\sqrt{\frac{abc}{a+b-c}}+\sqrt{\frac{abc}{b+c-a}}+\sqrt{\frac{abc}{a+c-b}}\geq a+b+c$, where a, b, c are the side lengths of some triangle.

Remark Note that $abc(a+b+c)\geq a^3(b+c-a)+b^3(a+c-b)+c^3(a+b-c)$.

33. $\sqrt[3]{ab}+\sqrt[3]{cd}\leq\sqrt[3]{(a+c+d)(a+c+b)}$, where $a>0, b>0, c>0,\ d>0$.

Remark Note that $\frac{ab}{(a+c)b} + \frac{cd}{d(a+c)} \geq \frac{(\sqrt[3]{ab}+\sqrt[3]{cd})^3}{(a+c+d)(b+a+c)}$.

34. $(a^5 - a^2 + 3)(b^5 - b^2 + 3)(c^5 - c^2 + 3) \geq (a+b+c)^3$, where $a > 0, b > 0, c > 0$.

Remark Note that $x^5 - x^2 + 3 \geq x^3 + 2$, where $x > 0$.

35. $3(a^3 + b^3 + c^3) \geq (a^2 + b^2 + c^2)^3$, where $a > 0,\ b > 0,\ c > 0$.

Remark Note that $3(a^3 + b^3 + c^3) \geq (a + b + c)(a^2 + b^2 + c^2)$.

36. $\left(1 + \frac{a_1^3}{a_2^2}\right) \cdot \left(1 + \frac{a_2^3}{a_3^2}\right) \cdots \left(1 + \frac{a_n^3}{a_1^2}\right) \geq (1 + a_1) \cdot (1 + a_2) \cdots (1 + a_n)$, where $a_1 > 0,\ a_2 > 0, \ldots, a_n > 0$.
37. $\left(a_1^3 + 1\right) \cdot \left(a_2^3 + 1\right) \cdots \left(a_n^3 + 1\right) \geq \left(a_1^2 a_2 + 1\right) \cdot \left(a_2^2 a_3 + 1\right) \cdots \left(a_n^2 a_1 + 1\right)$, where $a_1 > 0,\ a_2 > 0, \ldots, a_n > 0$.

Remark We have $\left(x^3 + 1\right)\left(x^3 + 1\right)\left(y^3 + 1\right) \geq \left(x^2 y + 1\right)^3$, where $x > 0,\ y > 0$.

38. $\frac{x_1}{1-x_1^2} + \frac{x_2}{1-x_2^2} + \cdots + \frac{x_n}{1-x_n^2} \geq \frac{(x_1+x_2+\cdots+x_n)^3}{(x_1+x_2+\cdots+x_n)^2-\left(x_1^2+x_2^2+\cdots+x_n^2\right)^2}$, where $0 < x_1 < 1,\ 0 < x_2 < 1, \ldots, 0 < x_n < 1$.
39. $n^n(x_1^n + 1) \cdot (x_2^n + 1) \cdots (x_n^n + 1) \geq \left(x_1 + x_2 + \cdots + x_n + \frac{1}{x_1} + \frac{1}{x_2} + \cdots + \frac{1}{x_n}\right)^n$, where $x_1 \cdot x_2 \cdots x_n = 1$
and $x_1 < 0,\ x_2 < 0, \ldots, x_n < 0$.

Remark Note that $(x_1^n + 1) \cdot (1 + x_2^n) \cdots (1 + x_n^n) \geq (x_1 \cdot 1 \cdots 1 + 1 \cdot x_2 \cdots x_n)^n = \left(x_1 + \frac{1}{x_1}\right)^n$.

40. $x_1 \sqrt[3]{1 + x_n - x_2} + x_2 \cdot \sqrt[3]{1 + x_1 - x_3} + \cdots + x_n \cdot \sqrt[3]{1 + x_{n-1} - x_1} \leq x_1 + x_2 + \cdots x_n$, where $x_1 > 0,\ x_2 > 0, \ldots, x_n > 0$.

Remark Note that $x_1 \cdot \sqrt[3]{1 + x_n - x_2} = \sqrt[3]{x_1} \cdot \sqrt[3]{x_1} \cdot \sqrt[3]{x_1 + x_n x_1 - x_1 x_2}$.

41. $(x_1^2 + x_2^2 + \cdots + x_n^2)\left(\frac{1}{x_1^2+x_1x_2} + \frac{1}{x_2^2+x_2x_3} + \cdots + \frac{1}{x_n^2+x_nx_1}\right) \geq \frac{n^2}{2}$, where $x_1 > 0,\ x_2 > 0, \ldots, x_n > 0$.

Remark Note that $\frac{1^3}{x_1^2+x_1x_2} + \frac{1^3}{x_n^2+x_2x_3} + \cdots + \frac{1^3}{x_n^2+x_nx_1} \geq \frac{n^3}{2(x_1+x_2+\cdots+x_n)^2}$.

42. $\frac{x_1^3}{(ax_1+bx_2)(ax_2+bx_1)} + \cdots + \frac{x_n^3}{(ax_n+bx_1)(ax_1+bx_n)} \geq \frac{x_1+\cdots+x_n}{(a+b)^2}$, where $a > 0,\ b > 0,\ x_1 > 0, x_2 > 0, \ldots, x_n > 0$.
43. $\frac{a_1^3}{b_1} + \cdots + \frac{a_n^3}{b_n} \geq 1$, where $a_i > 0,\ b_i > 0\ (i = 1, \ldots, n)$ and $\left(a_1^2 + \cdots + a_n^2\right)^3 = b_1^2 + \cdots + b_n^2$.

Remark Note that $\frac{a_1^3}{b_1} + \cdots + \frac{a_n^3}{b_n} = \frac{\left(a_1^2\right)^3}{\frac{a_1^3}{b_1} \cdot b_1^2} + \cdots + \frac{\left(a_n^2\right)^3}{\frac{a_n^3}{b_n} \cdot b_n^2}$.

44. $4\left(x^3 + y^3 + z^3 + xyz\right)^2 \geq \left(x^2 + y^2 + z^2 + t^2\right)^3$, where $x > 0,\ y > 0,\ z > 0$, and $t^2 = \frac{xyz}{\max(x,y,z)}$.

Remark Note that $x^3+y^3+z^3+xyz=\frac{(x^2)^2}{x}+\frac{(y^2)^2}{y}+\frac{(z^2)^2}{z}+\frac{\left(\sqrt[3]{x^2y^2z^2}\right)^2}{\sqrt[3]{xyz}}$.

45. $\frac{a_{1,1}^n}{a_{1,2}\cdot a_{1,3}\cdots a_{1,n}}+\frac{a_{2,1}^n}{a_{2,2}\cdot a_{2,3}\cdots a_{2,n}}+\dots+\frac{a_{m,1}^n}{a_{m,2}\cdot a_{m,3}\cdots a_{m,n}}\geq \frac{(a_{1,1}+a_{2,1}+\dots+a_{m,1})^n}{(a_{1,2}+a_{2,2}+\dots+a_{m,2})\cdot(a_{1,3}+a_{2,3}+\dots+a_{m,3})\cdots(a_{1,n}+a_{2,n}+\dots+a_{m,n})}$, where $a_{i,j}>0$, $i=1,2,\dots,m$, $j=1,2,\dots n$.

Remark Without loss of generality one can assume that $a_{1,1}+a_{2,1}+\dots+a_{m,1}=a_{1,2}+a_{2,2}+\dots+a_{m,2}=\dots=a_{1,n}+a_{2,n}+\dots+a_{m,n}=1$.

Note that $\frac{a_{k,1}^n}{a_{k,2}\cdot a_{k,3}\cdots a_{k,n}}+a_{k,2}+\dots+a_{k,n}\geq na_{k,1}$.

46. Find the smallest possible value of the expression $a_4+2b^4+3c^4$, if $a+b+c=1$.

Remark Note that

$$a^4+2b^4+3c^4=\frac{a^4}{1^3}+\frac{b^4}{\left(\sqrt[3]{\frac{1}{2}}\right)^3}+\frac{c^4}{\left(\sqrt[3]{\frac{1}{3}}\right)^3}\geq\frac{(|a|+|b|+|c|)^4}{\left(1+\sqrt[3]{\frac{1}{2}}+\sqrt[3]{\frac{1}{3}}\right)^3}\geq$$

$$\geq\frac{(|a+b+c|)^4}{\left(1+\sqrt[3]{\frac{1}{2}}+\sqrt[3]{\frac{1}{3}}\right)^3}=\frac{6}{\left(\sqrt[3]{2}+\sqrt[3]{3}+\sqrt[3]{6}\right)^3}.$$

47. $\frac{(b+c)^5}{a}+\frac{(c+a)^5}{b}+\frac{(a+b)^5}{c}\geq\frac{32}{9}(ab+bc+ca)$, where $a>0$, $b>0$, $c>0$ and $a+b+c=1$.

Remark Note that $\frac{(b+c)^5}{a}+\frac{(c+a)^5}{b}+\frac{(a+b)^5}{c}=\frac{(b+c)^5}{1\cdot1\cdot1\cdot a}+\frac{(c+a)^5}{1\cdot1\cdot1\cdot b}+\frac{(a+b)^5}{1\cdot1\cdot1\cdot c}\geq\frac{32}{27}$.

48. $\left(\sqrt[k-1]{2}-1\right)(a_1+a_2+\dots+a_n)<\sqrt[k]{2a_1^k+2^2a_2^k+\dots+2^na_n^k}$, where $a_1>0,\dots,a_n>0$ and $k\geq3$, $k\in\mathbb{N}$.

Remark Note that

$$2a_1^k+2^2a_2^k+\dots+2^na_n^k=\frac{a_1^k}{\left(1/\sqrt[k-1]{2}\right)^{k-1}}+\frac{a_2^k}{\left(1/\sqrt[k-1]{2^2}\right)^{k-1}}+\dots+\frac{a_n^k}{\left(1/\sqrt[k-1]{2^n}\right)^{k-1}}\geq$$

$$\geq\frac{(a_1+a_2+\cdots a_n)^k}{\left(1/\sqrt[k-1]{2}+1/\sqrt[k-1]{2^2}\right)^{k-1}}>\frac{(a_1+a_2+\dots+a_n)^k}{\left(1/\sqrt[k-1]{2}+1/\sqrt[k-1]{2^2}+\dots+1/\sqrt[k-1]{2^n}+\dots\right)^{k-1}}.$$

49. Let G be the intersection point of the medians of triangle $A_1A_2A_3$, and let C be the circumcircle of triangle $A_1A_2A_3$. Let the lines GA_1, GA_2, GA_3 intersect the circle C a second time at points B_1, B_2, B_3, respectively. Prove that $GB_1 + GB_2 + GB_3 \geq \sqrt{A_1A_2^2 + A_2A_3^2 + A_3A_1^2}$.

Remark See the proof of Problem 8.11.

Chapter 9
Using Derivatives and Integrals

Historical origins. Derivatives and integrals are foundational proof technique tools in mathematics. The derivative of a function $y = f(x)$ of a variable x is a measure of the rate at which the value y of the function changes with respect to the change of the variable x. In general, in the literature the following two distinct notations are commonly used for derivatives:

1. *Leibniz's notation,* introduced by German mathematician and philosopher *Gottfried Wilhelm von Leibniz,* born 1 July 1646 in Leipzig, Holy Roman Empire (now Leipzig, Germany), died 14 November 1716 in Hanover, Holy Roman Empire (now Hanover, Germany). According to Leibniz's notation, the derivative of y with respect to x is denoted by $\frac{dx}{dy}$, where dy represents the change in y and dx represents the change in x.
2. *Lagrange's notation,* introduced by the Italian–French mathematician *Joseph Louis Lagrange,* born 25 January 1736 in Turin, Kingdom of Sardinia (now Turin, Italy), died 10 April 1813 in Paris, France. According to Lagrange's notation, the derivative of a function $f(x)$ with respect to x is denoted by $f'(x)$, or sometimes it is denoted by $f'_x(x)$.

In this chapter, for simplicity and brevity we use Lagrange's notation for the derivative $f'(x)$.

Suppose we need to prove the inequality

$$f(x) \geq g(x) \tag{9.1}$$

on an interval $[a, b] = I$ or $[a, +\infty) = I$, where the functions $f(x)$ and $g(x)$ are defined and continuous on I.

Theorem 9.1 *If functions $f(x)$, $g(x)$ are differentiable on the domain I, $f(a) \geq g(a)$, and on I we have that $h'(x) \geq 0$, where $h(x) = f(x) - g(x)$, then $f(x) \geq g(x)$ holds for all $x \in I$.*

H. Sedrakyan and N. Sedrakyan, *Algebraic Inequalities*, Problem Books in Mathematics, https://doi.org/10.1007/978-3-319-77836-5_9

Proof If on the domain I we have $h'(x) \geq 0$, then the function $h(x)$ on this domain does not decrease, and consequently, $h(x) \geq h(a)$ at each point x of the domain I, that is, $f(x) - g(x) \geq f(a) - g(a) \geq 0$. Therefore, $f(x) \geq g(x)$.

This ends the proof.

Remark If on the domain I we have $h'(x) > 0 (x \neq a)$, then for $x \in I$ and $x > a$, one has $f(x) > g(x)$.

Now let us consider the following examples in order to see how Theorem 9.1 can be applied.

Example 9.1 Prove that $2^{x+1} > x + 2$ if $x \geq 1$.

Proof Let us consider the function $h(x) = 2^{x+1} - x - 2$ on $[1\ ,\ +\infty)$.

We have $h(1) = 1$ and $h'(x) = 2^{x+1}\ln 2 - 1$. Note that the function $y = 2^x$ is increasing on $[1\ ,+\infty)$, whence $h'(x) \geq 4\ln 2 - 1 > 0$.

Therefore, if $x \geq 1$, then $h(x) \geq h(1)$ or $2^{x+1} \geq x + 3$, and hence $2^{x+1} > x + 2$.

This ends the proof.

Example 9.2 Prove that $\sum\limits_{n=1}^{k} (\sum\limits_{m=1}^{k} \frac{a_m a_n}{m+n}) \geq 0$.

Proof Let us consider the function $h(x) = \sum\limits_{n=1}^{k} (\sum\limits_{m=1}^{k} \frac{x^m a_m . x^n a_n}{m+n})$ on $[0\ ,+\infty)$.

We have $xh'(x) = \sum\limits_{n=1}^{k} (\sum\limits_{m=1}^{k} x^m a_m \cdot x^n a^n) = (xa_1 + \cdots + x^k a_k)^2 \geq 0$, and therefore, for $x > 0$, we have $h'(x) \geq 0$.

Hence $h(1) \geq h(0) = 0$, that is, $\sum\limits_{n=1}^{k} (\sum\limits_{m=1}^{k} \frac{a_m a_n}{m+n}) \geq 0$.

This ends the proof.

Theorem 9.2 *If $f(b) \geq g(b)$ and on the domain I one has $h'(x) \leq 0, (h(x) = f(x) - g(x))$, then on I one has $f(x) \geq g(x)$, where $I = [a, b]$ or $I = (-\infty, b]$.*

Proof If $h'(x) \leq 0$ on I, then the function $h(x)$ in this domain is not increasing, and it attains its minimum value at the point $x = b$. On the other hand, $h(b) \geq 0$, and therefore for every x in I, one has $h(x) \geq 0$. Therefore, on I, one has $f(x) - g(x) \geq 0$, or equivalently, $f(x) \geq g(x)$.

This ends the proof.

Integrals are another crucial proof technique tool in mathematics. The notation $\int$ was also introduced by Gottfried Wilhelm von Leibniz in 1675. It represents the first letter, *s*, of the word sum (*summa* in Latin). Let us provide the following useful theorem as an application of integrals to prove inequalities.

Theorem 9.3 *If for every x in the domain I one has the inequality $f(x) \geq g(x)$, then*

$$\int_a^x f(t)dt \geq \int_a^x g(t)dt, \tag{9.2}$$

where $I = [a, b]$ *or* $I = [a, +\infty)$.

Proof If inequality (9.1) holds, then $F'(x) \geq G'(x)$, where $F'(x) = f(x)$, $G'(x) = g(x)$, and $F(a) = G(a) = 0$.

It follows that $F(x) \geq G(x)$. Hence, using that $F(x) = \int_a^x f(t)dt$, $G(x) = \int_a^x g(t)dt$, we obtain inequality (9.2).

This ends the proof.

Now let us consider the following example in order to see how Theorem 9.2 can be applied.

Example 9.3 Prove that $\ln(2\sin x) > \frac{1}{2}x(\pi - x) - \frac{5}{72}\pi^2$, if $x \in (\pi/6, \pi/2)$.

Proof Let us consider the inequality $\cot x > \frac{\pi}{2} - x$, which is seen to be valid if we use the fact that $\tan\alpha > \alpha$ for $0 < \alpha < \frac{\pi}{2}$ and substitute α by $\frac{\pi}{2} - x$.

Integrating the inequality under consideration, we obtain
$\int_{\frac{\pi}{6}}^x \cot t dt > \int_{\frac{\pi}{6}}^x \left(\frac{\pi}{2} - t\right) dt$, and hence
$\ln(\sin x) - \ln\frac{1}{2} > \left(\frac{\pi}{2} \cdot x - \frac{x^2}{2}\right) - \left(\frac{\pi}{2} \cdot \frac{\pi}{6} - \frac{1}{2} \cdot \left(\frac{\pi}{6}\right)^2\right)$, or
$\ln(2\sin x) > \frac{1}{2}x(\pi - x) - \frac{5}{72}\pi^2$.
This ends the proof.

Problems

Prove the following inequalities (9.1–9.22, 9.27–9.32)

9.1. $3a^3 + 7b^3 \geq 9ab^2$, where $a \geq 0,\ b \geq 0$.
9.2. $2^{n-1}(x^n + y^n) \geq (x + y)^n$, where $x > 0, y > 0, n \in \mathbb{N}$.
9.3. (a) $\cos x \geq 1 - \frac{x^2}{2}$,
(b) $\sin x \geq x - \frac{x^3}{3!}$, where $x \geq 0$,
(c) $\cos x \geq 1 - \frac{x^2}{2!} + \frac{x^4}{4!}$;
(d) $\sin x \geq x - \frac{x^3}{3!} + \frac{x^5}{5!}$, where $x \geq 0$ (use that $\sin x \leq x$).
9.4. $x - \sin x \leq 1 - \cos x \leq x\sqrt{2} - \sin x$, where $0 \leq x \leq \frac{\pi}{2}$.
9.5. $\tan x + \sin x \geq 2x$, where $0 \leq x < \frac{\pi}{2}$.
9.6. $x - \frac{x^2}{2} + \frac{x^3}{3} - \ldots + \frac{x^{2n+1}}{2n+1} - \frac{x^{2n+2}}{2n+2} \leq \ln(1 + x) \leq x - \frac{x^2}{2} + \frac{x^3}{3} - \ldots + \frac{x^{2n+1}}{2n+1}$, where $x \geq 0, n \in \mathbb{N}$.

9.7. $\ln(\cos x) \leq -\frac{x^2}{2}$, where $0 \leq x < \frac{\pi}{2}$.
9.8. $\sin x \leq \frac{x(\pi - x)}{2}$, where $0 \leq x \leq \frac{\pi}{2}$.
9.9. $\tan x - \frac{\tan^3 x}{3} \leq x$, where $0 \leq x < \frac{\pi}{2}$.
9.10. $(x + \frac{1}{x})\operatorname{arccot} x > 1$, where $x > 0$.
9.11. $\frac{1}{n+1} + \frac{1}{n+2} + \cdots + \frac{1}{3n} < \ln 3$, where $n \in \mathbb{N}$.
9.12. $\frac{3\cos x}{1+2\cos x} < \frac{\sin x}{x} < \frac{3}{4-\cos x}$, where $0 < x \leq \frac{\pi}{2}$.
9.13. *Young's inequality*: $ab \leq \frac{a^p}{p} + \frac{b^q}{q}$, where $a > 0,\ b > 0,\ p > 0,\ q > 0$ and $\frac{1}{p} + \frac{1}{q} = 1$.
9.14. $\frac{a^p - b^p}{a^p + b^p} > \frac{a^n - b^n}{a^n + b^n}$, where $a > b > 0,\ p > n$.
9.15. $(1 + x^t)^{\frac{1}{t}} - (1 + x^t)^{-\frac{1}{t}} \leq x$, where $x \geq 0, t \geq 2$.
9.16. $ab \leq e^a + b(\ln b - 1)$, where $b \geq 1$.
9.17. $\left(2 + \frac{(\ln x)^2}{3}\right) \ln x \leq \frac{x^2 - 1}{x}$, where $x \geq 1$.
9.18. $2^{\sin x} + 2^{\cos x} \geq 3$, where $0 \leq x \leq \frac{\pi}{2}$.
9.19. $\frac{n^{k+1}}{k+1} < 1^k + 2^k + \cdots + n^k < \left(1 + \frac{1}{n}\right)^{k+1} \cdot \frac{n^{k+1}}{k+1}$, where $n, k \in \mathbb{N}$.
9.20. (a) $\left(\frac{n}{e}\right)^n < n!$, where $n \in \mathbb{N}$,
(b) $n! < n\left(\frac{n}{e}\right)^n$, where $n \geq 7, n \in \mathbb{N}$.
9.21. $(a^\alpha + b^\alpha)^{\frac{1}{\alpha}} > (a^\beta + b^\beta)^{\frac{1}{\beta}}$, where $a > 0,\ b > 0,\quad 0 < \alpha < \beta$.
9.22. $\left(\frac{a+b}{c+d}\right)^{a+b} \leq \left(\frac{a}{c}\right)^a \cdot \left(\frac{b}{d}\right)^b$, where $a > 0,\ b > 0,\ c > 0,\ d > 0$.
9.23. Find the integer part of the expression $\frac{1}{\sqrt[3]{4}} + \frac{1}{\sqrt[3]{5}} + \cdots + \frac{1}{\sqrt[3]{10^6}}$.
9.24. Prove that $\frac{1}{2} + \frac{1}{3\sqrt{2}} + \frac{1}{4\sqrt{3}} + \cdots + \frac{1}{(n+1)\sqrt{n}} < 2$, where $n \geq 2,\quad n \in \mathbb{N}$.
9.25. Prove that

(a) *Bernoulli's inequality*: $(1 + x)^\alpha > 1 + \alpha x$, if $\alpha > 1, x > -1, x \neq 0$,
(b) *Bernoulli's inequality*: $(1 + x)^\alpha < 1 + \alpha x$, if $0 < \alpha < 1, x > -1, x \neq 0$.
(c) $(S - x_1)^{x_1} + \ldots + (S - x_n)^{x_n} > n - 1$, where $n \geq 2,\ x_1 > 0, \ldots, x_n > 0$, and $x_1 + \ldots + x_n = S$.

9.26. Let $f(x) = a_1 \sin x + a_2 \sin 2x + \ldots + a_n \sin nx$, where $a_1, \cdots, a_n$ are real numbers and n is a positive integer. We have that $|f(x)| \leq |sinx|$ for all real numbers x. Prove that $|a_1 + 2a_2 + \cdots + na_n| \leq 1$.
9.27. $x^2 \geq (1 + x)\ln^2(1 + x)$, where $x > -1$.
9.28. $\sqrt[n+1]{n+1} < \sqrt[n]{n}$, where $n \geq 3, n \in \mathbb{N}$.
9.29. $a_1 b_1^x + a_2 b_2^x + \cdots + a_n b_n^x \geq a_1 + \cdots + a_n$, where $a_i > 0,\ b_i > 0,\ i = 1, \ldots, n,\ x > 0$, and $b_1^{a_1} \cdot \ldots \cdot b_n^{a_n} = 1$.
9.30. $x^x > a\left(\frac{x+1}{a+1}\right)^{x+1}$, where $a > 0,\ x > \frac{1}{a}$.
9.31. $\left(\frac{a_1^\alpha + \cdots + a_n^\alpha}{n}\right)^{\frac{1}{\alpha}} \geq \left(\frac{a_1^\beta + \cdots + a_n^\beta}{n}\right)^{\frac{1}{\beta}}$, where $a_1 > 0, \cdots, a_n > 0,\ \alpha \geq \beta,\ \alpha \neq 0,\ \beta \neq 0. \alpha \geq \beta,\ \alpha \neq 0,\ \beta \neq 0$.
9.32. (a) $\sqrt{ab} < \frac{a-b}{\ln a - \ln b} < \frac{a+b}{2}$, where $a > 0,\ b > 0,\ a \neq b$,
(b) $\frac{2x}{x+2} < \ln(x + 1) < \frac{x(x+2)}{2(x+1)}$, where $x > 0$,
(c) $\frac{\ln x}{x-1} < \frac{1}{\sqrt{x}}$, where $x > 0, x \neq 1$,

(d) $\frac{2}{2x+1} < \ln(1+\frac{1}{x}) < \frac{2x+1}{2x(x+1)}$, where $x > 0$,
(e) $|x-y| \le |\ln x - \ln y|$, where $0 < x \le 1$, $0 < y \le 1$,
(f) $\ln\frac{1}{y} < \frac{\ln x - \ln y}{x-y}$, where $0 < y < x \le 1$,
(g) $\ln\left(1+\frac{1}{y}\right) < \frac{\ln x - \ln y}{x-y}$, where $x > 0$, $y > 0$ and $x + y \le 1$.

9.33. (a) Prove that $1 - x + \frac{x^2}{2!} - \frac{x^3}{3!} + \cdots + \frac{x^{2k}}{(2k!)} > 0$, where $k \in \mathbb{N}$.
(b) Prove that if a polynomial $P(x)$ of degree n is nonnegative for x, then $P(x) + P'(x) + P''(x) + \cdots + P^{(n)}(x) \ge 0$ for all values of x.

9.34. Prove that $a^r - b^r + c^r \ge (a-b+c)^r$, where $a \ge b \ge c \ge 0$ and $r \ge 1$.

9.35. Prove that $(1-x_1^k)^m + \ldots + (1-x_n^k)^m \ge n-1$, where $x_1 \ge 0, \ldots, x_n \ge 0$, $k, m \in \mathbb{N}$, $k \ge m$, and $x_1 + \ldots + x_n \le 1$.

9.36. Prove that $\left(\sum\limits_{i=1}^{n} a_i\right)^2 \le \sum\limits_{i,j=1}^{n} \frac{ij}{i+j-1} \cdot a_i a_j$.

Proofs

9.1. Let us consider the function $h(a) = 3a^3 + 7b^3 - 9ab^2$ on $[0, +\infty)$. If $b > 0$, then $h(0) = 7b^3 > 0$, $h'(a) = 9a^2 - 9b^2 = 9(a-b)(a+b)$.
Note that the function $h(a)$ is decreasing on $[0, b]$, and is increasing on $[b, +\infty)$, and since $h(b) = 3b^3 + 7b^3 - 9b^3 = b^3 > 0$, we have $h(a) > 0$ on $[0, +\infty)$. Therefore, $3a^3 + 7b^3 > 9ab^2$.
If $b = 0$, then obviously $3a^3 + 7b^3 \ge 9ab^2$.

9.2. Consider the function $h(y) = 2^{n-1}(x^n + y^n) - (x+y)^n$ on $[0, +\infty)$. If $x > 0$, then $h'(y) = n2^{n-1}y^{n-1} - n(x+y)^{n-1} = n((2y)^{n-1} - (x+y)^{n-1})$, $h'(y) = 0$ if $y = x$, $h'(y) < 0$ for $0 \le y < x$, and $h'(y) > 0$ for $y > x$. It follows that the function $h(y)$ is decreasing on $[0, x]$, and increasing on $[x, +\infty)$. On the other hand, $h(x) = 0$, and hence on $[0, +\infty)$ we have that $h(y) \ge 0$, that is, $2^{n-1}(x^n + y^n) \ge (x+y)^n$.

9.3. (a) If $x > 0$, then integrating the inequality $\sin x \le x$, we obtain $\int\limits_0^x \sin t\, dt \le \int\limits_0^x t\, dt$, or $-\cos x + 1 \le \frac{x^2}{2} - 0$, that is, $\cos x \ge 1 - \frac{x^2}{2}$.
If $x = 0$, then $\cos x = 1 = 1 - \frac{x^2}{2}$.
Since $\cos x = \cos|x| \ge 1 - \frac{|x|^2}{2} = 1 - \frac{x^2}{2}$, it follows that

$$\cos x \ge 1 - \frac{x^2}{2}. \tag{9.3}$$

(b) By Theorem 9.3, for inequality (9.3) we have $\int\limits_0^x \cos t dt \ge \int\limits_0^x \left(1 - \frac{t^2}{2}\right) dt$, and hence the given inequality holds.

(c) If $x > 0$, then by part (b) and using to Theorem 9.3, we obtain $\int_0^x \sin t dt \geq \int_0^x \left(t - \frac{t^3}{3!}\right) dt$, and thus the given inequality holds. If $x = 0$, then $\cos x = 1 = 1 - \frac{x^2}{2!} + \frac{x^4}{4!}$.
Since $\cos x = \cos |x| \geq 1 - \frac{|x|^2}{2!} + \frac{|x|^4}{4!} = 1 - \frac{x^2}{2!} + \frac{x^4}{4!}$, it follows that $\cos x \geq 1 - \frac{x^2}{2!} + \frac{x^4}{4!}$.

(d) The proof follows from part (c) and Theorem 9.3.

9.4. Let us evaluate $\sin x - \cos x$. We have $\sin x + \cos x = \sqrt{2} \sin\left(x + \frac{\pi}{4}\right)$, and note that for $0 \leq x \leq \frac{\pi}{2}$, on integrating the inequality $1 \leq \sin x + \cos x \leq \sqrt{2}$, we obtain $\int_0^x 1\, dt \leq \int_0^x (\cos t + \sin t)\, dt \leq \int_0^x \sqrt{2} dt$, therefore $x \leq \sin x - \cos x + 1 \leq x\sqrt{2}$.

9.5. According to inequality (3.2), we have $\cos x + \frac{1}{\cos^2 x} \geq 2\sqrt{\cos x \cdot \frac{1}{\cos^2 x}} = 2\sqrt{\frac{1}{\cos x}} \geq 2$, and from Theorem 9.3, it follows that $\int_0^x \left(\cos t + \frac{1}{\cos^2 t}\right) dt \geq \int_0^x 2dt$, and hence the given inequality holds.

9.6. Let us first us prove that if $x \geq 0$, then

$$1 - x + x^2 - \cdots + x^{2n} - x^{2n+1} \leq \frac{1}{1+x} \leq 1 - x + x^2 - \cdots + x^{2n}. \quad (9.4)$$

Using the formula for the sum of a geometric progression, we obtain $\frac{1-x^{2n+2}}{1+x} \leq \frac{1}{1+x} \leq \frac{1+x^{2n+1}}{1+x}$, or $1 - x^{2n+2} \leq 1 \leq 1 + x^{2n+1} \quad (x \geq 0)$.
From Theorem 9.3 and inequality (9.4), it follows that the given inequality holds.

9.7. Since $(\ln(\cos x))' = -\tan x$, $\left(-\frac{x^2}{2}\right)' = -x$, from the inequality $-\tan x \leq -x \ \left(0 \leq x < \frac{\pi}{2}\right)$ and Theorem 9.3 it follows that the given inequality holds.

9.8. Consider the function $F(x) = \sin x - \frac{x(\pi - x)}{2}$ on $\left[0, \frac{\pi}{2}\right]$ and the function $G(x) = F'(x) = \cos x + x - \frac{\pi}{2}$ on $[0, \pi/2]$. Since $G'(x) \geq 0$, it follows that for $x \leq \frac{\pi}{2}$, we have $G(x) \leq G(\pi/2)$, or $F'(x) = \cos x + x - \frac{\pi}{2} \leq 0$, and since $x \geq 0$, we have $F(x) \geq F(0) = 0$.
It follows that $\sin x - \frac{x(\pi - x)}{2} \geq 0$.

9.9. Consider the function $f(x) = x - \tan x + \frac{\tan^3 x}{3}$ on $\left[0, \frac{\pi}{2}\right)$. We have that $f'(x) = 1 - (\tan x)' + \tan^2 x \cdot (\tan x)' = \tan^4 x \geq 0$, and therefore, $f(x) \geq f(0) = 0$, that is, $x \geq \tan x - \frac{\tan^3 x}{3}$.

9.10. Let us prove that for $x > 0$, we have $\operatorname{arccot} x > \frac{x}{1+x^2}$.
Consider the function $f(x) = \operatorname{arccot} x - \frac{x}{1+x^2}$ on $(0, +\infty)$.
We have $f'(x) = -\frac{1}{1+x^2} - \frac{1-x^2}{(1+x^2)^2} = -\frac{2}{(1+x^2)^2} < 0$, and hence the function $f(x)$ is decreasing on $(0, +\infty)$. It follows that $f(x) > 0$ as $\lim\limits_{x \to +\infty} f(x) = 0$.

9.11. We have $\int_{n+k-1}^{n+k} \frac{1}{n+k} dx < \int_{n+k-1}^{n+k} \frac{1}{x} dx$, $k = 1, 2, \ldots, 2n$, and hence

$$\frac{1}{n+1}+\frac{1}{n+2}+\ldots+\frac{1}{3n}<\int\limits_{n}^{n+1}\frac{1}{x}dx+\int\limits_{n+1}^{n+2}\frac{1}{x}dx+\ldots+\int\limits_{3n-1}^{3n}\frac{1}{x}dx$$

$$=\int\limits_{n}^{3n}\frac{1}{x}dx=\ln x\,\Big|_{n}^{3n}=\ln 3.$$

9.12. Let us prove first that if $0<x\le\frac{\pi}{2}$, then $\frac{\sin x}{x}>\frac{3\cos x}{1+2\cos x}$, or for $0<x<\frac{\pi}{2}$, $\tan x+2\sin x>3x$.
Indeed, consider the function $f(x)=\tan x+2\sin x-3x$ on $\left(0,\frac{\pi}{2}\right)$. Since $f(0)=0$ and the derivative of the function $f(x)$ on $\left(0,\frac{\pi}{2}\right)$ is positive, we have $f'(x)=\frac{1}{\cos^2 x}+2\cos x-3=\frac{1}{\cos^2 x}+\cos x+\cos x-3>3\sqrt[3]{\frac{\cos x\cdot\cos x}{\cos^2 x}}-3=0$, since $\frac{1}{\cos^2 x}\ne\cos x$.
Therefore, the function $f(x)$ on $\left[0,\frac{\pi}{2}\right)$ is increasing, and hence $f(x)>f(0)$.
The inequality $\frac{\sin x}{x}<\frac{3}{4-\cos x}$ on $\left(0,\frac{\pi}{2}\right)$ is equivalent to the inequality $4\sin x-\sin x\cos x<3x$ In order to prove this, let us consider the function $F(x)=4\sin x-\sin x\cos x-3x$ in the given domain.
As for $0<x<\frac{\pi}{2}$, we have that

$$F'(x)=4\cos x-\cos 2x-3=4\cos x-2\cos^2 x-2$$
$$=-2(\cos x-1)^2<0.$$

Then on $\left[0,\frac{\pi}{2}\right)$, the function $F(x)$ is decreasing. It follows that $F(x)<F(0)=0$.

9.13. Consider the function $f(x)=\frac{x^{p-1}}{pb}+\frac{b^{q-1}}{qx}$ in $(0,+\infty)$. We have $f'(x)=\frac{p-1}{p}\cdot\frac{x^{p-2}}{b}-\frac{b^{q-1}}{qx^2}=\frac{x^p-b^q}{qbx^2}$, and therefore, $f'(x)>0$ for $x>b^{\frac{q}{p}}$ and $f'(x)<0$ for $0<x<b^{\frac{q}{p}}$.
Hence on $\left(0,b^{\frac{q}{p}}\right]$, the function $f(x)$ is decreasing, and on $\left[b^{\frac{q}{p}},+\infty\right)$ it is increasing.
Therefore, the function $f(x)$ attains its smallest value for $x=b^{\frac{q}{p}}$, that is, $\frac{x^{p-1}}{pb}+\frac{b^{q-1}}{qx}\ge\frac{\left(b^{\frac{q}{p}}\right)^{p-1}}{pb}+\frac{b^{q-1}}{qb^{\frac{q}{p}}}=\frac{1}{p}+\frac{1}{q}=1$, or $\frac{x^p}{p}+\frac{b^q}{q}\ge xb$. Thus for $x=a$ we obtain the given inequality.

9.14. Let us denote $\frac{a}{b}$ by c. In this case, we have $c>1$ and $\frac{a^p-b^p}{a^p+b^p}=\frac{c^p-1}{c^p+1}$, $\frac{a^n-b^n}{a^n+b^n}=\frac{c^n-1}{c^n+1}$. Consider the function $f(x)=\frac{c^x-1}{c^x+1}$ on $(-\infty,+\infty)$. We have $f'(x)=\frac{2c^x\ln c}{(c^x+1)^2}>0$, and therefore, the function $f(x)$ is increasing on the given domain. Hence, if $p>n$, then $f(p)>f(n)$, that is, $\frac{c^p-1}{c^p+1}>\frac{c^n-1}{c^n+1}$.

9.15. One can easily prove that the given inequality is equivalent to the following inequality: $(1+x^t)^{\frac{1}{t}}\le\frac{x+\sqrt{x^2+4}}{2}$. The proof follows from Problem 9.21 and inequality $(1+x^2)^{\frac{1}{2}}\le\frac{x+\sqrt{x^2+4}}{2}$.

9.16. Consider the function $f(x) = e^x + b(\ln b - 1) - xb$ in $(-\infty, +\infty)$.
We have $f'(x) = e^x - b$, and therefore, the function $f(x)$ is increasing on $[\ln b, +\infty)$ and decreasing on $(-\infty, \ln b]$. Hence for $x = \ln b$, the function attains its smallest value, that is, $f(x) \geq f(\ln b)$. Therefore, $f(a) \geq e^{\ln b} + b(\ln b - 1) - b \ln b = 0$.

9.17. Consider the function $f(x) = \frac{x^2-1}{x} - 2\ln x - \frac{\ln^3 x}{3}$ in $[1, +\infty)$.
We have $f'(x) = 1 + \frac{1}{x^2} - \frac{2}{x} - \frac{\ln^2 x}{x} = \frac{(x-1)^2}{x^2} - \frac{\ln^2 x}{x}$.
In order to find the sign of the function $f'(x)$, let us consider the sign of the function $g(x) = \frac{(x-1)^2}{x} - \ln^2 x$ in the given domain. We have $g'(x) = \frac{x^2-1}{x^2} - \frac{2\ln x}{x} = \frac{\frac{x^2-1}{x} - 2\ln x}{2x}$.
Now let us consider the function $F(x) = \frac{x^2-1}{x} - 2\ln x$ on $[1, +\infty)$.
Since $F'(x) = \frac{x^2+1}{x^2} - \frac{2}{x} = \frac{(x-1)^2}{x^2} \geq 0$, we have $x \geq 1$, $F(x) \geq F(1) = 0$, and therefore, $g'(x) \geq 0$, whence $g(x) \geq g(1) = 0$, and $f'(x) \geq 0$, that is, $f(x) \geq f(1) = 0$.

9.18. Consider the function $f(x) = 2^{\sin x} + 2^{\cos x} - 3$ on $\left[0, \frac{\pi}{4}\right]$.
We have $f'(x) = 2^{\cos x} \cdot \cos x \ln 2 (2^{\sin x - \cos x} - \tan x) \geq 0$, since $2^{\sin x - \cos x} - \tan x \geq 0$.
Indeed, let us consider the function $F(x) = sin\, x - cos\, x - log_2 \tan x$ in $\left(0, \frac{\pi}{4}\right]$. We obtain

$$F'(x) = \cos x + \sin x - \frac{1}{\sin x \cos x \ln 2} = \frac{1}{\sin x \cos x}\left(\frac{\sqrt{2}}{2}\sin(x + \pi/4)\sin 2x - \frac{1}{\ln 2}\right)$$

$$\leq \frac{1}{sin\, x cos\, x}\left(\frac{\sqrt{2}}{2} - \frac{1}{\ln 2}\right) < 0,$$

and therefore, $F(x) \geq F(\pi/4) = 0$ or $2^{\sin x - \cos x} - \tan x \geq 0$.
It follows that $f'(x) \geq 0$, and hence $f(x) \geq f(0) = 0$.
If $\frac{\pi}{4} \leq x \leq \frac{\pi}{2}$, then $f(\frac{\pi}{2} - x) \geq 0$, and using that $f(x) = f(\frac{\pi}{2} - x)$, we deduce that $f(x) \geq 0$.

9.19. We have

$$1^k + 2^k + \cdots + n^k < \int_1^2 x^k\, dx + \int_2^3 x^k\, dx + \cdots + \int_n^{n+1} x^k\, dx$$

$$= \int_1^{n+1} x^k dx = \frac{(n+1)^{k+1}}{k+1} - \frac{1}{k+1} < \frac{(n+1)^{k+1}}{k+1} = (1 + \frac{1}{n})^{k+1} \cdot \frac{n^{k+1}}{k+1}.$$

In a similar way, we obtain
$1^k + 2^k + \cdots + n^k > \int_0^1 x^k dx + \int_1^2 x^k dx + \cdots + \int_{n-1}^{n} x^k dx = \int_0^n x^k dx = \frac{x^{k+1}}{k+1}\Big|_0^n = \frac{n^{k+1}}{k+1}$.

9.20. (a) We have

$$\ln 2 + \ln 3 + \cdots + \ln n > \int_1^2 \ln x\, dx + \int_2^3 \ln x\, dx + \cdots + \int_{n-1}^n \ln x\, dx = \int_1^n \ln x\, dx$$

$$= (x \ln x - x)|_1^n = n \ln n - n + 1 > n \ln n - n,$$

and therefore, $n! > \left(\frac{n}{e}\right)^n$.

(b) One can easily verify that $7! < 7\left(\frac{7}{e}\right)^7$. Letting $n \geq 8$, we have

$$\ln 2 + \ln 3 + \cdots + \ln n < \ln 7! + \int_8^9 \ln x\, dx + \cdots + \int_n^{n+1} \ln x\, dx = \ln 7! + \int_8^{n+1} \ln x\, dx$$

$$= \ln 7! + (x \ln x - x)|_8^{n+1} < (n+1)(\ln(n+1) - 1) - 8(\ln 8 - 1) + 8 \ln 7 - 7 < (n+1)\ln n - n,$$

or $n! < n\left(\frac{n}{e}\right)^n$ (see Problem 3.16(b))

9.21. Consider the function $f(x) = (a^x + b^x)^{\frac{1}{x}}$ on $(0, +\infty)$.
We have

$$f'(x) = (a^x + b^x)^{\frac{1}{x}} \cdot \frac{a^x \ln a^x + b^x \ln b^x - \ln(a^x + b^x)^{a^x + b^x}}{x^2(a^x + b^x)}.$$

Consider the function $F(t) = t \ln t + c \ln c - (t + c)\ln(t + c)$ on $(0, c]$, where $c > 0$.
We have $F'(t) = 1 + \ln t - 1 - \ln(t + c) = \ln \frac{t}{t+c} < \ln 1 = 0$, and therefore, $F(t) \leq F(c) = 2c \ln c - 2c \ln 2c < 0$.
Let $a^x \geq b^x$. Taking $c = a^x$, $t = b^x$, we obtain $a^x \ln a^x + b^x \ln b^x - (a^x + b^x)\ln(a^x + b^x) < 0$, hence $f'(x) < 0$, and thus $f(x)$ is a decreasing function on $(0, +\infty)$.
Therefore, if $\beta > \alpha > 0$, then $(a^\alpha + b^\alpha)^{\frac{1}{\alpha}} > (a^\beta + b^\beta)^{\frac{1}{\beta}}$.

9.22. Consider the function $f(x) = \frac{x^a}{(x+d)^{a+b}}$ on $(0, +\infty)$.
We have $f'(x) = \frac{x^{a-1}}{(x+d)^{a+b+1}} \cdot (ad - bx)$. Therefore, if $0 < x < \frac{ad}{b}$, then $f'(x) > 0$, for otherwise, if $x > \frac{ad}{b}$, then $f'(x) < 0$. Hence, the function $f(x)$ is increasing on $\left(0, \frac{ad}{b}\right]$, and decreasing on $\left[\frac{ad}{b}, +\infty\right)$.
Thus, it follows that at the point $x = \frac{ad}{b}$, function $f(x)$ attains its greatest value.
We deduce that $f(x) \leq f\left(\frac{ad}{b}\right)$, and therefore $f(c) \leq f\left(\frac{ad}{b}\right)$, or $\frac{c^a}{(c+d)^{a+b}} \leq \frac{\left(\frac{ad}{b}\right)^a}{\left(d+\frac{ad}{b}\right)^{a+b}}$, and we obtain $\left(\frac{a+b}{c+d}\right)^{a+b} \leq \left(\frac{a}{c}\right)^a \cdot \left(\frac{b}{d}\right)^b$.

9.23. We have

$$\frac{1}{\sqrt[3]{4}}+\frac{1}{\sqrt[3]{5}}+\cdots+\frac{1}{\sqrt[3]{10^6}} < \int_3^4 \frac{1}{\sqrt[3]{x}}dx + \int_4^5 \frac{1}{\sqrt[3]{x}}dx + \cdots + \int_{10^6-1}^{10^6} \frac{1}{\sqrt[3]{x}}dx$$

$$= \int_3^{10^6} \frac{1}{\sqrt[3]{x}}dx = \frac{3}{2}\sqrt[3]{10^{12}} - \frac{3}{2}\sqrt[3]{9} = 14997 + 3\left(1 - \frac{\sqrt[3]{9}}{2}\right) < 14997.$$

In a similar way, we obtain

$$\frac{1}{\sqrt[3]{4}}+\frac{1}{\sqrt[3]{5}}+\cdots+\frac{1}{\sqrt[3]{10^6}} > \int_4^5 \frac{1}{\sqrt[3]{x}}dx + \int_5^6 \frac{1}{\sqrt[3]{x}}dx + \cdots + \int_{10^6}^{10^6+1} \frac{1}{\sqrt[3]{x}}dx$$

$$= \int_4^{10^6+1} \frac{1}{\sqrt[3]{x}}dx = \frac{3}{2}\sqrt[3]{(10^6+1)^2} - \frac{3}{2}\sqrt[3]{16} > \frac{3}{2}\sqrt[3]{10^{12}} - \frac{3}{2}\sqrt[3]{16}$$

$$= \frac{3}{2}10^4 - \frac{3}{2}\sqrt[3]{16} = 1500 - \frac{3}{2}\sqrt[3]{16} = 14996 + \frac{8-3\sqrt[3]{16}}{2} > 14996.$$

It follows that $14996 < \frac{1}{\sqrt[3]{4}} + \cdots + \frac{1}{\sqrt[3]{10^6}} < 14997$, and therefore, the integer part of the given number is equal to 14996.

9.24. We have

$$\frac{1}{2}+\frac{1}{3\sqrt{2}}+\frac{1}{4\sqrt{3}}+\cdots+\frac{1}{(n+1)\sqrt{n}}$$
$$< \frac{1}{2}+\frac{1}{3\sqrt{2}}+\frac{1}{4\sqrt{3}}+\frac{1}{5\sqrt{4}}+\frac{1}{5\sqrt{5}}+\frac{1}{6\sqrt{6}}+\cdots+\frac{1}{n\sqrt{n}}$$
$$< \frac{1}{2}+\frac{1}{3\sqrt{2}}+\frac{1}{4\sqrt{3}}+\frac{1}{5\sqrt{4}}+\int_4^5 \frac{1}{x\sqrt{x}}dx+\cdots+\int_{n-1}^n \frac{1}{x\sqrt{x}}dx$$
$$= \frac{1}{2}+\frac{1}{3\sqrt{2}}+\frac{1}{4\sqrt{3}}+\frac{1}{10}+\int_4^n \frac{1}{x\sqrt{x}}dx = \frac{1}{2}+\frac{1}{10}+\frac{1}{3\sqrt{2}}+\frac{1}{4\sqrt{3}}+1-\frac{2}{\sqrt{n}} < 1.6+\frac{1}{3\sqrt{2}}+\frac{1}{4\sqrt{3}} < 2,$$

since

$$\frac{1}{3\sqrt{2}}+\frac{1}{4\sqrt{3}} = \frac{2\sqrt{2}+\sqrt{3}}{12} < \frac{2\cdot 1.5+1.8}{12} = 0.4.$$

Therefore, $\frac{1}{2}+\frac{1}{3\sqrt{2}}+\cdots+\frac{1}{(n+1)\sqrt{n}} < 2$.

9.25. (a) Consider the function $f(x) = (1+x)^{\alpha} - 1 - \alpha x$ in $(-1, +\infty)$.
Since $f'(x) = \alpha(1+x)^{\alpha-1} - \alpha$ and $\alpha > 1$, it follows that for $-1 < x < 0$ we have $f'(x) < 0$. It then follows that $f(x) > f(0) = 0$, and for $x > 0$ we have $f'(x) > 0$, whence $f(x) > f(0) = 0$.

(b) See the proof of (a).

(c) Consider the following two cases.

(1) $\max(x_1, \ldots, x_n) \geq 1$. Let $\max(x_1, \ldots, x_n) = x_i, j \neq i$. Then we have $(S - x_j)^{x_j} \geq x_i^{x_j} \geq 1$,and therefore, $(S - x_1)^{x_1} + \ldots + (S - x_n)^{x_n} > n - 1$.

(2) $\max(x_1, \ldots, x_n) < 1$. Then from inequality (c) in Problem 9.25, it follows that

$$(S - x_k)^{x_k} = \frac{S - x_k}{(1 + S - x_k - 1)^{1-x_k}} \geq \frac{S - x_k}{1 + (1 - x_k)(S - x_k - 1)}$$
$$= \frac{S - x_k}{S - x_k(S - x_k)} > \frac{S - x_k}{S}, \text{ where } k = 1, \ldots, n.$$

Therefore, $(S - x_1)^{x_1} + \ldots + (S - x_n)^{x_n} > \frac{S-x_1}{S} + \ldots + \frac{S-x_n}{S} = n - 1$.

9.26. We have $-|\sin x| \leq f(x) \leq |\sin x|$. If $0 \leq x \leq \frac{\pi}{2}$, then $-\sin x \leq f(x) \leq \sin x$, whence $-\frac{\sin x}{x} \leq \frac{f(x)}{x} \leq \frac{\sin x}{x}$, and therefore, $-\lim\limits_{x\to 0} \frac{\sin x}{x} \leq \lim\limits_{x\to 0} \frac{f(x)}{x} \leq \lim\limits_{x\to 0} \frac{\sin x}{x} = 1\ (x > 0)$.
On the other hand,

$$\lim_{x\to 0} \frac{f(x)}{x} = \lim_{x\to 0} \frac{f(x) - f(0)}{x - 0} = f'(0) = a_1 + 2a_2 + \ldots + n \cdot a_n.$$

It follows that $-1 \leq a_1 + 2a_2 + \ldots + n \cdot a_n \leq 1$.

9.27. If $x \geq 0$, then the given inequality is equivalent to the following inequality: $\frac{x}{\sqrt{x+1}} \geq \ln(1+x)$. In order to prove this, let us consider the function $f(x) = \frac{x}{\sqrt{1+x}} - \ln(1+x)$ in $(-1, +\infty)$.
Since $f'(x) = \frac{(\sqrt{x+1}-1)^2}{2\sqrt{x+1}(x+1)} \geq 0$, we have for $x \geq 0$ that $f(x) \geq f(0) = 0$.
For $-1 < x \leq 0$, the given inequality is equivalent to the following inequality: $f(x) \leq f(0) = 0$.

9.28. Consider the function $f(x) = \frac{\ln x}{x}$ in $(0, +\infty)$. Since $f'(x) = \frac{1-\ln x}{x^2}$, in $(e, +\infty)$ the function $f(x)$ is decreasing, and hence if $n + 1 > n \geq 3$, then $f(n+1) < f(n)$, or $\frac{\ln(n+1)}{n+1} < \frac{\ln n}{n}$, and therefore $\sqrt[n+1]{n+1} < \sqrt[n]{n}$.

9.29. Let us first prove that for $x \geq 0$, one has $f(x) = b^x - 1 - x \ln b \geq 0$, since $f'(x) = \ln b(b^x - 1)$.
If $b \geq 1$, then $\ln b \geq 0$ and $b^x - 1 \geq 0$, and therefore $f'(x) \geq 0$.
If $0 < b < 1$, then $\ln b < 0$ and $b^x - 1 \leq 0$, and therefore $f'(x) \geq 0$.
Hence, we obtain that $f'(x) \geq 0$, that is $f(x) \geq f(0)$. It follows that $b^x \geq 1 + x \ln b$, and therefore

$$a_1b_1^x + a_2b_2^x + \cdots + a_nb_n^x \geq a_1(1 + x \ln b_1) + a_2(1 + x \ln b_2) + \cdots + a_n(1 + x \ln b_n)$$
$$= a_1 + a_2 + \cdots + a_n + x \ln(b_1^{a_1} \cdot b_2^{a_2} \cdot \cdots \cdot b_n^{a_n}) = a_1 + a_2 + \cdots + a_n.$$

9.30. The given inequality is equivalent to the following inequality: $x \ln x > \ln a + (x + 1) \ln \frac{x+1}{a+1}$.
Consider the function $f(x) = x \ln x - \ln a - (x + 1) \ln \frac{x+1}{a+1}$ in $\left[\frac{1}{a}, +\infty\right)$.
As for $x > \frac{1}{a}$, we have $f'(x) = \ln \frac{x(a+1)}{x+1} > 0$, and then the function $f(x)$ in the given domain is increasing. Hence for $x > \frac{1}{a}$, we have $f(x) > f\left(\frac{1}{a}\right) = 0$.

9.31. Consider the function $f(x) = \frac{\ln \frac{a_1^x + \cdots + a_n^x}{n}}{x}$ in $(-\infty, 0)$ and $(0, +\infty)$. We have

$$f'(x) = \frac{\frac{a_1^x \ln a_1 + \cdots + a_n^x \ln a_n}{a_1^x + \cdots + a_n^x} \cdot x - \ln \frac{a_1^x + \cdots + a_n^x}{n}}{x^2}$$
$$= \frac{n}{x^2\left(a_1^x + \ldots + a_n^x\right)} \left(\frac{a_1^x \ln a_1^x + \cdots + a_n^x \ln a_n^x}{n} - \frac{a_1^x + \cdots + a_n^x}{n} \cdot \ln \frac{a_1^x + \cdots + a_n^x}{n} \right) \geq 0,$$

since the function $f(t) = t \ln t$ is convex in $(0, +\infty)$.
Indeed, $f'(t) = \ln t + 1$ and $f''(t) = \frac{1}{t} > 0$.
Therefore, if $\alpha \geq \beta > 0$ or $0 > \alpha \geq \beta$, then $\left(\frac{a_1^\alpha + \cdots + a_n^\alpha}{n}\right)^{1/\alpha} \geq \left(\frac{a_1^\beta + \cdots + a_n^\beta}{n}\right)^{1/\beta}$.
If $\alpha > 0 > \beta$, then $\left(\frac{a_1^\alpha + \cdots + a_n^\alpha}{n}\right)^{1/\alpha} \geq \left(\sqrt[n]{a_1^\alpha \cdots a_n^\alpha}\right)^{1/\alpha} = \sqrt[n]{a_1 \cdots a_n}$,
$\left(\frac{a_1^\beta + \cdots + a_n^\beta}{n}\right)^{1/\beta} \leq \left(\sqrt[n]{a_1^\beta \cdots a_n^\beta}\right)^{1/\beta} = \sqrt[n]{a_1 \cdots a_n}$, and therefore

$$\left(\frac{a_1^\alpha + \cdots + a_n^\alpha}{n}\right)^{1/\alpha} \geq \left(\frac{a_1^\beta + \cdots + a_n^\beta}{n}\right)^{1/\beta}.$$

9.32. (a) Without loss of generality one can assume that $a \geq b$.
In this case, the given inequalities are equivalent to the following inequalities:

$$\frac{2\left(\frac{a}{b} - 1\right)}{\frac{a}{b} + 1} < \ln \frac{a}{b} < \sqrt{\frac{a}{b}} - \sqrt{\frac{b}{a}}.$$

Consider the function $f(x) = \ln x - 2 \cdot \frac{x-1}{x+1}$ in $[1, +\infty)$.
Note that $f'(x) = \frac{(x-1)^2}{x(x+1)^2} > 0 \quad (x > 1)$, and therefore, the function $f(x)$ is increasing on $[1, +\infty)$ and for $x > 1$, we have $f(x) > f(1) = 0$.
Taking $x = \frac{a}{b}$, we obtain $\frac{2\left(\frac{a}{b}-1\right)}{\frac{a}{b}+1} < \ln \frac{a}{b}$. In order to prove the second inequality, let us define $\sqrt{\frac{a}{b}} = x$ and consider the function $g(x) = 2 \ln x - x + \frac{1}{x}$ in $[1, +\infty)$

We have $g'(x) = -\frac{(x-1)^2}{x^2} < 0 \quad (x > 1)$, and therefore, the function $g(x)$ is decreasing on $[1, +\infty)$, and then for $x > 1$, we have $g(x) < g(1) = 0$, that is, $\ln \frac{a}{b} < \sqrt{\frac{a}{b}} - \sqrt{\frac{b}{a}}$.

(b) Let us take $a = x + 1$ and $b = 1$. From Problem 9.32(a), it follows that $\frac{2x}{x+2} < \ln(x + 1) < \frac{x}{\sqrt{x+1}}$. Note that $\frac{x}{\sqrt{x+1}} < \frac{x(x+2)}{2(x+1)}$.

(c) Taking $a = x$, $b = 1$ and using that $\sqrt{ab} < \frac{a-b}{\ln a - \ln b}$, we obtain $\sqrt{x} < \frac{x-1}{\ln x}$, and hence $\frac{\ln x}{x-1} < \frac{1}{\sqrt{x}}$.

(d) Taking $a = x + 1$, $b = x$ and using Problem 9.32(a), we obtain $\sqrt{x(x+1)} < \frac{1}{\ln \frac{x+1}{x}} < \frac{x+(x+1)}{2}$, and therefore $\frac{2}{2x+1} < \ln \frac{x+1}{x} < \frac{1}{\sqrt{x(x+1)}}$. Note that $\frac{1}{\sqrt{x(x+1)}} < \frac{2x+1}{2x(x+1)}$.

(e) Using that $\frac{a-b}{\ln a - \ln b} < \frac{a+b}{2}$ for the numbers x, y and $\frac{a-b}{\ln a - \ln b} > 0$, we see that $\left|\frac{x-y}{\ln x - \ln y}\right| < \frac{x+y}{2} \quad (x \neq y)$, or $|\ln x - \ln y| > \frac{2}{x+y} \cdot |x - y| \geq |x - y|$, since $0 < x, y \leq 1$, and for $x = y$, we obtain that the given inequality holds.

(f) Let $I = (0, 1]$, $x_1 = y$, $x_2 = x$, $x_3 = 1$, and $f(x) = -\ln x$. Then from Problem 7.12(a), it follows that $x \ln y - \ln y \geq y \ln x - \ln x > y \ln y - \ln x$.

(g) From Problem 9.25(b), it follows that $\left(1 + \frac{1}{y}\right)^{x-y} < 1 + \frac{x-y}{y} = \frac{x}{y}$; hence we obtain $\ln\left(1 + \frac{1}{y}\right) < \frac{\ln x - \ln y}{x-y}$.

9.33. (a) Let us define

$$p_k(x) = 1 - x + \frac{x^2}{2!} - \ldots + \frac{x^{2k}}{(2k)!}, \quad k = 0, 1, 2, \ldots.$$

If $x \leq 0$, then $p_k(x) > 0$. Now let us prove that if $x > 0$, then $p_k(x) - e^{-x} > 0$. For $k = 0$, we have $p_0(x) - e^{-x} = 1 - e^{-x} > 0$.

Assume that for $k = n$ (if $x > 0$), we have $p_n(x) - e^{-x} > 0$, and let us prove that it holds for $k = n+1$. (That is, the following inequality holds: $p_{n+1}(x) - e^{-x} > 0$.)

Indeed, let $f(x) = p_{n+1}(x) - e^{-x}$. Then for $x > 0$ we have $f''(x) = p_n(x) - e^{-x} > 0$, and therefore, $f'(x) > f'(0) = 0$, whence for $x > 0$ it follows that $f(x) > f(0) = 0$.

Therefore, we have obtained $p_{n+1}(x) > e^{-x}$ if $x > 0$, and hence for $x > 0$ it follows that $p_k(x) > 0$.

(b) Since for every value of x we have $p(x) \geq 0$, it follows that n is an even number. On the other hand, the polynomial $F(x) = P(x) + P'(x) + \cdots + P^{(n)}(x)$ has degree equal to n.

So the polynomial $F(x)$ has the smallest value.

Let $\min\limits_{(-\infty,+\infty)} F(x) = F(x_0)$, in which case $F'(x_0) = 0$. Thus, it follows that $F'(x_0) = P'(x_0) + P''(x_0) + \cdots + P^{(n)}(x_0) + P^{(n+1)}(x_0) = P'(x_0) + P''(x_0) + \cdots + P^{(n)}(x_0) = 0$, and $F(x_0) = P(x_0) + P'(x_0) + P''(x_0) + \cdots + P^{(n)}(x_0) = P(x_0) \geq 0$. Therefore, $F(x) \geq F(x_0) \geq 0$, and hence $F(x) \geq 0$ for all values of x.

Taking $P(x) = \frac{x^{2n}}{(2n)!}$, we obtain $\frac{x^{2n}}{(2n)!} + \frac{x^{2n-1}}{(2n-1)!} + \cdots + 1 \geq 0$ (see Problem 9.33(a)).

9.34. Consider the function $f(x) = a^r - b^r + x^r - (a - b + x)^r$, in $[0, b]$.
Then $f'(x) = rx^{r-1} - r(a-b+x)^{r-1} = rx^{r-1}(1 - (1 + \frac{a-b}{x})^{r-1}) \leq 0$, since $\frac{a-b}{x} \geq 0$ and $r - 1 \geq 0$.
Hence, in $[0, b]$ the function $f(x)$ is nonincreasing, and hence $f(x) \geq f(b) = 0$, that is, $f(c) \geq 0$ or $a^r - b^r + c^r \geq (a - b + c)^r$.

9.35. Note that for $0 \leq x \leq 1$ we have $0 \leq x^k \leq x^m \leq 1$, and then $(1 - x^k)^m \geq (1-x^m)^m$, and so it is sufficient to prove that $(1-x_1^m)^m + \ldots + (1-x_n^m)^m \geq n-1$.
Let $n \geq 3$ and $x_1 \leq \ldots \leq x_n$. Then $x_1 + 2x_2 \leq 1$. Let us prove that

$$(1 - x_1^m)^m + (1 - x_2^m)^m \geq 1 + (1 - (x_1 + x_2)^m)^m.$$

Consider the function $f(x) = (1 - x^m)^m - (1 - (x + x_2)^m)^m$ in $[0, h]$, where $h = \min(x_2, 1 - 2x_2)$ (if $h = 0$, then $x_1 = 0$, and hence (1) holds).
Since $f'(x) = m^2 \left((x + x_2 - (x + x_2)^{m+1})^{m-1} - (x - x^{m+1})^{m-1}\right)$ and $x_2 \geq x_2((x + x_2) + x_2)^m \geq x_2((x + x_2)^m + \ldots + x_2^m) \geq x_2((x + x_2)^m + \ldots + x^m)$, we have $x + x_2 - (x + x_2)^{m+1} \geq x - x^{m+1}$.
It follows that $f'(x) \geq 0$, and hence $f(x_1) \geq f(0)$.
Therefore, $(1-x_1^m)^m - (1-(x_1+x_2)^m)^m \geq 1-(1-x_2^m)^m$. From inequality (1) it follows that it is enough to prove that $(1 - x_1^m)^m + \ldots + (1 - x_n^m)^m \geq n - 1$, for $n = 2$.
Let $0 \leq x_1 \leq x_2$ and $x_1 + x_2 \leq 1$. Let us prove that $(1-x_1^m)^m + (1-x_2^m)^m \geq 1$.
We have $(1-x_1^m)^m + (1-x_2^m)^m \geq (1-x_1^m)^m + (1-(1-x_1)^m)^m$ and $0 \leq x_1 \leq \frac{1}{2}$.
Consider the function $g(x) = (1 - x^m)^m + (1 - (1 - x)^m)^m$ in $\left[0, \frac{1}{2}\right]$.
Since $g'(x) = m^2 \left((1 - x - (1 - x)^{m+1})^{m-1} - (x - x^{m+1})^{m-1}\right)$ and $1 - 2x = (1-2x)(1-x+x)^m \geq (1-2x)((1-x)^m + \ldots + x^m)$, we have $1-x-(1-x)^{m+1} \geq x - x^{m+1}$.
Hence $g'(x) \geq 0$, and therefore, $g(x_1) \geq g(0)$, that is, $(1 - x_1^m)^m + (1 - (1 - x_1)^m)^m \geq 1$.
It follows that $(1 - x_1^m)^m + (1 - x_2^m)^m \geq 1$.

9.36. If we set $a_i = \frac{b_i}{i}$, $i = 1, \ldots, n$, then we need to prove that $p(1) \geq 0$, where $p(x) = \sum\limits_{i,j=1}^{n} \frac{b_i b_j}{i+j-1} \cdot x^{i+j-1} - x\left(\sum\limits_{i=1}^{n} \frac{b_i}{i} \cdot x^{i-1}\right)^2$.
We have

$$p'(x) = \sum_{i,j=1}^{n} b_i b_j x^{i+j-2} - \left(\frac{\left(\sum\limits_{i=1}^{n} \frac{b_i}{i} \cdot x^i\right)^2}{x}\right)'$$

$$= \sum_{i,j=1}^{n} b_i b_j x^{i+j-2} - \frac{2\left(\sum_{i=1}^{n} \frac{b_i}{i} \cdot x^i\right)\left(\sum_{i=1}^{n} b_i \cdot x^{i-1}\right)x - \left(\sum_{i=1}^{n} \frac{b_i}{i} \cdot x^i\right)^2}{x^2}$$

$$= \left(\frac{\sum_{i=1}^{n} b_i \cdot x^i - \sum_{i=1}^{n} \frac{b_i}{i} \cdot x^i}{x}\right)^2 = \left(\sum_{i=1}^{n} \left(b_i - \frac{b_i}{i}\right) \cdot x^{i-1}\right)^2,$$

for $x \neq 0$.
Therefore, for $x > 0$, we have $p'(x) \geq 0$, and hence $p(x)$ is a nondecreasing function on $[0, +\infty)$. It follows that $p(1) \geq p(0) = 0$.

Problems for Independent Study

Prove the following inequalities (1–13, 16–20).

1. $2\sin x \geq \frac{\pi}{2} + (x - \frac{\pi}{2})\cos x$, where $0 \leq x \leq \frac{\pi}{2}$.
2. $\frac{x^p-1}{p} > \frac{x^q-1}{q}$, where $p > q > 0$.
3. $\sqrt{1} + \sqrt{2} + \cdots + \sqrt{n} > \frac{2}{3}n\sqrt{n}$, where $n \in \mathbb{N}$.
4. $e^x \geq x^e$, where $x > 0$.
5. $\frac{\sin\alpha}{\alpha} > 1 - \frac{\alpha^2}{6}$, where $0 < \alpha < \frac{\pi}{2}$.
6. $2\sin\alpha + \tan\alpha > 3\alpha$, where $0 < \alpha < \frac{\pi}{2}$.
7. $\sin x + \frac{\sin 2x}{2} + \frac{\sin 3x}{3} > 0$, where $0 < x < \pi$.
8. $\left(\frac{a+1}{b+1}\right)^{b+1} \geq \left(\frac{a}{b}\right)^b$, where $a > 0,\ b > 0$.
9. $\frac{e^b-e^a}{b-a} < \frac{e^c-1}{2c}(b+a)+1$, where $0 \leq a < b \leq c$.
10. (a) $\frac{\sin\alpha}{\alpha} > \frac{\sin\beta}{\beta}$, where $0 < \alpha < \beta < \frac{\pi}{2}$;
    (b) $\frac{\tan\alpha}{\alpha} < \frac{\tan\beta}{\beta}$, where $0 < \alpha < \beta < \frac{\pi}{2}$.
11. $\frac{2}{\pi}x < \sin x < x$, where $0 < x < \pi/2$.
12. $(\sin x)^{-2} \leq x^{-2} + 1 - \frac{4}{\pi^2}$, where $0 < x < \frac{\pi}{2}$.
13. (a) $\tan x > x + \frac{x^3}{3}$, where $0 < x < \frac{\pi}{2}$,
    (b) $x\cos x < 0,6$, where $0 < x < \frac{\pi}{2}$,
    (c) $\left(\frac{\sin x}{x}\right)^3 \geq \cos x$, where $0 < x < \frac{\pi}{2}$.
14. Let p_n and q_n be the respective perimeters of regular n-gons inscribed in and circumscribed about a circle with radius $\frac{1}{2}$. Let us divide the interval (p_n, q_n) into three equal parts. To which part does the number π belong?
15. Find real values of x such that $a^x \geq x^a$, where $x \geq 0$ and $a \geq 1$.
16. (a) $\frac{a+b}{a+b+2ab} + \frac{a+c}{a+c+2} \geq \frac{2(b+1)(c+1)}{2bc+3b+c+2}$, where $a > 0,\ b > 0,\ c > 0$;
    (b) $\frac{a+b}{a+b+2ab} + \frac{c+d}{c+d+2cd} + \frac{a+c}{a+c+2} + \frac{b+d}{b+d+2} \geq 2$, where $a > 0,\ b > 0,\ c > 0,\ d > 0$.
17. $x^p + x^{-p} + 2^p \leq (x + x^{-1})^p + 2$, where $x > 0$ and $p \geq 2$.

18. (a) $\ln x > \frac{3(x^2-1)}{x^2+4x+1}$, where $x > 1$,
    (b) $\frac{a-b}{\ln a - \ln b} < \frac{1}{3}\left(2\sqrt{ab} + \frac{a+b}{2}\right)$, where $a > 0$, $b > 0$, $a \neq b$.
19. $(\cos x)^{\cos^2 x} > (\sin x)^{\sin^2 x}$, where $0 < x < \frac{\pi}{4}$.

Remark According to the Bernoulli's inequality (see Problem 9.25(a)), we have $(\cos x)^{\frac{\cos^2 x}{\sin^2 x}} > 1 + (\cos x - 1) \cdot \frac{\cos^2 x}{\sin^2 x}$.

20. $\left(a_1^p + \ldots + a_n^p\right)^{\frac{1}{p}} \left(b_1^q + \ldots + b_n^q\right)^{\frac{1}{q}} \geq a_1 b_1 + \ldots + a_n b_n$, where $a_i > 0$, $b_i > 0$, $i = 1, \ldots, n$, $p > 0$, $q > 0$ and $\frac{1}{p} + \frac{1}{q} = 1$.

Remark Use the inequality of Problem 9.13 with the condition $a_1^p + \ldots + a_n^p = 1$ and $b_1^q + \ldots + b_n^q = 1$.

Chapter 10
Using Functions

Consider functions $f : \mathbb{R}^n \to \mathbb{R}$ and $g : \mathbb{R}^n \to \mathbb{R}$. Assume that we need to prove the inequality $f(x_1, x_2, \ldots, x_n) \geq g(x_1, x_2, \ldots, x_n)$ for some values of variables $x_1, x_2, \ldots, x_n \in \mathbb{R}$.

Rewriting this inequality as $f(x_1, x_2, \ldots, x_n) - g(x_1, x_2, \ldots, x_n) \geq 0$, we study the dependence on $x_i (1 \leq i \leq n)$ of the function $F(x) = f(x_1, x_2, \ldots, x_n) - g(x_1, x_2, \ldots, x_n)$, where the variables $x_1, x_2, \ldots, x_{i-1}, x_{i+1}, \ldots, x_n$ are considered constants. Let us consider the following example.

Example 10.1 Prove that

$$\left(x_1^2 + x_2^2 + \cdots + x_n^2\right) \cos \frac{\pi}{n+1} \geq x_1x_2 + x_2x_3 + \cdots + x_{n-1}x_n,$$

where $n \geq 2$.

Proof Let us consider the following function:

$$F(x_1) = \cos \frac{\pi}{n+1} x_1^2 - x_2x_1 + \left(x_2^2 + \cdots + x_n^2\right) \cos \frac{\pi}{n+1} - x_2x_3 - \cdots - x_{n-1}x_n.$$

This is a quadratic expression in x_1. Note that the quadratic expression ax^2+bx+c for $a > 0$ attains its smallest value at the point $x = -\frac{b}{2a}$. Hence since $\cos \frac{\pi}{n+1} > 0$, it follows that $F(x_1)$ attains its smallest value at the point $x_1 = \frac{x_2}{2 \cos \frac{\pi}{n+1}}$, and therefore,

$$F(x_1) = (x_1^2 + x_2^2 + \cdots + x_n^2) \cos \frac{\pi}{n+1} - x_1x_2 - \cdots - x_{n-1}x_n \geq F\left(\frac{x_2}{2 \cos \frac{\pi}{n+1}}\right).$$

One can easily prove that

$$F\left(\frac{x_2}{2 \cos \frac{\pi}{n+1}}\right) = \left(\frac{\sin \frac{3\pi}{n+1}}{2 \sin \frac{2\pi}{n+1} \cos \frac{\pi}{n+1}} x_2^2 + x_3^2 + \cdots + x_n^2\right) \cos \frac{\pi}{n+1} - x_2x_3 - \cdots - x_{n-1}x_n.$$

H. Sedrakyan and N. Sedrakyan, *Algebraic Inequalities*, Problem Books in Mathematics, https://doi.org/10.1007/978-3-319-77836-5_10

Let us consider the following quadratic function: $G(x_2) = F\left(\frac{x_2}{2\cos\frac{\pi}{n+1}}\right)$. We obtain

$$G(x_2) \geq G\left(\frac{\sin\frac{2\pi}{n+1}}{\sin\frac{3\pi}{n+1}}x_3\right) = \left(\frac{\sin\frac{4\pi}{n+1}}{2\sin\frac{3\pi}{n+1}\cos\frac{\pi}{n+1}}x_3^2 + \cdots + x_n^2\right)\cos\frac{\pi}{n+1} - x_3x_4 - \cdots - x_{n-1}x_n.$$

In a similar way, for variables $x_3, \ldots, x_n$, we obtain $F(x_1) \geq G(x_2) \geq \cdots \geq \frac{\sin\frac{(n+1)\pi}{n+1}}{2\sin\frac{n\pi}{n+1}\cos\frac{\pi}{n+1}}x_n^2\cos\frac{\pi}{n+1} = 0$, and therefore, $(x_1^2 + x_2^2 + \cdots + x_n^2)\cos\frac{\pi}{n+1} \geq x_1x_2 + \cdots + x_{n-1}x_n$.

One can use the following properties of functions in proving a large number of inequalities.

Property 1 If a function $f(x)$ is defined on $[a, b]$ and is decreasing on $[a, c]$ and increasing on $[c, b]$, then on $[d, e]$, then the function $f(x)$ attains its maximum value at one of the endpoints of $[d, e]$ $(a \leq d < e \leq b)$.

Property 2 If the function $f(x)$ is defined on $[a, b]$ and is increasing on $[a, c]$, and is decreasing on $[c, b]$, then on $[d, e]$, the function $f(x)$ attains its smallest value at one of the endpoints $[d, e]$ $(a \leq d < e \leq b)$.

One can easily note that if the function $f'(x)$ is increasing on $[a, b]$, then the function $f(x)$ attains its greatest value on $[a, b]$ at the point a or b, and if $f'(x)$ is decreasing on $[a, b]$, then $f(x)$ attains its greatest value on $[a, b]$ at the point a or b.

These properties of functions can be used to prove a large number of inequalities.

Example 10.2 Prove that $a^2 + b^2 + c^2 \leq a^2b + b^2c + c^2a + 1$, where $0 \leq a \leq 1$, $0 \leq b \leq 1$, $0 \leq c \leq 1$.

Proof Consider the function $f(a) = a^2(1 - b) - c^2a + b^2 + c^2 - bc^2 - 1$ in $[0, 1]$.

If $b \neq 1$, then $f(a)$ is a quadratic trinomial in a, and the branches of its graph are directed upward. Hence, the function $f(a)$ attains its greatest value at one of the endpoints in $[0, 1]$. Since $f(0) = b^2 + c^2 - b^2c - 1 = (1 - c)\big(b^2 - (1 + c)\big) \leq 0$, $f(1) = 1 - b - c^2 + b^2 + c^2 - b^2c - 1 = b(b - 1) - b^2c \leq 0$, it follows that $f(a) \geq 0$ on $[0, 1]$.

This ends the proof.

If $b = 1$, then the proof can be done similarly.

Example 10.3 Prove that $x_1 + x_2 + x_3 - x_1x_2 - x_2x_3 - x_3x_1 \leq 1$, where $0 \leq x_i \leq 1$, i = 1, 2, 3.

Proof Let us consider the following monotonic function $f(x) = x + x_2 + x_3 - xx_2 - x_2x_3 - xx_3 = x(1 - x_2 - x_3) + x_2 + x_3 - x_3x_2$, which attains its greatest value at one of the endpoints $[0, 1]$. We have $f(0) = x_2 + x_3 - x_2x_3 = 1 + (1 - x_3)(x_2 - 1) \leq 1$, $f(1) = 1 - x_3 - x_2 + x_2 + x_3 - x_2x_3 = 1 - x_2x_3 \leq 1$. Therefore, in $[0, 1]$ one has $f(x) \leq 1$ or $f(x_1) = x_1 + x_2 + x_3 - x_1x_2 - x_2x_3 - x_3x_1 \leq 1$.

This ends the proof.

Problems

Prove the following inequalities (10.2–10.14).

10.1. Prove that, if $\left|ax^2 + bx + c\right| \leq 1$ holds for all numbers x belonging to $[-1, 1]$, then for those x one has the following inequality: $\left|cx^2 - bx + a\right| \leq 2$.

10.2. $(a+b+c)^2 < 4(ab+bc+ac)$, where a, b, and c are the side lengths of some triangle.

10.3. $\frac{a}{b+c+1} + \frac{b}{c+a+1} + \frac{c}{a+b+1} + (1-a)(1-b)(1-c) \leq 1$, where $0 \leq a \leq 1$, $0 \leq b \leq 1$, $0 \leq c \leq 1$.

10.4. *Schweitzer's inequality*: $(a+b+c+d+e)\left(\frac{1}{a}+\frac{1}{b}+\frac{1}{c}+\frac{1}{d}+\frac{1}{e}\right) \leq 25 + 6\left(\sqrt{\frac{p}{q}} - \sqrt{\frac{q}{p}}\right)^2$, where $0 < p \leq a, b, c, d, e \leq q$.

10.5. (a) *Chebyshev's inequality*: $\left(\sum\limits_{i=1}^{n} m_i a_i\right)\left(\sum\limits_{i=1}^{n} m_i b_i\right) \leq \sum\limits_{i=1}^{n} m_i a_i b_i$, where $a_1 \leq a_2 \leq \cdots \leq a_n$, $b_1 \leq b_2 \leq \cdots \leq b_n$, $\sum\limits_{i=1}^{n} m_i = 1$, $m_i > 0 (i = 1, \ldots, n)$.

(b) $\left(\sum\limits_{i=1}^{n} m_i a_i\right)\left(\sum\limits_{i=1}^{n} m_i b_i\right) \geq \sum\limits_{i=1}^{n} m_i a_i b_i$, where $a_1 \leq a_2 \leq \cdots \leq a_n$, $b_1 \geq b_2 \geq \cdots \geq b_n$, $\sum\limits_{i=1}^{n} m_i = 1$, $m_i > 0 (i = 1, \ldots, n)$.

10.6. $(1+a_1)(1+a_2)\cdots(1+a_n) \geq 1 + a_1 + a_2 + \cdots + a_n$, where the numbers $a_i > -1$, $i = 1, \ldots, n$, and the numbers $a_1, \ldots, a_n$ have the same sign.

10.7. $(x_1 + \cdots + x_n)\left(\frac{1}{x_1} + \cdots + \frac{1}{x_n}\right) \leq \frac{(a+b)^2}{4ab} \cdot n^2$, where $x_1, \ldots, x_n \in [a, b]$, $0 < a < b$.

10.8. $\frac{(1-a)(1-b)}{ab} \geq \frac{(1-c)(1-a-b+c)}{c(a+b-c)}$, where $0 < a \leq c \leq b$, $a + b < 1$.

10.9. $\frac{1}{1+a} + \frac{1}{1+b} \geq \frac{1}{1+m} + \frac{1}{1+\frac{ab}{m}}$, where $1 < a < m < b$.

10.10. $abc \leq \frac{1}{8}(pa+qb+rc)$, where $a, b, c > 0$, $p, q, r \in \left[0; \frac{1}{2}\right]$, $a+b+c = p+q+r = 1$.

10.11. $a_i + a_j > a_k$, where $n \geq 3$, $i < j < k$, $0 < a_1 < \ldots < a_n$ and $\left(a_1^2 + \cdots + a_n^2\right)^2 > (n-1)\left(a_1^4 + \cdots + a_n^4\right)$.

10.12. $x+y+z \geq xy+yz+zx$, where $x \geq 0$, $y \geq 0$, $z \geq 0$ and $xy+yz+zx+xyz \leq 4$.

10.13. $\left|b^2 - 4ac\right| \leq \left|B^2 - 4AC\right|$, if $\left|ax^2 + bx + c\right| \leq \left|Ax^2 + Bx + C\right|$, for all real values of x.

10.14. $\left(1 + \frac{2}{\sqrt{3}}\right)(x^2 - x + 1)(y^2 - y + 1)(z^2 - z + 1) \geq (xyz)^2 - xyz + 1$.

10.15. (a) $(1 - x_1 \cdots x_n)^\lambda + (1 - y_1^\lambda) \cdots (1 - y_n^\lambda) \geq 1$, where $\lambda > 1$, $0 < x_i < 1$, $0 < y_i < 1$, $x_i + y_i = 1$, $i = 1, \ldots, n$.

(b) $(1-p^n)^m + (1-q^m)^n \geq 1$, where $p+q = 1$, $p > 0$, $q > 0$, $m, n \in \mathbb{N}$.

Proofs

10.1. We have $|cx^2 - bx + a| \leq |cx^2 - c| + |c - bx + a| = |c|(1 - x^2) + |c - bx + a| \leq |c| + |c - bx + a|$.
If $x = 0$, then from $|ax^2 + bx + c| \leq 1$ it follows that $|c| \leq 1$.
We deduce that $\left|cx^2 - bx + a\right| \leq 1 + |c - bx + a|$.
Note that the function $f(x) = |c - bx + a|$ in $[-1, 1]$ attains its greatest value at one of the endpoints. Therefore, $f(x) \leq \max(f(-1), f(1)) = \max(|c + b + a|, |c - b + a|) \leq 1$, whence $\left|cx^2 - bx + a\right| \leq 1 + 1 = 2$.

10.2. Since a, b and c are the side lengths of some triangle, we have $|b - c| < a < b + c$. Without loss of generality, one can assume that $b \geq c$.
Consider the function $f(a) = (a + b + c)^2 - 4(ab + bc + ac)$ in $[b - c; b + c]$.
We have $f(a) = a^2 - 2a(b + c) + b^2 + c^2 - 2bc$. The graph of this function is a parabola with branches pointing upward. Consequently, in the given domain, $f(a)$ attains its largest value at one of the points $b - c$, $b + c$, that is, $f(b - c) = 4c(c - b) \leq 0$, $f(b + c) = -4bc < 0$. Hence in $(b - c, b + c)$ we have $f(a) < 0$.

10.3. Consider the function $f(a) = \frac{a}{b+c+1} + \frac{b}{c+a+1} + \frac{c}{a+b+1} + (1 - a)(1 - b)(1 - c)$ in $[0, 1]$. Obviously, the derivative is increasing:

$$f'(a) = \frac{1}{b + c + 1} - \frac{b}{(c + a + 1)^2} - \frac{c}{(a + b + 1)^2} - (1 - b)(1 - c)$$

in $[0, 1]$, and therefore, it satisfies Property 1 above, that is, $f(a)$ attains its greatest value at one of the points 0, 1.
Substituting a by 0 or 1 and considering the obtained function as a function depending on b, we obtain that it also attains its greatest value at one of the points $b = 0$, $b = 1$.
Since the given inequality is symmetric with respect to a, b, and c, we obtain that the expression in the left-hand side attains its greatest value at one of the following triples: (0, 0, 0), (0, 0, 1), (0, 1, 1), (1, 1, 1). Note that in all these cases, the left-hand side is equal to 1.
Alternative proof According to Problem 1.26, we have

$$\frac{a}{1 + b + c} + \frac{b}{1 + a + c} + \frac{c}{1 + a + b} + (1 - a)(1 - b)(1 - c) \leq a - a\frac{b + c}{2} +$$
$$+ \frac{abc}{3} + b - b\frac{a + c}{2} + \frac{abc}{3} + c - c\frac{a + b}{2} + \frac{abc}{3} + (1 - a)(1 - b)(1 - c) = 1.$$

10.4. Consider the function $f(a) = (a + b + c + d + e)\left(\frac{1}{a} + \frac{1}{b} + \frac{1}{c} + \frac{1}{d} + \frac{1}{e}\right)$, or $f(a) = Aa + \frac{B}{a} + AB + 1$, where $A = \frac{1}{b} + \frac{1}{c} + \frac{1}{d} + \frac{1}{e}$, $B = b + c + d + e$.
Since $f'(a) = A - \frac{B}{a^2}$ (B>0), the function $f(a)$ is increasing in $[p, q]$, and therefore, it attains its greatest value at one of the points p, q.
It follows that $f(a) \leq \max(f(p), f(q))$.

Consider the function $g(b) = (p+b+c+d+e)\left(\frac{1}{p}+\frac{1}{b}+\frac{1}{c}+\frac{1}{d}+\frac{1}{e}\right)$ or $h(b) = (q+b+c+d+e)\left(\frac{1}{q}+\frac{1}{b}+\frac{1}{c}+\frac{1}{d}+\frac{1}{e}\right)$ in $[p, q]$.
In a similar way we obtain that $g(b) \leq \max(g(p), g(q))$ and $h(b) \leq \max(h(p), h(q))$.
Continuing these arguments with respect to variables c, d, e, we get that the expression attains its greatest value when some of the variables are equal to p and the others are equal to q. Therefore, $(a+b+c+d+e)\left(\frac{1}{a}+\frac{1}{b}+\frac{1}{c}+\frac{1}{d}+\frac{1}{e}\right) \leq (mp+nq)\left(m\frac{1}{p}+n\frac{1}{q}\right)$, where $m \geq 0,\ n \geq 0,\ m+n=5,\ m, n \in \mathbb{Z}$.
Now let us prove that $(mp+nq)\left(\frac{m}{p}+\frac{n}{q}\right) \leq 25+6\left(\sqrt{\frac{p}{q}}-\sqrt{\frac{q}{p}}\right)^2$.

$$m^2+n^2+\frac{mnp}{q}+\frac{mnq}{p} \leq 25\left(\sqrt{\frac{p}{q}}-\sqrt{\frac{q}{p}}\right)^2,$$

$$(m+n)^2+mn\left(\frac{p}{q}+\frac{q}{p}-2\right) \leq 25+6\left(\frac{p}{q}+\frac{q}{p}-2\right),$$

$$25+mn\left(\frac{p}{q}+\frac{q}{p}-2\right) \leq 25+6\left(\frac{p}{q}+\frac{q}{p}-2\right),$$

since one can easily note that $mn \leq 6$.

Remark Problem 10.4 can be generalized in the following way.
Prove that if $0 < p < a_1, a_2, \ldots, a_n < q$, then
$(a_1+\cdots+a_n)\left(\frac{1}{a_1}+\cdots+\frac{1}{a_n}\right) \leq n^2+\psi(n)\left(\sqrt{\frac{p}{q}}-\sqrt{\frac{q}{p}}\right)^2$ (see Problem 10.7), where

$$\psi(n) = \begin{cases} \frac{n^2}{4}, & n = 2k (k = 0, 1, 2, \ldots), \\ \frac{n^2-1}{4}, & n = 2k+1 (k = 0, 1, 2, \ldots) \end{cases}$$

10.5. (a) Consider the function

$$f(x) = \left(m_2 + x\sum_{\substack{i=1\\ i\neq 2}}^{n} m_i a_i\right)\left(\sum_{i=1}^{n} m_i b_i\right) - \sum_{\substack{i=1\\ i\neq 2}}^{n} m_i a_i b_i - m_2 x b_2$$

in $[a_1, a_3]$. Since $f(x)$ is a linear function, it attains its greatest value at one of the points a_1, a_3. In the expression $M = f(a_2)$, substituting a_2 by a_1 or a_3 and in the obtained expression substituting a_3 by x, in a similar

way, one can prove that this expression attains its greatest value at one of the points a_1, a_4.
Continuing these assumptions for the numbers $a_4, \dots, a_{n-1}, b_2, \dots, b_{n-1}$, we obtain $M \le (\alpha a_1 + \beta a_n)(\gamma b_1 + \delta b_n) - k a_1 b_1 - (\alpha - k) a_1 b_n - (\delta - \alpha + k) b_n a_n - (\beta + \alpha - \delta - k) a_n b_1$, where α is the sum of the m_i whose corresponding a_i were substituted by a_1, and γ is the sum of the m_i whose corresponding b_i were substituted by b_1 and $\alpha + \beta = 1, \quad \gamma + \delta = 1, \quad k = \min(\alpha, \gamma)$.
We have

$$M \le \alpha\gamma a_1 b_1 + (\alpha\delta - \alpha) a_1 b_n + (\beta\gamma - \beta - \alpha + \delta) a_n b_1 + (\beta\delta - \delta + \alpha) a_n b_n - \\ - k a_1 b_1 + k a_1 b_n + k a_n b_1 - k a_n b_n = \alpha\gamma a_1 b_1 - \alpha\gamma a_1 b_n - \alpha\gamma a_n b_1 + \alpha\gamma a_n b_n - k(a_1 - a_n)(b_1 - b_n) = \\ = (\alpha\gamma - k)(a_1 - a_n)(b_1 - b_n) \le 0,$$

as $k = \alpha$ or $k = \gamma$ and $0 \le \alpha \le 1, \quad 0 \le \gamma \le 1$.

(b) Let us rewrite (a) for numbers $a_1 \le a_2 \le \dots \le a_n$ and $-b_1 \le -b_2 \le \dots \le -b_n$.
It follows that $\left(\sum\limits_{i=1}^{n} m_i a_i\right)\left(\sum\limits_{i=1}^{n} m_i(-b_i)\right) \le \sum\limits_{i=1}^{n} m_i a_i(-b_i)$, and on multiplying both sides by -1, we obtain the given inequality.

10.6. If $a_1 > 0, \dots, a_n > 0$, then in this case it is sufficient to expand the left-hand side of the inequality.
If $-2 \le a_1 < 0, \dots, -2 \le a_n < 0$, then let us consider the function $f(x) = (1+x)(1+a_2)\cdots(1+a_n) - 1 - x - a_2 - \dots - a_n$ in $[-2, 0]$. Since it is a linear function, it attains its smallest value at one of the points $-2, 0$.
One can easily prove that the expression $(1 + a_1) \cdots (1 + a_n) - 1 - a_1 - \dots - a_n$ attains its smallest value if some of the numbers are equal to -2 and the others are equal to 0.
Hence, if $a_1 = \dots = a_n = 0$, then we have $(1 + a_1) \cdots (1 + a_n) - 1 - a_1 - \dots - a_n = 1 - 1 = 0$.
If some of the numbers are equal to -2, then $(1 + a_1) \cdots (1 + a_n) - 1 - a_1 - \dots - a_n = (-1)^k - 1 + 2k \ge 0$, where k is the number of -2's.

10.7. Consider the function $f(x) = (x + x_2 + \dots + x_n)\left(\frac{1}{x} + \frac{1}{x_2} + \dots + \frac{1}{x_n}\right)$ in $[a, b]$.
Note that one can rewrite this function in the following way: $f(x) = 1 + \frac{A}{x} + Bx + AB$, where $A = x_2 + \dots + x_n, \ B = \frac{1}{x_2} + \dots + \frac{1}{x_n}$.
Since $f'(x) = B - \frac{A}{x^2} (B > 0)$, it follows that $f'(x)$ is increasing in $[a, b]$, and therefore it attains its greatest value at one of the points a, b.
It follows that $f(x_1) \le \max(f(a), f(b))$.
In a similar way, let us consider the following function:
$g(x) = (b + x + x_3 + \dots + x_n)\left(\frac{1}{b} + \frac{1}{x} + \frac{1}{x_3} + \dots + \frac{1}{x_n}\right)$, or $g(x) = (a + x + x_3 + \dots + x_n)\left(\frac{1}{a} + \frac{1}{x} + \frac{1}{x_3} + \dots + \frac{1}{x_n}\right)$ in $[a, b]$. We obtain $g(x_2) \le \max(g(a), g(b))$.

Repeating the same for the variables $x_3, \ldots, x_n$, we obtain that the expression $(x_1+\cdots+x_n)\left(\frac{1}{x_1}+\cdots+\frac{1}{x_n}\right)$ attains its greatest value if some of the variables equal a and the others equal b.
Let us take $x_1=\cdots=x_k=a, \quad x_{k+1}=\cdots=x_n=b$.
It is sufficient to prove that $(ka+(n-k)b)\left(\frac{k}{a}+\frac{n-k}{b}\right) \le \frac{(a+b)^2}{4ab}\cdot n^2$.
This inequality holds because $((n-2k)a-(n-2k)b)^2 \ge 0$.

10.8. Consider the function $f(x)=\frac{(1-x)(1-a-b+x)}{x(a+b-x)}$ in $[a,b]$.
One can easily prove that $f(x)=\left(\frac{1}{a+b}-1\right)\left(\frac{1}{x}+\frac{1}{a+b-x}\right)+1$.
We have $f'(x)=\frac{(2x-(a+b))(a+b)}{x^2(a+b-x)^2}\left(\frac{1}{a+b}-1\right)$, and hence the function $f(x)$ is decreasing on $\left[a,\frac{a+b}{2}\right]$ and increasing in $\left[\frac{a+b}{2},b\right]$. It follows that the function $f(x)$ attains its greatest value on $[a,b]$ at one of the points a, b.
Since $f(a)=f(b)=\frac{(1-a)(1-b)}{ab}$, we have $f(x)\le\frac{(1-a)(1-b)}{ab}$, where by taking $x=c$, we obtain the given inequality.

10.9. Consider the function $f(x)=\frac{1}{1+x}+\frac{1}{1+\frac{ab}{x}}=\frac{1}{1+x}+1-\frac{ab}{x+ab}$ in $[a,b]$.
Since $f'(x)=\frac{(ab-1)(x-\sqrt{ab})(x+\sqrt{ab})}{(x+1)^2(x+ab)^2}$, we see that the function $f(x)$ is decreasing on $\left[a,\sqrt{ab}\right]$ and increasing on $\left[\sqrt{ab},b\right]$. Therefore, $f(x)\le\max(f(a),\ f(b))$.
Since $f(a)=f(b)=\frac{1}{1+a}+\frac{1}{1+b}$ and $a\le m\le b$, we have $f(m)\le\frac{1}{1+a}+\frac{1}{1+b}$.

10.10. Note that $\frac{1}{2}-q\le p\le\frac{1}{2}$, and so for the function $f(p)=pa+qb+(1-p-q)c$, we have

$$f(p)\ge\min\left(f\left(\frac{1}{2}-q\right),f\left(\frac{1}{2}\right)\right)=\min\left(\left(\frac{1}{2}-q\right)a+qb+\frac{1}{2}c,\ \frac{1}{2}a+qb+\left(\frac{1}{2}-q\right)c\right)$$

$$\ge\min\left(\frac{1}{2}a+\frac{1}{2}c,\ \frac{1}{2}b+\frac{1}{2}c,\ \frac{1}{2}a+\frac{1}{2}b\right).$$

Let $\min\left(\frac{1}{2}a+\frac{1}{2}c,\frac{1}{2}b+\frac{1}{2}c,\frac{1}{2}a+\frac{1}{2}b\right)=\frac{1}{2}(a+c)$. Therefore,

$$pa+qb+rc\ge\frac{1}{2}(a+c)\ge(a+c)2(a+c)(1-(a+c))=2(a+c)^2b\ge 8abc.$$

10.11. Letting $m\in\{1,\ldots,n\}$, we see that the given inequality is equivalent to the following inequality:

$$(n-2)a_m^4-2a_m^2\left(a_1^2+\cdots+a_n^2-a_m^2\right)+(n-1)\left(a_1^4+\cdots+a_n^4-a_m^4\right)-\left(a_1^2+\cdots+a_n^2-a_m^2\right)^2<0,$$

which means that the quadratic function

$$(n-2)x^2-2x\left(a_1^2+\cdots+a_n^2-a_m^2\right)+(n-1)\left(a_1^4+\cdots+a_n^4-a_m^4\right)-\left(a_1^2+\cdots+a_n^2-a_m^2\right)^2$$

for $x=a_m^2$ attains a negative value, whence $D>0$, and so we have $\left(a_1^2+\cdots+a_n^2-a_m^2\right)^2>(n-2)\left(a_1^4+\cdots+a_n^4-a_m^4\right)$.

Repeating the same process, we obtain $\left(a_i^2+a_j^2+a_k^2\right)^2 > 2\left(a_i^4+a_j^4+a_k^4\right)$, or $(a_i+a_j+a_k)(a_i+a_j-a_k)(a_i-a_j+a_k)(-a_i+a_j+a_k) > 0$, and therefore, $a_i+a_j > a_k$.

10.12. If $\max(xy, yz, zx) \le 1$, then $x+y+z \ge \sqrt{xy}+\sqrt{yz}+\sqrt{zx} \ge xy+yz+zx$, and therefore, $x+y+z \ge xy+yz+zx$.
If $xy > 1$, let us consider the following two cases.

(a) If $x+y \ge 4$, then $x+y+z \ge 4 \ge xy+yz+zx$.
(b) If $x+y < 4$, then $z \le \frac{4-xy}{x+y+xy}$, and one needs to prove that $z \le \frac{x+y-xy}{x+y-1}$.

Prove that $\frac{4-xy}{x+y+xy} \le \frac{x+y-xy}{x+y-1}$, or $f(b) = b^2-(a-1)b+4a-4-a^2 \le 0$, where $a = x+y,\ b = xy$.
Since $1 \le b \le \frac{a^2}{4}$, we have $f(b) \le \max\left(f(1), f\left(\frac{a^2}{4}\right)\right)$.
Note that $f(1) = (a-1)(2-a) \le 0$, since $a \ge 2\sqrt{xy} \ge 2$, and $f\left(\frac{a^2}{4}\right) = \frac{(a-2)^2(a+4)(a-4)}{16} \le 0$, and therefore $f(b) \le 0$.

10.13. Note that if $x \ne 0$, then $\left|a+\frac{b}{x}+\frac{c}{x^2}\right| \le \left|A+\frac{B}{x}+\frac{C}{x^2}\right|$, and hence letting $x \to +\infty$, we obtain $|a| \le |A|$. (1)
Consider the following cases.

(a) If $A = 0$, then from (1) it follows that $a = 0$, and therefore, $\left|b+\frac{c}{x}\right| \le \left|B+\frac{C}{x}\right|$, whence $|b| \le |B|$. It follows that $\left|b^2-4ac\right| = \left|b^2\right| \le \left|B^2\right| = \left|B^2-4AC\right|$.
(b) If $A \ne 0$, let $B^2-4AC > 0$, and let x_1, x_2 be the roots of the polynomial Ax^2+Bx+C. Then from $\left|ax^2+bx+c\right| \le \left|Ax^2+Bx+C\right|$ it follows that $ax_i^2+bx_i+c = 0,\ i = 1,2$, and hence $a = b = c = 0$ or $(x_2-x_1)^2 = \frac{b^2-4ac}{a^2} = \frac{B^2-4AC}{A^2}$. Therefore, $\left|b^2-4ac\right| \le \left|B^2-4AC\right|$.
(c) If $A \ne 0$, let $B^2-4AC \le 0$, and without loss of generality one can assume that $A > 0$. Therefore, $\left|ax^2+bx+c\right| \le Ax^2+Bx+C$, and hence, $(A-a)x^2+(B-b)x+C-c \ge 0$ and $(A+a)x^2+(B+b)x+C+c \ge 0$, for all real values of x.

We have $(B-b)^2-4(A-a)(C-c) \le 0$ and $(B+b)^2-4(A+a)(C+c) \le 0$, and hence$(B-b)^2-4(A-a)(C-c)+(B+b)^2-4(A+a)(C+c) \le 0$.

Therefore, $b^2-4ac \le 4AC-B^2 = \left|B^2-4AC\right|$.
In order to complete the proof, it is left to consider the case $b^2-4ac < 0$. Without loss of generality one can assume that $a > 0$, and thus $ax^2+bx+c \le Ax^2+Bx+C$, and it follows that $\frac{4ac-b^2}{4a} \le \frac{4AC-B^2}{4A}$, and hence $\left|b^2-4ac\right| = 4ac-b^2 \le 4AC-B^2 = \left|B^2-4AC\right|$.

10.14. If $xyz = 0$, then $\left(1+\frac{2}{\sqrt{3}}\right)(x^2-x+1)(y^2-y+1)(z^2-z+1) \ge \left(1+\frac{2}{\sqrt{3}}\right)\cdot\frac{3}{4}\cdot\frac{3}{4} > 1 = (xyz)^2-xyz+1$.
If $xyz \ne 0$, then without loss of generality one can assume that $yz > 0$.

Let $k = 1 + \frac{2}{\sqrt{3}}$, $A = (y^2 - y + 1)(z^2 - z + 1)$, $B = yz$. Then one needs to prove that

$$(kA - B^2)x^2 - (kA - B)x + kA - 1 \geq 0. \qquad (1)$$

Note that $kA \geq k \cdot \frac{3}{4}y^2 \cdot \frac{3}{4}z^2 > y^2z^2 = B^2$, and then it suffices to prove that $D = (kA - B)^2 - 4(kA - B^2)(kA - 1) \leq 0$, or $3k^2A^2 - 2k(2B^2 - B + 2)A + 3B^2 \geq 0$.
Let us prove that

$$A \geq \frac{2B^2 - B + 2 + 2\sqrt{(B-1)^2(B^2 + B + 1)}}{3k}. \qquad (2)$$

Note that

$$A = (y + z)^2 - (B + 1)(y + z) + B^2 - B + 1. \qquad (3)$$

Let $t = \sqrt{B} + \frac{1}{\sqrt{B}}$. Consider the following two cases.

(a) If $t > 4$, then from (3), it follows that $A \geq \frac{3}{4}(B-1)^2$, and let us prove that $\frac{9k}{4}(B-1)^2 \geq 2B^2 - B + 2 + 2\sqrt{(B-1)^2(B^2 + B + 1)}$, $\frac{9k}{4}\left(B + \frac{1}{B} - 2\right) \geq 2\left(B + \frac{1}{B}\right) - 1 + 2\sqrt{\left(B + \frac{1}{B} - 2\right)\left(B + \frac{1}{B} + 1\right)}$, or $\frac{9k}{4}\left(t^2 - 4\right) \geq 2(t^2 - 2) - 1 + 2\sqrt{(t^2 - 4)(t^2 - 1)}$. Then the obtained inequality holds because $\frac{9k}{4}\left(t^2 - 4\right) - (4t^2 - 10) > \left(\frac{9k}{4} - 4\right)16 - 9k + 10 > 0$, and therefore, $2(t^2 - 2) - 1 + 2\sqrt{(t^2 - 4)(t^2 - 1)} \leq 2(t^2 - 2) - 1 + (t^2 - 4) + (t^2 - 1) = 4t^2 - 10 < \frac{9k}{4}\left(t^2 - 4\right)$.

(b) If $t \leq 4$, we have $y + z \geq 2\sqrt{yz} \geq \frac{B+1}{2}$, and then from (3) it follows that $A \geq \left(2\sqrt{B}\right)^2 - 2(B + 1)\sqrt{B} + B^2 - B + 1$, and let us prove that $3k\left(B^2 + 3B + 1 - 2(B + 1)\sqrt{B}\right) \geq 2B^2 - B + 2 + 2\sqrt{(B-1)^2(B^2 + B + 1)}$, $3k(t - 1)^2 \geq 2t^2 - 5 + 2\sqrt{(t^2 - 4)(t^2 - 1)}$, or $\left(t - 1 - \sqrt{3}\right)^2\left(\left(9 + 4\sqrt{3}\right)t^2 - \left(18 + 6\sqrt{3}\right)t + 12 + 2\sqrt{3}\right) \geq 0$.
This inequality holds because
$D = \left(18 + 6\sqrt{3}\right)^2 - 4\left(9 + 4\sqrt{3}\right)\left(12 + 2\sqrt{3}\right) < 0$.

10.15. (a) We proceed by induction.
For $n = 1$ we have $(1 - x_1)^\lambda + 1 - y_1^\lambda = 1$.
Let $n \geq 2$ and suppose that the inequality holds for $n - 1$ numbers.
Consider the function $f(x) = (1 - (1 - x) \cdot x_2 \cdots x_n)^\lambda + (1 - x^\lambda) \cdot (1 - y_2^\lambda) \cdots (1 - y_n^\lambda)$ on $[0, 1]$.
Since $f'(x) = \lambda x_2 \cdots x_n(1 - (1 - x) \cdot x_2 \cdots x_n)^{\lambda-1} - \lambda x^{\lambda-1} \cdot (1 - y_2^\lambda) \cdots (1 - y_n^\lambda)$, we have $0 \leq x < x_0$, where $x_0 =$

$\frac{1-x_2\cdots x_n}{\left(\frac{(1-y_2^\lambda)\cdots(1-y_n^\lambda)}{x_2\cdots x_n}\right)^{\frac{1}{\lambda-1}}-x_2\cdots x_n}$, and then we obtain $f'(x) > 0$, and if $x_0 < x \leq 1$, then we obtain $f'(x) < 0$.
(Note that $(1-y_2^\lambda)\cdots(1-y_n^\lambda) > (1-y_2)\cdots(1-y_n) = x_2\cdots x_n$, whence $0 < x_0 < 1$.)
It follows that $f(y_1) \geq \min(f(0); f(1))$.
Since $f(0) = (1-x_2\cdots x_n)^\lambda + (1-y_2^\lambda)\cdots(1-y_n^\lambda) \geq 1$ and $f(1) = 1$, we have $f(y_1) \geq 1$.
Hence, the given inequality holds for every positive integer n.

(b) If $m = 1$, then $(1-p^n)^m + (1-q^m)^n = 1$.
If $m > 1$, then in Problem 10.15(a), taking $\lambda = m$, $x_1 = \cdots = x_n = p$, we obtain $(1-p^n)^m + (1-q^m)^n \geq 1$.
This ends the proof.

Problems for Independent Study

Prove the following inequalities (2–12, 14–20).

1. Given that $0 \leq p \leq 1$, $0 \leq r \leq 1$, and the identity $(px+(1-p)y)(rx+(1-r)y) = ax^2+bxy+cy^2$, prove that one of the numbers a, b, c is not less than 4/9.
2. $x_1+\cdots+x_n-x_1x_2-x_2x_3-\cdots-x_{n-1}x_n-x_nx_1 \leq \left[\frac{n}{2}\right]$, where $0 \leq x_i \leq 1$ $(i=1,\ldots,n)$ and $n \geq 3, n \in \mathbb{N}$.
3. $(x_1+x_2+\cdots+x_n+1)^2 \geq 4(x_1^2+\cdots+x_n^2)$, where $x_1,\ldots,x_n \in [0,1]$.
4. $\frac{\left(\sum\limits_{i=1}^n a_i^2\right)\left(\sum\limits_{i=1}^n b_i^2\right)}{\left(\sum\limits_{i=1}^n a_ib_i\right)^2} \leq \left(\frac{\sqrt{\frac{AB}{ab}}+\sqrt{\frac{ab}{AB}}}{2}\right)^2$, where $0 < a \leq a_i \leq A, 0 < b \leq b_i \leq B$, $i = 1,\ldots,n$.
5. $a^2b^2+b^2c^2+c^2a^2 > \frac{1}{2}(a^4+b^4+c^4)$, where a, b, c are the side lengths of some triangle.
6. (a) $x^2+y^2+z^2+xy+yz+zx+x+y+z+\frac{3}{8} \geq 0$,
   (b) $xy(1-z)+yz(1-x)+xz(1-y) \leq 1$, where $0 \leq x, y, z \leq 1$.
7. $2^n \geq (1+a_1)\cdots(1+a_n)+(1-a_1)\cdots(1-a_n) \geq 2$, where $0 \leq a_i \leq 1, i = 1,\ldots,n$.
8. $2^{a+b}+2^{b+c}+2^{c+a} < 1+2^{a+b+c+1}$, where $a > 0$, $b > 0$, $c > 0$.
9. $\min\left((a-b)^2,(b-c)^2,(a-c)^2\right) \leq \frac{a^2+b^2+c^2}{2}$.
10. $a^4+ab+b^2 \geq 4a^2b+a^3b$, where $a, b \geq 0$.
11. $aB+bC+cA < k^2$, where $a > 0$, $b > 0$, $c > 0$, $A > 0$, $B > 0$, $C > 0$ and $a+A = b+B = c+C = k$.
12. $\frac{a}{bc+1}+\frac{b}{ca+1}+\frac{c}{ab+1} \leq 2$, where $a, b, c \in [0,1]$.
13. Prove that the perimeter of the quadrilateral section of a regular tetrahedron with edge length a is smaller than $3\,a$.

14. $\dfrac{\sum_{i=1}^{n} a_i x_i \sum_{i=1}^{n} \frac{a_i}{x_i}}{\left(\sum_{i=1}^{n} a_i\right)^2} \leq \frac{1}{4}\left(\sqrt{\frac{M}{m}} + \sqrt{\frac{m}{M}}\right)^2$, where $a_1, \ldots, a_n > 0$ and $0 < m \leq x_i \leq M$, $i = 1, \ldots, n$.
15. $8ac + 8bd \leq 3a^2 + 3b^2 + 3c^2 + 3d^2 + ab + bc + cd + da$, where $0 \leq a \leq b \leq c \leq d$.
16. (a) $\frac{a^2+b^2-c^2}{a+b-c} \leq a + b - 2$, where $2 \leq a, b, c \leq 3$.
    (b) $\frac{a_1^2+a_2^2-a_3^2}{a_1+a_2-a_3} + \frac{a_2^2+a_3^2-a_4^2}{a_2+a_3-a_4} + \cdots + \frac{a_n^2+a_1^2-a_2^2}{a_n+a_1-a_2} \leq 2a_1 + 2a_2 + \cdots + 2a_n - 2n$, where $n \geq 3, 2 \leq a_i \leq 3, i = 1, \ldots, n$.
17. $\frac{1}{2} < \frac{a(c-d)+2d}{b(d-c)+2c} \leq 2$, where $1 \leq a \leq 2,\ \ 1 \leq b \leq 2,\ \ 1 \leq c \leq 2,\ \ 1 \leq d \leq 2$.
18. $\alpha\gamma - \beta^2 \leq 0$, where $a\gamma - 2b\beta + c\alpha = 0$ and $ac - b^2 > 0$.
19. $(a_1 + \cdots + a_n - b_1 - \cdots - b_n)^2 + 2|a_1 b_2 - a_2 b_1| + \cdots + 2|a_1 b_n - a_n b_1| + \cdots + 2|a_{n-1} b_n - a_n b_{n-1}| \leq |a_1^2 - b_1^2| + \cdots + |a_n^2 - b_n^2| + 2|a_1 a_2 - b_1 b_2| + \cdots + 2|a_1 a_n - b_1 b_n| + \cdots + 2|a_{n-1} a_n - b_{n-1} b_n|$, where $n \geq 2$ and $a_i > 0,\ \ b_i > 0,\ i = 1, \ldots, n$.
20. $\frac{a_1}{1+a_1+\cdots+a_n-a_1} + \cdots + \frac{a_n}{1+a_1+\cdots+a_n-a_n} + (1-a_1)\cdots(1-a_n) \leq 1$, where $n \geq 2$, $0 \leq a_i \leq 1,\ i = 1, \ldots, n$.

Chapter 11
Jensen's Inequality

Historical origins. *Jensen's inequality* is named after the Danish mathematician *Johan Ludwig William Valdemar Jensen,* born 8 May 1859 in Nakskov, Denmark, died 5 March 1925 in Copenhagen, Denmark. This inequality was proved in a paper that Jensen published in 1906. He never received an academic degree in mathematics. In 1876 he entered *Copenhagen College of Technology* and studied mathematics among various other subjects. Mathematics was his favorite subject, and he published his first paper in mathematics when he was still a student. Jensen studied advanced topics of mathematics later by himself. He never held an academic position. In 1881 he began work as an engineer for a *Copenhagen division of the international Bell company* (from 1882 *Copenhagen telephone company*) and did mathematics in his spare time. From 1890 to 1924 he was head of the technical department of the Copenhagen telephone company, and from 1892 to 1903 he was president of the *Danish Mathematical Society*.

Theorem 11.1 *(Jensen's inequality) Suppose the following inequalities hold for all a and b in $D(f) = I$:*

$$\frac{f(a)+f(b)}{2} \geq f\left(\frac{a+b}{2}\right), \tag{11.1}$$

$$\left[\frac{f(a)+f(b)}{2} \leq f\left(\frac{a+b}{2}\right)\right]. \tag{11.2}$$

Then for all $\alpha_1, \ldots, \alpha_n \in Q_+$[1], where $\alpha_1 + \cdots + \alpha_n = 1$, and all numbers $x_1, \ldots, x_n \in I$, one has the following inequalities:

$$\alpha_1 f(x_1) + \cdots + \alpha_n f(x_n) \geq f(\alpha_1 x_1 + \cdots + \alpha_n x_n), \tag{11.3}$$

$$[\alpha_1 f(x_1) + \cdots + \alpha_n f(x_n) \leq f(\alpha_1 x_1 + \cdots + \alpha_n x_n)]. \tag{11.4}$$

[1] Q_+ is the set of the positive rational numbers

H. Sedrakyan and N. Sedrakyan, *Algebraic Inequalities*, Problem Books in Mathematics, https://doi.org/10.1007/978-3-319-77836-5_11

Proof See Problem 7.8.

Note that if on I the function $f(x)$ satisfies (11.1), then $f(x)$ is said to be *Jensen convex* or *midconvex* on I, and if $f(x)$ satisfies (11.2), then $f(x)$ is said to be *Jensen concave* or *midconcave* on I.

Example 11.1 Prove that $\frac{\cos x_1+\cdots+\cos x_n}{n} \le \cos\left(\frac{x_1+\cdots+x_n}{n}\right)$, where $x_1, \ldots, x_n \in \left[0, \frac{\pi}{2}\right]$.

Proof Prove that for all numbers $a, b \in \left[0, \frac{\pi}{2}\right]$ inequality (11.2) holds, that is, $\frac{\cos a+\cos b}{2} \le \cos\left(\frac{a+b}{2}\right)$. Indeed, $\frac{\cos a+\cos b}{2} = \cos\frac{a+b}{2}\cos\frac{a-b}{2} \le \cos\frac{a+b}{2}$.

Assume that $\alpha_1 = \cdots = \alpha_n = \frac{1}{n}$. From (11.4), it follows that

$$\frac{\cos x_1 + \cdots + \cos x_n}{n} \le \cos\frac{x_1 + \cdots + x_n}{n}.$$

This ends the proof.

Theorem 11.2 *If for all numbers a and b from $D(f) = I$ and for all α, β such that $\alpha \ge 0,\ \beta \ge 0,\ \ \alpha + \beta = 1$, one has*

$$\alpha f(a) + \beta f(b) \ge f(\alpha a + \beta b), \tag{11.5}$$

$$[\alpha f(a) + \beta f(b) \le f(\alpha a + \beta b)], \tag{11.6}$$

then for all $\alpha_1, \ldots, \alpha_n, x_1, \ldots, x_n$, one has

$$\alpha_1 f(x_1) + \cdots + \alpha_n f(x_n) \ge f(\alpha_1 x_1 + \cdots + \alpha_n x_n), \tag{11.7}$$

$$[\alpha_1 f(x_1) + \cdots + \alpha_n f(x_n) \le f(\alpha_1 x_1 + \cdots + \alpha_n x_n)], \tag{11.8}$$

where $\alpha_1 \ge 0,\ \ldots,\alpha_n \ge 0,\ \ \alpha_1 + \cdots + \alpha_n = 1$, and $x_1, \ldots, x_n \in I$.

Proof See Problem 7.9.

Note that if on I the function $f(x)$ satisfies (11.5), then $f(x)$ is said to be *convex* in I, and if $f(x)$ satisfies (11.6), then $f(x)$ is said to be *concave* in I.

Theorem 11.3 *If for a function $f(x)$ in $D(f) = I$ one has*

$$f''(x) \ge 0, \tag{11.9}$$

$$\left[or\ f''(x) \le 0\right], \tag{11.10}$$

then (11.7) or (11.8) holds.

Proof Assume that (11.9) holds. Let us first prove that (11.5) holds.

Indeed, from the finite increments formula, it follows that

$$\begin{aligned}\alpha f(x_1)+\beta f(x_2)-f(\alpha x_1+\beta x_2) &= \alpha(f(x_1)-f(\alpha x_1+\beta x_2))+\beta(f(x_2)-f(\alpha x_1+\beta x_2))\\ &= \alpha f'(c_1)(x_1-\alpha x_1-\beta x_2)+\beta f'(c_2)(x_2-\alpha x_1-\beta x_2)\\ &= \alpha\beta\big(f'(c_2)-f'(c_1)\big)(x_2-x_1)=\alpha\beta f''(c)(c_2-c_1)(x_2-x_1),\end{aligned}$$

where $x_1 < c_1 < \alpha x_1 + \beta x_2 < c_2 < x_2$ and $c_1 < c < c_2$.

Hence, the sign of the left-hand side is the same as the sign of $f''(c)$, and thus (11.5) holds.

According to Theorem 11.2, we have that (11.7) holds.

This ends the proof.

Example 11.2 Prove that $\left(\sum\limits_{i=1}^{n} a_i^k\right)\left(\sum\limits_{i=1}^{n} c_i^k\right)^{k-1} \geq \left(\sum\limits_{i=1}^{n} a_i c_i^{k-1}\right)^k$, where $a_i > 0,\ c_i > 0\ i = 1, \ldots, n$, and $k \notin (0, 1)$.

Proof Consider the function $f(x) = x^k$ in $(0, +\infty)$.

Since $f''(x) = k(k-1)x^{k-2} \geq 0$, then taking $\alpha_i = \frac{c_i^k}{\sum\limits_{j=1}^{n} c_j^k}$, $x_i = \left(\frac{a_i}{c_i}\right)^k$, where $i = 1, \ldots, n$, we obtain by Theorem 11.3 that

$$\sum_{i=1}^{n} \frac{c_i^k}{\sum\limits_{j=1}^{n} c_j^k} \cdot \left(\frac{a_i}{c_i}\right)^k \geq \left(\sum_{i=1}^{n} \frac{c_i^k}{\sum\limits_{j=1}^{n} c_j^k} \cdot \frac{a_i}{c_i}\right)^k.$$

This ends the proof.

Problems

Prove the following inequalities (11.1–11.21)

11.1. $\frac{\sin x_1+\cdots+\sin x_n}{n} \leq \sin\frac{x_1+\cdots+x_n}{n}$, where $x_1, \ldots, x_n \in [0, \pi]$.

11.2. $\frac{a+b+c}{3} \leq a^{a/(a+b+c)} \cdot b^{b/(a+b+c)} \cdot c^{c/(a+b+c)} \leq \frac{a^2+b^2+c^2}{a+b+c}$, where $a, b, c \in \mathbb{N}$.

11.3. $\left(1+\frac{b-c}{a}\right)^a \cdot \left(1+\frac{c-a}{b}\right)^b \cdot \left(1+\frac{a-b}{c}\right)^c \leq 1$, where a, b, c are the side lengths of some triangle and $a, b, c \in \mathbb{Q}$.

11.4. $a^2b(a-b)+b^2c(b-c)+c^2a(c-a) \geq 0$, where a, b, c are the side lengths of some triangle.

11.5. $a^3b + b^3c + c^3a \geq a^2bc + b^2ca + c^2ab$, where $a > 0,\ b > 0,\ c > 0$.

11.6. $\left(1+\frac{1}{x}\right)\left(1+\frac{1}{y}\right)\left(1+\frac{1}{z}\right) \geq 64$, where $x > 0,\ y > 0,\ z > 0$, and $x+y+z = 1$.

11.7. $\frac{12}{\sqrt{a}+\sqrt{b}+\sqrt{c}+\sqrt{d}} \leq \frac{1}{\sqrt{a}+\sqrt{b}} + \frac{1}{\sqrt{a}+\sqrt{c}} + \frac{1}{\sqrt{a}+\sqrt{d}} + \frac{1}{\sqrt{b}+\sqrt{c}} + \frac{1}{\sqrt{b}+\sqrt{d}} + \frac{1}{\sqrt{c}+\sqrt{d}}$, where $a > 0,\ b > 0,\ c > 0,\ d > 0$.

11.8. $\frac{m^p}{p} + \frac{n^q}{q} \geq mn$, where $m > 0,\ n > 0,\ p > 0,\ q > 0$, and $\frac{1}{p} + \frac{1}{q} = 1$.

11.9. $\sqrt{a}(a+c-b) + \sqrt{b}(a+b-c) + \sqrt{c}(b+c-a) \leq \sqrt{(a^2+b^2+c^2)(a+b+c)}$, where a, b, c are the side lengths of some triangle.

11.10. $\left(\sum\limits_{i=1}^{n} a_i^2\right)\left(\sum\limits_{i=1}^{n} c_i^2\right) \geq \left(\sum\limits_{i=1}^{n} a_i c_i\right)^2$.

11.11. $\left(\sum\limits_{i=1}^{n} a_i^k\right)\left(\sum\limits_{i=1}^{n} c_i^k\right)^{k-1} \leq \left(\sum\limits_{i=1}^{n} a_i c_i^{k-1}\right)^k$, where $a_1, a_2, \ldots, a_n, c_1, \ldots, c_n > 0$, and $0 < k < 1$.

11.12. (a) $\left(\sum\limits_{i=1}^{n} a_i^p\right)^{\frac{1}{p}}\left(\sum\limits_{i=1}^{n} b_i^q\right)^{\frac{1}{q}} \geq \sum\limits_{i=1}^{n} a_i b_i$, where $a_i > 0,\ b_i > 0,\ i = 1, \ldots, n,\ \ p > 0,\ q > 0$, and $\frac{1}{p} + \frac{1}{q} = 1$;

(b) $\sum\limits_{i=1}^{n} \frac{a_i^p}{b_i^{p-1}} \geq \frac{\left(\sum\limits_{i=1}^{n} a_i\right)^p}{\left(\sum\limits_{i=1}^{n} b_i\right)^{p-1}}$, where $a_i > 0,\ b_i > 0,\ i = 1, \ldots, n$, and $p > 1$.

11.13. $\frac{a_1}{-a_1+a_2+\cdots+a_n} + \frac{a_2}{a_1-a_2+\cdots+a_n} + \cdots + \frac{a_n}{a_1+a_2+\cdots+a_{n-1}-a_n} \geq \frac{n}{n-2}$, where $a_1, a_2, \ldots, a_n$ are the side lengths of some n-gon.

11.14. $\frac{\sum\limits_{i=1}^{n} d_i^m}{n} \cdot \frac{\sum\limits_{i=1}^{n} b_i^m}{n} \geq \left(\frac{\sum\limits_{i=1}^{n} d_i b_i}{n}\right)^m$, where $d_1 > 0, \ldots, d_n > 0,\ b_1 > 0, \ldots, b_n > 0$, and $m \geq 2$.

11.15. (a) $\frac{x_1}{x_2+x_3} + \frac{x_2}{x_3+x_1} + \frac{x_3}{x_1+x_3} \geq \frac{3}{2}$, where $x_1 > 0,\ x_2 > 0,\ x_3 > 0$;

(b) $\frac{x_1}{x_2+x_n} + \frac{x_2}{x_3+x_1} + \cdots + \frac{x_{n-1}}{x_n+x_{n-2}} + \frac{x_n}{x_1+x_{n-1}} \geq 2$, where $n \geq 4,\ x_1 > 0, \cdots, x_n > 0$.

11.16. $\left|\frac{x}{y} + \frac{y}{z} + \frac{z}{x} - \frac{y}{x} - \frac{z}{y} - \frac{x}{z}\right| \leq 2\left(1 - \frac{xy+yz+xz}{x^2+y^2+z^2}\right)$, where x, y, z are the side lengths of some triangle.

11.17. (a) $\frac{a}{b+c+d} + \frac{b}{a+c+d} + \frac{c}{a+b+d} + \frac{d}{a+b+c} \geq \frac{4}{3}$, where $a > 0,\ b > 0,\ c > 0,\ d > 0$;

(b) $\frac{a_1}{a_2+a_3+\cdots+a_n} + \frac{a_2}{a_1+a_3+\cdots+a_n} + \cdots + \frac{a_n}{a_1+a_2+\cdots+a_{n-1}} \geq \frac{n}{n-1}$, where $n \geq 2,\ \ a_1 > 0, \ldots, a_n > 0$.

11.18. $\frac{a}{b+c} + \frac{b}{c+d} + \frac{c}{a+d} + \frac{d}{a+b} \geq 2$, where $a > 0,\ b > 0,\ c > 0,\ d > 0$.

11.19. $a+b+c \leq \frac{a^2+b^2}{2c} + \frac{b^2+c^2}{2a} + \frac{c^2+a^2}{2b} \leq \frac{a^3}{bc} + \frac{b^3}{ca} + \frac{c^3}{ab}$, where $a > 0,\ b > 0,\ c > 0$.

11.20. $\frac{a^3}{a^2+ab+b^2} + \frac{b^3}{b^2+bc+c^2} + \frac{c^3}{c^2+ca+a^2} \geq \frac{a+b+c}{3}$, where $a > 0,\ b > 0,\ c > 0$.

11.21. $\left(\frac{a+b}{c+d}\right)^{a+b} \leq \left(\frac{a}{c}\right)^a \cdot \left(\frac{b}{d}\right)^b$, where $a > 0,\ b > 0,\ c > 0,\ d > 0$.

11.22. Let (x_n) be a sequence with positive terms such that $1 = x_0 \geq x_1 \geq x_2 \geq \cdots \geq x_n \geq \cdots$.

Prove that there exists n such that for every such sequence (x_n), one has

$$\frac{x_0^2}{x_1} + \frac{x_1^2}{x_2} + \cdots + \frac{x_{n-1}^2}{x_n} \geq 3.999.$$

11.23. Perpendiculars $MA_1,\ MB_1,\ MC_1$ are drawn from the inner point M of the triangle ABC to the sides $BC,\ CA,\ AB$, respectively. For which point M of the triangle ABC is the value of the expression $\frac{BC}{MA_1} + \frac{CA}{MB_1} + \frac{AB}{MC_1}$ the smallest?

11.24. Let $a_1, \ldots, a_n, b_1, \ldots, b_n > 0$ and $J \subseteq \{1, \ldots, n\}$. Prove that if for every $\mathfrak{J}$ one has
$\sum_{i\in\mathfrak{J}} b_i \geq \left(\sum_{i\in\mathfrak{J}} a_i\right)^m$, where $m \in \mathbb{N}$, then $\left(\sum_{i=1}^n a_i b_i\right)^2 \geq \left(\sum_{i=1}^n a_i^2\right)^{m+1}$.

11.25. Let $a_1 > 0, \ldots, a_n > 0, b_1 > 0, \ldots, b_n > 0$. Prove
(a) *Radon's inequality*:

$$\frac{a_1^p}{b_1^{p-1}} + \cdots + \frac{a_n^p}{b_n^{p-1}} \geq \frac{(a_1 + \cdots + a_n)^p}{(b_1 + \cdots + b_n)^{p-1}},$$

where $p(p-1) \geq 0$,
(b) $\frac{a_1^p}{b_1^{p-1}} + \cdots + \frac{a_n^p}{b_n^{p-1}} \leq \frac{(a_1+\cdots+a_n)^p}{(b_1+\cdots+b_n)^{p-1}}$,

$$\frac{a_1^p}{b_1^{p-1}} + \cdots + \frac{a_n^p}{b_n^{p-1}} \leq \frac{(a_1 + \cdots + a_n)^p}{(b_1 + \cdots + b_n)^{p-1}},$$

where $p(p-1) < 0$,
(c) $\left(\frac{p+1}{p}\right)^p \cdot \left(\frac{a_1^{p+1}}{b_1^p} + \cdots + \frac{a_n^{p+1}}{b_n^p}\right) > a_1 \cdot \left(\frac{a_1}{b_1}\right)^p + \ldots + a_n \cdot \left(\frac{a_1+\cdots+a_n}{b_1+\cdots+b_n}\right)^p \geq$
$\frac{p}{p+1}\left(b_1 \cdot \left(\frac{a_1}{b_1}\right)^{p+1} + \cdots + b_n\left(\frac{a_1+\cdots+a_n}{b_1+\cdots+b_n}\right)^{p+1}\right) + \frac{1}{p+1} \cdot \frac{(a_1+\cdots+a_n)^{p+1}}{(b_1+\cdots+b_n)^p}$,
where $p > 0$.

11.26. (a) Given the interval G and a function $f(x)$ defined on the interval I, suppose that for all numbers x_1 and x_2 from the interval I, where $x_1 + x_2 \in G$, one has $f\left(\frac{x_1+x_2}{2}\right) \leq \frac{f(x_1)+f(x_2)}{2}$.
Prove that if the $x_1, \ldots, x_n$ are numbers from the interval I such that $x_1 + x_1', \ldots, x_n + x_n' \in G$, where the numbers $x_1', \ldots, x_n'$ are the numbers $x_1, \ldots, x_n$ written in some other order, then $f\left(\frac{x_1+\cdots+x_n}{n}\right) \leq \frac{f(x_1)+\cdots+f(x_n)}{n}$.
Prove that
(b) $\frac{a_1}{(n-1)a_1+a_2} + \cdots + \frac{a_n}{(n-1)a_n+a_1} \leq 1$, where $n \geq 2$, $a_i > 0$, $i = 1, \ldots, n$,
(c) $\frac{1}{\sqrt{1+x_1}} + \cdots + \frac{1}{\sqrt{1+x_n}} \geq \min\left(1, \frac{n}{\sqrt{1+\lambda}}\right)$, where $n \geq 2$, $x_i > 0$, $i = 1, \ldots, n$, and $x_1 \ldots x_n = \lambda^n$ $(\lambda > 0)$,
(d) $\frac{(1-x_1)\cdots(1-x_n)}{x_1\cdots x_n} \geq \left(\frac{1-\frac{x_1+\ldots+x_n}{n}}{\frac{x_1+\cdots+x_n}{n}}\right)^n$, where $n \geq 2$, $x_i > 0$, $i = 1, \ldots, n$, and $x_i + x_j \leq 1$, $i, j \in \{1, \ldots, n\}$.

11.27. Prove that $\frac{1}{1+x^2} + \frac{1}{1+y^2} + \frac{1}{1+z^2} \leq \frac{27}{10}$, where $x + y + z = 1$.

11.28. Prove that $\frac{1}{(1+a)^\alpha} + \frac{1}{(1+b)^\alpha} + \frac{1}{(1+c)^\alpha} + \frac{1}{(1+d)^\alpha} \geq 2^{2-\alpha}$, where $\alpha \geq 2$, $a > 0$, $b > 0$, $c > 0$, $d > 0$, and $abcd = 1$.

Proofs

11.1. For all $a, b \in [0, \pi]$ we have
$\frac{\sin a+\sin b}{2} \leq \sin \frac{a+b}{2}$.
We have $\frac{\sin a+\sin b}{2} = \sin \frac{a+b}{2} \cos \frac{a-b}{2} \leq \sin \frac{a+b}{2}$.
Taking $\alpha_1 = \cdots = \alpha_n = \frac{1}{n}$ and using inequality (11.4), we deduce that $\frac{\sin x_1+\cdots+\sin x_n}{n} \leq \sin \frac{x_1+\cdots+x_n}{n}$.

11.2. Since for all $a > 0$, $b > 0$ one has $\frac{\log a+\log b}{2} \leq \log \frac{a+b}{2}$, it follows that if we set $\alpha_1 = \frac{a}{a+b+c}$, $\alpha_2 = \frac{b}{a+b+c}$, $\alpha_3 = \frac{c}{a+b+c}$, then by inequality (11.4), we have $\alpha_1 \log a + \alpha_2 \log b + \alpha_3 \log c \leq \log(\alpha_1 a + \alpha_2 b + \alpha_3 c)$, thus $a^{a/(a+b+c)} \cdot b^{b/(a+b+c)} \cdot c^{c/(a+b+c)} \leq \frac{a^2+b^2+c^2}{a+b+c}$.
Let us consider the function $f(x) = x \ln x$ in $(0, +\infty)$.
Since $f''(x) = \frac{1}{x} > 0$, we have that (11.7) holds for the function $f(x)$. Taking $\alpha_1 = \alpha_2 = \alpha_3 = \frac{1}{3}$, it follows that $\frac{1}{3}(a \ln a + b \ln b + c \ln c) \geq \frac{1}{3}(a+b+c) \cdot \ln \frac{a+b+c}{3}$. Therefore, $a^a \cdot b^b \cdot c^c \geq \left(\frac{a+b+c}{3}\right)^{a+b+c}$, or $a^{a/(a+b+c)} \cdot b^{b/(a+b+c)} \cdot c^{c/(a+b+c)} \geq \frac{a+b+c}{3}$.

11.3. Let $f(x) = \log x$, $x_1 = 1 + \frac{b-c}{a}$, $x_2 = 1 + \frac{c-a}{b}$, $x_3 = 1 + \frac{a-b}{c}$, $\alpha_1 = \frac{a}{a+b+c}$, $\alpha_2 = \frac{b}{a+b+c}$, $\alpha_3 = \frac{c}{a+b+c}$, hence $\frac{1}{a+b+c}\left(a \log\left(1 + \frac{b-c}{a}\right) + b \log\left(1 + \frac{c-a}{b}\right) + c \log\left(1 + \frac{a-b}{c}\right)\right) \leq \log\left(\frac{a}{a+b+c}\left(1 + \frac{b-c}{a}\right) + \frac{b}{a+b+c}\left(1 + \frac{c-a}{b}\right) + \frac{c}{a+b+c}\left(1 + \frac{a-b}{c}\right)\right)$,
and therefore $\left(1 + \frac{b-c}{a}\right)^a \cdot \left(1 + \frac{c-a}{b}\right)^b \cdot \left(1 + \frac{a-b}{c}\right)^c \leq 1$.

11.4. Consider the function $f(x) = x^2$. One can easily prove that for all numbers m and n, one has $\alpha f(m) + \beta f(n) \geq f(\alpha m + \beta n)$, where $\alpha \geq 0$, $\beta \geq 0$, and $\alpha + \beta = 1$.
According to Theorem 11.2 inequality (11.7) holds, and taking $x_1 = c$, $x_2 = b$, $x_3 = a$, $\alpha_1 = \frac{a+c-b}{a+b+c}$, $\alpha_2 = \frac{b+c-a}{a+b+c}$, $\alpha_3 = \frac{a+b-c}{a+b+c}$, we obtain

$$c^2 \frac{a+c-b}{a+b+c} + b^2 \frac{b+c-a}{a+b+c} + a^2 \frac{a+b-c}{a+b+c} \geq \left(\frac{ac + c^2 - bc + b^2 + bc - ab + a^2 + ab - ac}{a+b+c}\right)^2,$$

whence

$$c^2\left((a+c)^2 - b^2\right) + b^2\left((b+c)^2 - a^2\right) + a^2\left((a+b)^2 - c^2\right) \geq \left(a^2 + b^2 + c^2\right)^2,$$

or

$$a^2 b(a-b) + b^2 c(b-c) + c^2 a(c-a) \geq 0.$$

11.5. **Hint**. Take $f(x) = x^2$, $x_1 = a + b - c$, $x_2 = a + c - b$, $x_3 = b + c - a$,

$$\alpha_1 = \frac{a}{a+b+c}, \ \alpha_2 = \frac{b}{a+b+c}, \ \alpha_3 = \frac{c}{a+b+c}.$$

Alternative proof. Note that

$$a^3b + b^3c + c^3a - \left(a^2bc + b^2ca + c^2ab\right)$$
$$= ab(a-c)^2 + bc(b-a)^2 + ac(b-c)^2 \geq 0.$$

11.6. Let us consider the function $f(t) = \ln\left(1 + \frac{1}{t}\right)$ on $(0, +\infty)$.
Since $f''(t) = \frac{1}{t^2} - \frac{1}{(t+1)^2} > 0$, inequality (11.7) holds. Taking $\alpha_1 = \alpha_2 = \alpha_3 = \frac{1}{3}, t_1 = x, \ t_2 = y, \ t_3 = z$, we obtain $\frac{1}{3}\left(\ln\left(1 + \frac{1}{x}\right) + \ln\left(1 + \frac{1}{y}\right) + \ln\left(1 + \frac{1}{z}\right)\right) \geq \ln\left(1 + \frac{3}{x+y+z}\right)$; hence from $x + y + z = 1$, it follows that $\left(1 + \frac{1}{x}\right)\left(1 + \frac{1}{y}\right)\left(1 + \frac{1}{z}\right) \geq \left(1 + \frac{3}{x+y+z}\right)^3 = 64$.

11.7. Let us consider the function $f(x) = \frac{1}{x}$ $(0, +\infty)$. We have $\frac{1}{2}\left(\frac{1}{a} + \frac{1}{b}\right) \geq \frac{2}{a+b}$, and so by Theorem 11, taking $x_1 = \sqrt{a} + \sqrt{b}, \ x_2 = \sqrt{a} + \sqrt{c}$,

$$x_3 = \sqrt{a} + \sqrt{d}, \ x_4 = \sqrt{b} + \sqrt{c}, \ x_5 = \sqrt{b} + \sqrt{d},$$
$$x_6 = \sqrt{c} + \sqrt{d}, \quad \alpha_1 = \cdots = \alpha_6 = \frac{1}{6},$$

we obtain that the given inequality holds.

11.8. Let us consider the function $f(x) = \ln x$ in $(0, +\infty)$. Since $f''(x) = \frac{-1}{x^2} < 0$, inequality (11.6) holds. Taking $\alpha = \frac{1}{p}, \ \beta = \frac{1}{q}, \quad a = m^p, \quad b = n^q$, we obtain that the given inequality holds.

11.9. Let us consider the function $f(x) = \sqrt{x}$ in $(0, +\infty)$. One can easily prove that for the function $f(x)$, the condition (11.10) holds.
Taking $\alpha_1 = \frac{a+c-b}{a+b+c}, \ \alpha_2 = \frac{a+b-c}{a+b+c}, \alpha_3 = \frac{c+b-a}{a+b+c}, \ x_1 = a, \ x_2 = b, \ x_3 = c$, we deduce inequality (11.8) in the given form.
Hint. For $a = 9, b = 4, c = 1$ the inequality does not hold.

11.10. Let $c_1^2 + \cdots + c_n^2 \neq 0$. Consider the function $f(x) = x^2$. Since $f''(x) > 0$, on taking

$$\alpha_i = \frac{c_i^2}{c_1^2 + \cdots + c_n^2}, \ x_i = \left(\frac{a_i}{c_i}\right)^2,$$

where $i = 1, \ldots, n$, using Theorem 11.3, we obtain

$$\sum_{i=1}^{n} \frac{c_i^2}{c_1^2 + \cdots + c_n^2} \cdot \left(\frac{a_i}{c_i}\right)^2 \geq \left(\sum_{i=1}^{n} \frac{c_i^2}{c_1^2 + \cdots + c_n^2} \cdot \frac{a_i}{c_i}\right)^2;$$

hence the given inequality holds.
If $c_1 = \cdots = c_n = 0$, then we obtain an obvious inequality.

11.11. The proof is similar to Problem 11.2, using $f''(x) < 0$.

11.12. (a) The proof is similar to Problem 11.2, on substituting k by p, and c_i by $b_i^{\frac{q}{p}}$.

(b) According to Problem 11.12a, we have

$$\left(\sum_{i=1}^{n}\left(\frac{a_i}{b_i^{\frac{p-1}{p}}}\right)^p\right)^{\frac{1}{p}}\left(\sum_{i=1}^{n}\left(b_i^{\frac{p-1}{p}}\right)^{\frac{p}{p-1}}\right)^{\frac{p-1}{p}} \geq \sum_{i=1}^{n} a_i,$$

and therefore,

$$\left(\sum_{i=1}^{n}\frac{a_i^p}{b_i^{p-1}}\right)\left(\sum_{i=1}^{n} b_i\right)^{p-1} \geq \left(\sum_{i=1}^{n} a_i\right)^p,$$

or

$$\sum_{i=1}^{n}\frac{a_i^p}{b_i^{p-1}} \geq \frac{\left(\sum\limits_{i=1}^{n} a_i\right)^p}{\left(\sum\limits_{i=1}^{n} b_i\right)^{p-1}}.$$

11.13. Consider the function $f(x) = \frac{1}{x}$ in $(0, +\infty)$.
Taking $x_i = a_1 + \cdots + a_{i-1} - a_i + \cdots + a_n$, $\alpha_i = \frac{a_i}{a_1 + \cdots + a_n}$, where $i = 1, \ldots, n$ (11.7), we obtain

$$\frac{1}{a_1 + \cdots + a_n} \cdot \sum_{i=1}^{n} \frac{a_i}{a_1 + \cdots + a_{i-1} - a_i + \cdots + a_n} \geq \frac{a_1 + \cdots + a_n}{\sum\limits_{i=1}^{n} a_i\left(a_1 + \cdots + a_{i-1} - a_i + \cdots + a_n\right)},$$

whence

$$\sum_{i=1}^{n} \frac{a_i}{a_1 + \cdots + a_{i-1} - a_i + \cdots + a_n} \geq \frac{(a_1 + \cdots + a_n)^2}{(a_1 + \cdots + a_n)^2 - 2(a_1^2 + \cdots + a_n^2)} \geq \frac{n}{n-2},$$

since
$n\left(a_1^2 + \cdots + a_n^2\right) \geq (a_1 + \cdots + a_n)^2$ (see Problem 2.2).

11.14. In Problem 11.2, taking $c_i = 1$, we obtain $n^{k-1}\sum\limits_{i=1}^{n} a_i^k \geq \left(\sum\limits_{i=1}^{n} a_i\right)^k$, or $\frac{\sum\limits_{i=1}^{n} a_i^k}{n} \geq \left(\frac{\sum\limits_{i=1}^{n} a_i}{n}\right)^k$. Now taking $k = \frac{m}{2}$, $a_i = d_i^2$, it follows that

$$\frac{\sum\limits_{i=1}^{n} d_i^m}{n} \geq \left(\frac{\sum\limits_{i=1}^{n} d_i^2}{n}\right)^{\frac{m}{2}}.$$

In a similar way, we deduce that $\frac{\sum\limits_{i=1}^{n} b_i^m}{n} \geq \left(\frac{\sum\limits_{i=1}^{n} b_i^2}{n}\right)^{\frac{m}{2}}$.

Multiplying the obtained inequalities term by term and using the inequality of Exercise 11.10,

we obtain $\frac{\sum_{i=1}^{n} b_i^m}{n} \cdot \frac{\sum_{i=1}^{n} d_i^m}{n} \geq \left(\frac{\sum_{i=1}^{n} b_i^2 \cdot \sum_{i=1}^{n} d_i^2}{n^2} \right)^{m/2} \geq \left(\frac{\sum_{i=1}^{n} b_i d_i}{n} \right)^m.$

11.15. Consider the function $f(x) = \frac{1}{x}$ in $(0, +\infty)$.
(a) Taking $\alpha_i = \frac{x_i}{x_1+x_2+x_3}$, $i = 1, 2, 3$, we obtain

$$\frac{x_1}{x_1+x_2+x_3} \cdot \frac{1}{x_2+x_3} + \frac{x_2}{x_1+x_2+x_3} \cdot \frac{1}{x_1+x_3} + \frac{x_3}{x_1+x_2+x_3} \cdot \frac{1}{x_1+x_2}$$
$$\geq \frac{1}{\frac{x_1}{x_1+x_2+x_3} \cdot (x_2+x_3) + \frac{x_2}{x_1+x_2+x_3}(x_1+x_3) + \frac{x_3}{x_1+x_2+x_3}(x_1+x_2)},$$

and hence.

$$\frac{x_1}{x_2+x_3} + \frac{x_2}{x_1+x_3} + \frac{x_3}{x_1+x_2} \geq \frac{(x_1+x_2+x_3)^2}{2(x_1x_2+x_2x_3+x_1x_3)}.$$

From the inequality $\frac{x_1^2+x_2^2+x_3^2}{3} \geq \left(\frac{x_1+x_2+x_3}{3}\right)^2$, or $x_1^2+x_2^2+x_3^2 \geq x_1x_2+x_2x_3+x_1x_3$, we have

$$\frac{(x_1+x_2+x_3)^2}{2(x_1x_2+x_2x_3+x_1x_3)} = \frac{1}{2} \cdot \frac{x_1^2+x_2^2+x_3^2}{x_1x_2+x_2x_3+x_1x_3} + 1 \geq \frac{3}{2},$$

and therefore $\frac{x_1}{x_2+x_3} + \frac{x_2}{x_1+x_3} + \frac{x_3}{x_1+x_2} \geq \frac{3}{2}$.
(b) $\alpha_i = \frac{x_i}{(x_1+\cdots+x_n)}$, $i = 1, \ldots, n$, and we have that

$$\frac{x_1}{x_1+\cdots+x_n} \cdot \frac{1}{x_2+x_n} + \frac{x_2}{x_1+\cdots+x_n} \cdot \frac{1}{x_1+x_3} + \cdots + \frac{x_n}{x_1+\cdots+x_n} \cdot \frac{1}{x_1+x_{n-1}}$$
$$\geq \frac{1}{\frac{x_1(x_2+x_n)+x_2(x_3+x_1)+\cdots+x_n(x_1+x_{n-1})}{x_1+\cdots+x_n}},$$

whence

$$\frac{x_1}{x_2+x_n} + \frac{x_2}{x_1+x_3} + \cdots + \frac{x_n}{x_1+x_{n-1}} \geq \frac{(x_1+\cdots+x_n)^2}{x_1x_2+x_1x_n+x_2x_3+x_2x_1+\cdots+x_n(x_1+x_{n-1})}$$
$$= \frac{(x_1+\cdots+x_n)^2}{2(x_1x_2+x_2x_3+\cdots+x_{n-1}x_n+x_nx_1)} \geq 2$$

(see the solution to Problem 7.6).

Hence, it follows that

$$\frac{x_1}{x_2 + x_n} + \frac{x_2}{x_1 + x_3} + \cdots + \frac{x_n}{x_1 + x_{n-1}} \geq 2.$$

11.16. Let $A = \frac{x}{y} + \frac{y}{z} + \frac{z}{x} - \frac{y}{x} - \frac{z}{y} - \frac{x}{z} \geq 0$. Consider the function $f(x) = \frac{1}{x}$ in $(0, +\infty)$.
Let us take $\alpha_1 = \frac{y+z-x}{y+z+x}$, $\quad \alpha_2 = \frac{y+x-z}{y+z+x}$, $\quad \alpha_3 = \frac{z+x-y}{y+z+x}$.
Hence, $\frac{1}{x+y+z}\left(\frac{y+z-x}{y} + \frac{x+y-z}{x} + \frac{z+x-y}{z}\right) \geq \frac{1}{\frac{x^2+y^2+z^2}{x+y+z}}$, or $3 - A \geq \frac{(x+y+z)^2}{x^2+y^2+z^2}$,
whence $A \leq 3 - \frac{(x+y+z)^2}{x^2+y^2+z^2} = 2\left(1 - \frac{xy+xz+yz}{x^2+y^2+z^2}\right)$.
One can easily prove that $xy+xz+yz > \frac{1}{2}(x^2 + y^2 + z^2)$, and thus it follows that $A < 1$.

11.17. (b) We have

$$\frac{a_1}{a_1 + \cdots + a_n} \cdot \frac{1}{a_2 + \cdots + a_n} + \frac{a_2}{a_1 + \cdots + a_n} \cdot \frac{1}{a_1 + a_3 + \cdots + a_n} + \cdots + \frac{a_n}{a_1 + \cdots + a_n} \cdot \frac{1}{a_1 + \cdots + a_{n-1}}$$
$$\geq \frac{1}{\frac{a_1(a_2+\cdots+a_n)+a_2(a_1+a_3+\cdots+a_n)+\cdots+a_n(a_1+\cdots+a_{n-1})}{a_1+\cdots+a_n}};$$

therefore,

$$\frac{a_1}{a_2 + \cdots + a_n} + \cdots + \frac{a_n}{a_1 + \cdots + a_{n-1}}$$
$$\geq \frac{(a_1 + \cdots + a_n)^2}{(a_1 + \cdots + a_n)^2 - \left(a_1^2 + \cdots + a_n^2\right)} \geq \frac{n}{n-1},$$

since $n\left(a_1^2 + \cdots + a_n^2\right) \geq (a_1 + \cdots + a_n)^2$.

11.18. We have

$$\frac{a}{a+b+c+d} \cdot \frac{1}{b+c} + \frac{b}{a+b+c+d} \cdot \frac{1}{c+d} + \frac{c}{a+b+c+d} \cdot \frac{1}{d+a}$$
$$+ \frac{d}{a+b+c+d} \cdot \frac{1}{a+b} \geq \frac{1}{\frac{a(b+c)+b(c+d)+c(d+a)+d(a+b)}{a+b+c+d}}.$$

Therefore,

$$\frac{a}{b+c} + \frac{b}{c+d} + \frac{c}{d+a} + \frac{d}{a+b} \geq \frac{(a+b+c+d)^2}{ab+ac+bc+bd+cd+ac+ad+db} \geq 2,$$

since $(a+b+c+d)^2 - 2(ab+ac+bc+bd+cd+ac+ad+bd) = (a-c)^2 + (b-d)^2 \geq 0$.

11.19. Consider the function $f(x) = \frac{1}{x}$ in $(0, +\infty)$.

Taking

$$\alpha_1 = \alpha_6 = \tfrac{a}{2(a+b+c)}, \quad \alpha_2 = \alpha_3 = \tfrac{b}{2(a+b+c)}, \quad \alpha_4 = \alpha_5 = \tfrac{c}{2(a+b+c)},$$
$$x_1 = \tfrac{c}{a}, \quad x_2 = \tfrac{c}{b}, \quad x_3 = \tfrac{a}{b}, \quad x_4 = \tfrac{a}{c}, \quad x_5 = \tfrac{b}{c}, \quad x_6 = \tfrac{b}{a},$$

we obtain

$$\frac{a}{2(a+b+c)} \cdot \frac{1}{\frac{c}{a}} + \frac{b}{2(a+b+c)} \cdot \frac{1}{\frac{c}{b}} + \frac{b}{2(a+b+c)} \cdot \frac{1}{\frac{a}{b}} + \frac{c}{2(a+b+c)}$$
$$\cdot \frac{1}{\frac{a}{c}} + \frac{c}{2(a+b+c)} \cdot \frac{1}{\frac{b}{c}} + \frac{a}{2(a+b+c)} \cdot \frac{1}{\frac{b}{a}} \geq \frac{1}{\frac{2c+2a+2b}{2(a+b+c)}} = 1.$$

Similarly, one can obtain the second inequality:

$$\begin{aligned}
\frac{a^3}{bc} + \frac{b^3}{ca} + \frac{c^3}{ab} &= \frac{1}{2}\left(a^2 \cdot \frac{1}{\frac{bc}{a}} + b^2 \cdot \frac{1}{\frac{ac}{b}}\right) + \frac{1}{2}\left(a^2 \cdot \frac{1}{\frac{bc}{a}} + c^2 \cdot \frac{1}{\frac{ab}{c}}\right) + \frac{1}{2}\left(c^2 \cdot \frac{1}{\frac{ab}{c}} + b^2 \cdot \frac{1}{\frac{ac}{b}}\right) \\
&\geq \frac{1}{2} \cdot \frac{\left(a^2+b^2\right)^2}{2abc} + \frac{1}{2} \cdot \frac{\left(a^2+c^2\right)^2}{2abc} + \frac{1}{2} \cdot \frac{\left(c^2+b^2\right)^2}{2abc} \\
&= \frac{a^2+b^2}{2c} \cdot \frac{a^2+b^2}{2ab} + \frac{a^2+c^2}{2b} \cdot \frac{a^2+c^2}{2ac} + \frac{b^2+c^2}{2a} \cdot \frac{b^2+c^2}{2bc} \\
&\geq \frac{a^2+b^2}{2c} + \frac{a^2+c^2}{2b} + \frac{b^2+c^2}{2a}.
\end{aligned}$$

11.20. Consider the function $f(x) = \frac{1}{x}$ in $(0, +\infty)$. Taking

$$\alpha_1 = \tfrac{a^2}{a^2+b^2+c^2},\ \alpha_2 = \tfrac{b^2}{a^2+b^2+c^2},\ \alpha_3 = \tfrac{c^2}{a^2+b^2+c^2},$$
$$x_1 = \tfrac{a^2+ab+b^2}{a},\ x_2 = \tfrac{b^2+bc+c^2}{b},\ x_3 = \tfrac{c^2+ca+a^2}{c},$$

we obtain

$$\frac{a^2}{a^2+b^2+c^2} \cdot \frac{1}{\frac{a^2+ab+b^2}{a}} + \frac{b^2}{a^2+b^2+c^2} \cdot \frac{1}{\frac{b^2+bc+c^2}{b}} + \frac{c^2}{a^2+b^2+c^2} \cdot \frac{1}{\frac{c^2+ac+a^2}{c}}$$
$$\geq \frac{1}{\frac{a^3+a^2b+ab^2+b^3+b^2c+bc^2+c^3+c^2a+a^2c}{a^2+b^2+c^2}},$$

or

$$\frac{a^3}{a^2+ab+b^2} + \frac{b^3}{b^2+bc+c^2} + \frac{c^3}{c^2+ac+a^2} \geq \frac{\left(a^2+b^2+c^2\right)^2}{(a+b+c)\left(a^2+b^2+c^2\right)} \geq \frac{a+b+c}{3}.$$

11.21. Consider the function $f(x) = -\ln x$ on $(0, +\infty)$. It is a convex function, and therefore,

$$-\ln\left(\frac{a}{a+b}\cdot\frac{c}{a}+\frac{b}{a+b}\cdot\frac{d}{b}\right)\leq -\frac{a}{a+b}\ln\frac{c}{a}-\frac{b}{a+b}\ln\frac{d}{b},$$

or

$$\left(\frac{a+b}{c+d}\right)^{a+b}\leq\left(\frac{a}{c}\right)^{a}\cdot\left(\frac{b}{d}\right)^{b}.$$

Remark If $a_1 > 0, \ldots, a_n > 0, b_1 > 0, \ldots, b_n > 0$, then

$$\left(\frac{a_1+\cdots+a_n}{b_1+\cdots+b_n}\right)^{a_1+\cdots+a_n}\leq\left(\frac{a_1}{b_1}\right)^{a_1}\cdot\ldots\cdot\left(\frac{a_n}{b_n}\right)^{a_n}.$$

11.22. Consider the function $f(x) = \frac{1}{x}$ on $(0, +\infty)$. Taking $\alpha_i = \frac{x_{i-1}}{x_0+x_1+\cdots+x_{n-1}}$, $i = 1, \ldots, n$, we obtain

$$\frac{x_0^2}{x_1}+\frac{x_1^2}{x_2}+\cdots+\frac{x_{n-1}^2}{x_n}=\left(\frac{x_0}{x_0+x_1+\cdots+x_{n-1}}\cdot\frac{1}{\frac{x_1}{x_0}}+\cdots+\frac{x_{n-1}}{x_0+x_1+\cdots+x_{n-1}}\cdot\frac{1}{\frac{x_n}{x_{n-1}}}\right)\cdot(x_0+x_1+\cdots+x_{n-1})$$
$$\geq\frac{(x_0+x_1+\cdots+x_{n-1})^2}{x_1+x_2+\cdots+x_{n-1}+x_n}=k_n.$$

Now let us prove that there exists a positive integer n_0, such that for $n > n_0$ we have $k_n \geq 3.999$.
Indeed, if $n > \frac{3,999}{0,001} + 1$, we have

$$(x_0+x_1+\cdots+x_{n-1})^2-3,999(x_1+x_2+\cdots+x_{n-1}+x_n)\geq 0,001(n-1)x_n-3,999x_n\geq 0,$$

and one can take $n_0 = 4000$.

11.23. Consider the function $f(x) = \frac{1}{x}$ in $(0, +\infty)$. Taking $\alpha_1 = \frac{BC}{AB+BC+AC}$, $\alpha_2 = \frac{AC}{AB+BC+AC}$, $\alpha_3 = \frac{AB}{AB+BC+AC}$, $x_1 = MA_1$, $x_2 = MB_1$, $x_3 = MC_1$, we obtain

$$\frac{BC}{MA_1}+\frac{CA}{MB_1}+\frac{AB}{MC_1}=\left(\frac{BC}{BC+AC+AB}\cdot\frac{1}{MA_1}+\frac{CA}{BC+AC+AB}\cdot\frac{1}{MB_1}+\frac{AB}{BC+AC+AB}\cdot\frac{1}{MC_1}\right)$$
$$(BC+AC+AB)\geq\frac{(BC+AC+AB)^2}{BC\cdot MA_1+AC\cdot MB_1+AB\cdot MC_1}=\frac{4p^2}{2S}=\frac{2p^2}{S},$$

and thus the smallest value of the expression $\frac{BC}{MA_1}+\frac{CA}{MB_1}+\frac{AB}{MC_1}$ is equal to $\frac{2p^2}{S}$, and equality holds when $MA_1 = MB_1 = MC_1$.

11.24. Without loss of generality one can assume that $a_1 \geq a_2 \geq \cdots \geq a_n$. According to Problem 14.10a, we have

$$\sum_{i=1}^{n}a_ib_i=b_1(a_1-a_2)+(b_1+b_2)(a_2-a_3)+\cdots$$
$$+(b_1+\cdots+b_{n-1})(a_{n-1}-a_n)+(b_1+\cdots+b_n)a_n.$$

It follows that

$$\sum_{i=1}^{n} a_i b_i \geq a_1^m (a_1 - a_2) + (a_1 + a_2)^m (a_2 - a_3) + \cdots$$
$$+ (a_1 + \cdots + a_{n-1})^m (a_{n-1} - a_n) + (a_1 + \cdots + a_n)^m a_n$$
$$= A.$$

Since the function $f(x) = x^m$ is convex on $[0, +\infty)$, we have

$$\geq \left(\frac{a_1(a_1 - a_2)}{a_1} + \frac{(a_1 + a_2)(a_2 - a_3)}{a_1} + \cdots + \frac{(a_1 + \cdots + a_{n-1})(a_{n-1} - a_n)}{a_1} + \frac{(a_1 + \cdots + a_n)a_n}{a_1}\right)^m \left(\frac{a_1^2 + a_2^2 + \cdots + a_n^2}{a_1}\right)^m.$$

Hence, we deduce that

$$\left(\sum_{i=1}^{n} a_i b_i\right)^2 \geq A^2 \geq \frac{\left(a_1^2 + a_2^2 + \cdots + a_n^2\right)^{2m}}{a_1^{2m-2}} = \left(\frac{a_1^2 + a_2^2 + \cdots + a_n^2}{a_1^2}\right)^{m-1} \left(a_1^2 + a_2^2 + \cdots + a_n^2\right)^{m+1} \geq$$

$$\geq \left(a_1^2 + a_2^2 + \cdots + a_n^2\right)^{m+1}.$$

Thus, it follows that $\left(\sum_{i=1}^{n} a_i b_i\right)^2 \geq \left(a_1^2 + a_2^2 + \cdots + a_n^2\right)^{m+1}$.

11.25. (a) Consider the function $f(x) = x^{1-p}$ in $(0, +\infty)$. Since $f''(x) = p(p-1)x^{-p-1} \geq 0$, it follows that for the function $f(x)$ one has inequality (11.7). Taking $\alpha_i = \frac{a_i}{a_1 + \cdots + a_n}$, $\quad i = 1, \ldots, n$, we obtain

$$\frac{a_1}{a_1 + \cdots + a_n} \cdot \left(\frac{b_1}{a_1}\right)^{1-p} + \cdots + \frac{a_n}{a_1 + \cdots + a_n} \cdot \left(\frac{b_n}{a_n}\right)^{1-p}$$
$$\geq \left(\frac{a_1}{a_1 + \cdots + a_n} \cdot \frac{b_1}{a_1} + \cdots + \frac{a_n}{a_1 + \cdots + a_n} \cdot \frac{b_n}{a_n}\right)^{1-p},$$

or

$$\frac{a_1^p}{b_1^{p-1}} + \cdots + \frac{a_n^p}{b_n^{p-1}} \geq \frac{(a_1 + \cdots + a_n)^p}{(b_1 + \cdots + b_n)^{p-1}}.$$

(b) The proof is similar to Problem 11.25a, using $f''(x) < 0$.
(c) Defining $B_i = \frac{b_1 + \cdots + b_i}{a_1 + \cdots + a_i}$, $i = 1, \ldots, n$, then if $i = 2, \ldots n$, we obtain

$b_i = B_i(a_1 + \cdots + a_i) - B_{i-1}(a_1 + \cdots + a_{i-1})$, and therefore,

$$\frac{b_i}{B_i^{p+1}} = \frac{a_1 + \cdots + a_i}{B_i^p} - (a_1 + \cdots + a_{i-1}) \cdot \frac{B_{i-1}}{B_i^{p+1}}. \tag{11.11}$$

By to the Problem 9.13, we have

$$\frac{\left(\frac{B_{i-1}^{\frac{p}{p+1}}}{B_i^p}\right)^{\frac{p+1}{p}}}{\frac{p+1}{p}} + \frac{\left(\frac{1}{B_{i-1}^{\frac{p}{p+1}}}\right)^{p+1}}{p+1} \geq \frac{1}{B_i^p},$$

and thus it follows that

$$\frac{B_{i-1}}{B_i^{p+1}} \geq \frac{p+1}{pB_i^p} - \frac{1}{pB_i^{p-1}}. \tag{11.12}$$

From (11.11) and (11.12) we obtain $\frac{p+1}{p} \cdot \frac{a_i}{B_i^p} - \frac{b_i}{B_i^{p+1}} \geq \frac{a_1 + \cdots + a_i}{pB_1^p} - \frac{a_1 + \cdots + a_{i-1}}{pB_{1-1}^p}$, $i = 2, \ldots, n$, and $\frac{p+1}{p} \cdot \frac{a_1}{B_1^p} - \frac{b_1}{B_1^{p+1}} = \frac{a_1}{pB_1^p} \geq \frac{a_1}{pB_1^p}$, and summing all n inequalities, we obtain

$$\frac{p+1}{p} \cdot \left(\frac{a_1}{B_1^p} + \cdots + \frac{a_n}{B_n^p}\right) - \left(\frac{b_1}{B_1^{p+1}} + \cdots + \frac{b_n}{B_n^{p+1}}\right) = \frac{a_1}{pB_1^p} \geq \frac{a_1 + \cdots + a_n}{pB_n^p}. \tag{11.13}$$

Therefore, from (11.13) it follows that

$$\frac{p+1}{p} \cdot \left(\frac{a_1}{B_1^p} + \cdots + \frac{a_n}{B_n^p}\right) > \frac{b_1}{B_1^{p+1}} + \cdots + \frac{b_n}{B_n^{p+1}}, \tag{11.14}$$

and according to Problem 11.25a, it follows that

$$\frac{b_1}{B_1^{p+1}} + \cdots + \frac{b_n}{B_n^{p+1}} = \frac{\left(\frac{a_1}{B_1^p}\right)^{\frac{p+!}{p}}}{\left(\frac{a_1^{p+1}}{b_1^p}\right)^{\frac{1}{p}}} + \cdots + \frac{\left(\frac{a_n}{B_n^p}\right)^{\frac{p+!}{p}}}{\left(\frac{a_n^{p+1}}{b_n^p}\right)^{\frac{1}{p}}} \geq \frac{\left(\frac{a_1}{B_1^p} + \cdots + \frac{a_n}{B_n^p}\right)^{\frac{p+!}{p}}}{\left(\frac{a_1^{p+1}}{b_1^p} + \cdots + \frac{a_n^{p+1}}{b_n^p}\right)^{\frac{1}{p}}}. \tag{11.15}$$

From (11.14) and (11.15) we obtain

$$a_1 \cdot \left(\frac{a_1}{b_1}\right)^p + \cdots + a_n \cdot \left(\frac{a_1 + \cdots + a_n}{b_1 + \cdots + b_n}\right)^p < \left(\frac{p+1}{p}\right)^p \left(\frac{a_1^{p+1}}{b_1^p} + \cdots + \frac{a_n^{p+1}}{b_n^p}\right). \tag{11.16}$$

From (11.13), it follows that

$$a_1 \cdot \left(\frac{a_1}{b_1}\right)^p + \cdots + a_n \cdot \left(\frac{a_1 + \cdots + a_n}{b_1 + \cdots + b_n}\right)^p$$
$$\geq \frac{p}{p+1}\left(b_1 \cdot \left(\frac{a_1}{b_1}\right)^{p+1} + \cdots + b_n \left(\frac{a_1 + \cdots + a_n}{b_1 + \cdots + b_n}\right)^{p+1}\right)$$
$$+ \frac{1}{p+1} \cdot \frac{(a_1 + \cdots + a_n)^{p+1}}{(b_1 + \cdots + b_n)^p}. \tag{11.17}$$

Remark

1. If $p < -1$ and $a_1,\ldots, a_n, b_1,\ldots, b_n$ are arbitrary positive numbers, then
$$a_1 \cdot \left(\frac{a_1}{b_1}\right)^p + \cdots + a_n \cdot \left(\frac{a_1 + \cdots + a_n}{b_1 + \cdots + b_n}\right)^p < \left(\frac{p+1}{p}\right)^p \left(\frac{a_1^{p+1}}{b_1^p} + \cdots + \frac{a_n^{p+1}}{b_n^p}\right).$$
2. If $p > 1$ and $a_1,\ldots, a_n$ are arbitrary positive numbers, then
$$a_1^p + \cdots + \left(\frac{a_1 + \cdots + a_n}{n}\right)^p + \frac{n}{p-1} \cdot \left(\frac{a_1 + \cdots + a_n}{n}\right)^p$$
$$< \left(\frac{p}{p-1}\right)^p \cdot (a_1^p + \cdots + a_n^p).$$
3. If $p > 1$ and $b_1,\ldots, b_n$ are arbitrary positive numbers, then
$$b_1^{\frac{1}{1}} + \cdots + (b_1 \cdot \ldots \cdot b_n)^{\frac{1}{n}} < \left(\frac{p}{p-1}\right)^p \cdot (b_1 + \cdots + b_n).$$
4. If $b_1,\ldots, b_n$ are arbitrary positive numbers, then
$$b_1^{\frac{1}{1}} + \cdots + (b_1 \cdot \ldots \cdot b_n)^{\frac{1}{n}} < e \cdot (b_1 + \cdots + b_n).$$

11.26. (a) Consider the numbers $x_i'' = \frac{x_i + x_i'}{2}$, $i = 1, \ldots, n$. Note that $\min(x_i, x_i') \leq x_i'' \leq \max(x_i, x_i')$ and $\min(x_i + x_j, x_i' + x_j') \leq x_i'' + x_j'' \leq \max(x_i + x_j, x_i' + x_j')$, for all $i, j \in \{1, \ldots, n\}$. Therefore, $x_i'' \in I$ and $x_i'' + x_j'' \in G$.
If we prove that

$$f\left(\frac{x_1'' + \cdots + x_n''}{n}\right) \leq \frac{f(x_1'') + \cdots + f(x_n'')}{n}, \tag{11.18}$$

then since $f(x_i'') \leq \frac{f(x_i) + f(x_i')}{2}$, $i = 1, \ldots, n$, and $f(x_1) + \cdots + f(x_n) = f(x_1') + \cdots + f(x_n')$, $\frac{x_1'' + \cdots + x_n''}{n} = \frac{x_1 + \cdots + x_n}{n}$, we have $f\left(\frac{x_1 + \cdots + x_n}{n}\right) \leq \frac{f(x_1) + \cdots + f(x_n)}{n}$.
Assume that inequality (11.18) holds for m numbers, and let us prove that it holds for $2m$ numbers.

Indeed, we have

$$f\left(\frac{x_1''+x_2''+\cdots+x_{2m-1}''+x_{2m}''}{2m}\right)=f\left(\frac{\frac{x_1''+x_2''}{2}+\cdots+\frac{x_{2m-1}''+x_{2m}''}{2}}{m}\right)$$
$$\leq\frac{f\left(\frac{x_1''+x_2''}{2}\right)+\cdots+f\left(\frac{x_{2m-1}''+x_{2m}''}{2}\right)}{m}$$
$$\leq\frac{\frac{f(x_1'')+f(x_2'')}{2}+\cdots+\frac{f(x_{2m-1}'')+f(x_{2m}'')}{2}}{m}$$
$$=\frac{f(x_1'')+f(x_2'')+\ldots+f(x_{2m-1}'')+f(x_{2m}'')}{2m}.$$

Now let us prove that if inequality (11.18) holds for k ($k>2$) numbers, then it holds for $k-1$ numbers.
Let $x_1'',\ldots,x_{k-1}''\in I$ and $x_i''+x_j''\in G$, where $i,j\in\{1,\ldots,k-1\}$.
Note that $\min(x_1'',\ldots,x_{k-1}'')\leq\frac{x_1''+\cdots+x_{k-1}''}{k-1}=x''\leq\max(x_1'',\ldots,x_{k-1}'')$, and therefore, $x''\in I$, $2x'', x''+x_i''\in G$, for all $i=1,\ldots,k-1$, whence $f\left(\frac{x_1''+\cdots+x_{k-1}''+x''}{k}\right)\leq\frac{f(x_1'')+\cdots+f(x_{k-1}'')+f(x'')}{k}$. Thus it follows that $f\left(\frac{x_1''+\cdots+x_{k-1}''}{k-1}\right)\leq\frac{f(x_1'')+\cdots+f(x_{k-1}'')}{k-1}$. Since inequality (11.18) for two numbers holds, from this statement it follows that it holds for n numbers.
Remark If $G=R$, then we obtain Jensen's inequality.
(b) Let us prove that if $n\geq 2$, $y_i>0$, $i=1,\ldots,n$, and $y_1\cdot\ldots\cdot y_n=\frac{1}{(n-1)^n}$, then

$$\frac{1}{1+y_1}+\cdots+\frac{1}{1+y_n}\leq n-1. \tag{11.19}$$

Indeed, let us consider the function $f(x)=\frac{1}{1+3^x}$ on R, and $G=(-\infty,0]$. Note that if $x_1+x_2\in G$, then $-\frac{2}{1+3^{\frac{x_1+x_2}{2}}}-\left(-\frac{1}{1+3^{x_1}}-\frac{1}{1+3^{x_2}}\right)=\frac{\left(3^{\frac{x_1+x_2}{2}}-1\right)\left(3^{\frac{x_1}{2}}-3^{\frac{x_2}{2}}\right)^2}{(1+3^{x_1})(1+3^{x_2})\left(1+3^{\frac{x_1+x_2}{2}}\right)}\leq 0$, and hence

$$f\left(\frac{x_1+x_2}{2}\right)\leq\frac{f(x_1)+f(x_2)}{2}.$$

If $y_iy_j\leq 1$ for all i and j, then $x_i+x_j\in G$, where $y_i=3^{x_i}$, $i=1,\ldots,n$, and thus for $x_1'=x_2,\ \ldots\ ,\ x_{n-1}'=x_n,\ x_n'=x_1$, the condition of Problem 11.25a holds. Therefore,

$$\frac{1}{1+y_1}+\cdots+\frac{1}{1+y_n}\leq\frac{n}{1+\sqrt[n]{y_1\cdot\ldots\cdot y_n}}=n-1.$$

If there exist numbers i and j such that $y_i y_j > 1$, then

$$\frac{1}{1+y_1}+\cdots+\frac{1}{1+y_n} \leq \frac{1}{1+y_i}+\frac{1}{1+y_j}+\sum_{k\neq i, k\neq j}\frac{1}{1+y_k} < 1+\sum_{k\neq i, k\neq j}\frac{1}{1+y_k} < 1+(n-2) = n-1.$$

For $y_i = \frac{a_{i+1}}{(n-1)a_i}$, $i = 1, \ldots, n$, $a_{n+1} = a_1$, from inequality (11.19), it follows that

$$\frac{a_1}{(n-1)a_1+a_2}+\cdots+\frac{a_n}{(n-1)a_n+a_1} \leq 1.$$

(c) Consider the function $f(x) = \frac{1}{\sqrt{1+3^x}}$ in R, and $G = [2, +\infty)$. Note that if $x_1 + x_2 \in G$, then

$$\begin{aligned}\left(\sqrt{1+3^{x_1}}+\sqrt{1+3^{x_2}}\right)^2 &= 2+3^{x_1}+3^{x_2}+2\sqrt{(1+3^{x_1})(1+3^{x_2})}\\ &\geq 2+3^{x_1}+3^{x_2}+2\left(1+3^{\frac{x_1+x_2}{2}}\right),\end{aligned}$$

whence

$$\begin{aligned}&\left(\sqrt{1+3^{x_1}}+\sqrt{1+3^{x_2}}\right)^2\left(1+3^{\frac{x_1+x_2}{2}}\right)\\ &\geq \left(4+3^{x_1}+3^{x_2}+2\cdot 3^{\frac{x_1+x_2}{2}}\right)\left(1+3^{\frac{x_1+x_2}{2}}\right)\\ &\geq 4(1+3^{x_1})(1+3^{x_2}),\end{aligned}$$

since

$$\left(4+3^{x_1}+3^{x_2}+2\cdot 3^{\frac{x_1+x_2}{2}}\right)\left(1+3^{\frac{x_1+x_2}{2}}\right)-4(1+3^{x_1})(1+3^{x_2}) = \left(3^{\frac{x_1+x_2}{2}}-3\right)\left(3^{\frac{x_1}{2}}-3^{\frac{x_2}{2}}\right)^2 \geq 0.$$

It follows that

$$f\left(\frac{x_1+x_2}{2}\right) \leq \frac{f(x_1)+f(x_2)}{2}.$$

If $y_i y_j \geq 9$ for all i and j, then $x_i + x_j \in G$, where $y_i = 3^{x_i}$, $i = 1, \ldots, n$, and therefore for $x_1' = x_2, \ldots, x_{n-1}' = x_n$, $x_n' = x_1$, the conditions of Problem 11.25a hold.
Hence

$$\frac{1}{\sqrt{1+y_1}}+\cdots+\frac{1}{\sqrt{1+y_n}} \geq \frac{n}{1+\sqrt[n]{y_1\cdot\ldots\cdot y_n}} = \frac{n}{\sqrt{1+\lambda}}.$$

If there exist numbers i and j such that $y_i y_j < 9$, then

$$\frac{1}{\sqrt{1+y_1}}+\cdots+\frac{1}{\sqrt{1+y_n}} \geq \frac{1}{\sqrt{1+y_1}}+\frac{1}{\sqrt{1+y_2}} > 1,$$

since

$$\left(\sqrt{1+y_1}+\sqrt{1+y_2}\right)^2 \geq 4+y_1+y_2+2\sqrt{y_1y_2} > (1+y_1)(1+y_2).$$

(d) Consider the function $f(x) = \ln\frac{1-x}{x}$ on $(0,1)$, and $G = (0,1]$. Note that if $x_1, x_2 \in (0,1)$ and $x_1+x_2 \in G$, then $f\left(\frac{x_1+x_2}{2}\right) = \ln\frac{1-\frac{x_1+x_2}{2}}{\frac{x_1+x_2}{2}} \leq \frac{\ln\frac{1-x_1}{x_1}+\ln\frac{1-x_2}{x_2}}{2} = \frac{f(x_1)+f(x_2)}{2}$, since $\left(\frac{2-(x_1+x_2)}{x_1+x_2}\right)^2 \leq \frac{(1-x_1)(1-x_2)}{x_1x_2}$, or $(x_1-x_2)^2(1-x_1-x_2) > 0$, whence

$$f\left(\frac{x_1+x_2}{2}\right) \leq \frac{f(x_1)+f(x_2)}{2}.$$

Thus, from Problem 11.25a, it follows that

$$f\left(\frac{x_1+\cdots+x_n}{n}\right) \leq \frac{f(x_1)+\cdots+f(x_n)}{n},$$

or

$$\frac{(1-x_1)\cdot\ldots\cdot(1-x_n)}{x_1\cdot\ldots\cdot x_n} \geq \left(\frac{1-\frac{x_1+\cdots+x_n}{n}}{\frac{x_1+\cdots+x_n}{n}}\right)^n.$$

11.27. Without loss of generality one can assume that $x \leq y \leq z$. Let us consider three cases.

(a) If $x \leq -\frac{1}{\sqrt{3}}$, then $y+z \geq 1+\frac{1}{\sqrt{3}} > \frac{3}{2}$. Therefore, $z > \frac{3}{4}$. It follows that $\frac{1}{1+x^2}+\frac{1}{1+y^2}+\frac{1}{1+z^2} < 1+1+\frac{16}{25} < \frac{27}{10}$.
(b) If $y \geq \frac{1}{\sqrt{3}}$, then $z \geq \frac{1}{\sqrt{3}}$, whence, $\frac{1}{1+x^2}+\frac{1}{1+y^2}+\frac{1}{1+z^2} \leq 1+\frac{3}{4}+\frac{3}{4} < \frac{27}{10}$.
(c) If $x, y \in \left(-\frac{1}{\sqrt{3}}, \frac{1}{\sqrt{3}}\right)$, then consider the function $f(t) = \frac{1}{1+t^2}$ in $\left(-\frac{1}{\sqrt{3}}, \frac{1}{\sqrt{3}}\right)$.

Note that $f''(t) = \frac{2(1+t^2)(3t^2-1)}{(1+t^2)^2} < 0$, and thus by Jensen's inequality, it follows that $\frac{1}{1+x^2}+\frac{1}{1+y^2} \leq \frac{2}{1+\left(\frac{x+y}{2}\right)^2} = \frac{2}{1+\left(\frac{1-z}{2}\right)^2}$. Therefore, $\frac{1}{1+x^2}+\frac{1}{1+y^2}+\frac{1}{1+z^2} \leq \frac{2}{1+\left(\frac{1-z}{2}\right)^2}+\frac{1}{1+z^2} \leq \frac{27}{10}$, since the last inequality is equivalent to the inequality $(3z-1)^2(3z^2-4z+5) \geq 0$.

11.28. Let us prove that if $\alpha \geq 2$ and $a > 0,\ b > 0$, then

$$\frac{1}{(1+a)^\alpha}+\frac{1}{(1+b)^\alpha} \geq \frac{2^{2-\alpha}}{1+(ab)^{\frac{\alpha}{2}}} \tag{11.20}$$

On setting $x = \frac{1}{1+a},\ y = \frac{1}{1+b}$, we see that inequality (11.20) can be rewritten as

$$\left(x^{\frac{\alpha}{2}}y^{-\frac{\alpha}{2}}+x^{-\frac{\alpha}{2}}y^{\frac{\alpha}{2}}\right)\left(x^{\frac{\alpha}{2}}y^{\frac{\alpha}{2}}+(1-x)^{\frac{\alpha}{2}}(1-y)^{\frac{\alpha}{2}}\right)\geq 2^{2-\alpha}.$$

Without loss of generality one can assume that $y^{\frac{\alpha}{2}}(1-y)^{\frac{\alpha}{2}}\geq x^{\frac{\alpha}{2}}(1-x)^{\frac{\alpha}{2}}$. Thus by to the Cauchy–Bunyakovsky–Schwarz inequality, it follows that

$$\begin{aligned}&\left(x^{\frac{\alpha}{2}}y^{-\frac{\alpha}{2}}+x^{-\frac{\alpha}{2}}y^{\frac{\alpha}{2}}\right)\left(x^{\frac{\alpha}{2}}y^{\frac{\alpha}{2}}+(1-x)^{\frac{\alpha}{2}}(1-y)^{\frac{\alpha}{2}}\right)\\&\geq\left(x^{\frac{\alpha}{2}}+\sqrt{y^{\frac{\alpha}{2}}(1-y)^{\frac{\alpha}{2}}x^{-\frac{\alpha}{2}}(1-x)^{\frac{\alpha}{2}}}\right)^2\\&\geq\left(x^{\frac{\alpha}{2}}+(1-x)^{\frac{\alpha}{2}}\right)^2,\end{aligned}$$

and by Jensen's inequality, it follows that $\left(x^{\frac{\alpha}{2}}+(1-x)^{\frac{\alpha}{2}}\right)^2\geq\left(2\left(\frac{x+(1-x)}{2}\right)^{\frac{\alpha}{2}}\right)^2=2^{2-\alpha}$.
Therefore, $(x^{\frac{\alpha}{2}}y^{-\frac{\alpha}{2}}+x^{-\frac{\alpha}{2}}y^{\frac{\alpha}{2}})(x^{\frac{\alpha}{2}}y^{\frac{\alpha}{2}}+(1-x)^{\frac{\alpha}{2}}(1-y)^{\frac{\alpha}{2}})\geq 2^{2-\alpha}$.
Note that

$$\begin{aligned}\frac{1}{(1+a)^\alpha}+\frac{1}{(1+b)^\alpha}+\frac{1}{(1+c)^\alpha}+\frac{1}{(1+d)^\alpha}&\geq\frac{2^{2-\alpha}}{1+(ab)^{\frac{\alpha}{2}}}+\frac{2^{2-\alpha}}{1+(cd)^{\frac{\alpha}{2}}}\\&=\frac{2^{2-\alpha}}{1+(ab)^{\frac{\alpha}{2}}}+\frac{2^{2-\alpha}}{1+\left(\frac{1}{ab}\right)^{\frac{\alpha}{2}}}=2^{2-\alpha}.\end{aligned}$$

Problems for Independent Study

1. Prove that if α, β, γ are the angles of some triangle, then

   (a) $\cos\alpha+\cos\beta+\cos\gamma\leq\frac{3}{2}$,
   (b) $\frac{1}{\sin^2\alpha/2}+\frac{1}{\sin^2\beta/2}+\frac{1}{\sin^2\gamma/2}\geq 12$,
   (c) $\frac{1}{\cos\alpha/2}+\frac{1}{\cos\beta/2}+\frac{1}{\cos\gamma/2}\geq 2\sqrt{3}$,
   (d) $\frac{1}{\sin\alpha/2}+\frac{1}{\sin\beta/2}+\frac{1}{\sin\gamma/2}\geq 6$.

2. Prove that if α, β, γ are the angles of some acute triangle, then

   (a) $\frac{\sqrt{1+8\cos^2\alpha}}{\sin\alpha}+\frac{\sqrt{1+8\cos^2\beta}}{\sin\beta}+\frac{\sqrt{1+8\cos^2\gamma}}{\sin\gamma}\geq 6$,
   (b) $\frac{1}{\cos\alpha}+\frac{1}{\cos\beta}+\frac{1}{\cos\gamma}\geq 6$.

3. Prove that if for a convex quadrilateral $ABCD$ one has $\sin\frac{\angle A}{2}\sin\frac{\angle B}{2}\sin\frac{\angle C}{2}\sin\frac{\angle D}{2}=\frac{1}{4}$, then $ABCD$ is a rectangle.
4. Prove that $\frac{x_1}{x_2+x_3}+\frac{x_2}{x_3+x_4}+\cdots+\frac{x_{n-1}}{x_n+x_1}+\frac{x_n}{x_1+x_2}\geq\frac{n}{2}$, where $3\leq n\leq 6$ and $x_i>0,\ i=1,\cdots,n$.
5. Prove that among all convex n-gons inscribed in a given circle, the one with greatest area is the regular n-gon.

6. Prove that among all convex n-gons inscribed in a given circle, the one with the greatest perimeter is the regular n-gon.
7. Prove that $\frac{a_1}{1+xa_1}+\frac{a_2}{1+xa_2}+\cdots+\frac{a_n}{1+xa_n} \leq \frac{n}{n+x}$, where $x > 0,\ a_1 \geq 0,\ \cdots,\ a_n \geq 0$, and $a_1+\cdots+a_n = 1$.
8. Prove that $\sin\alpha\ \sin\beta + \sin\beta\ \sin\gamma + \sin\gamma\ \sin\delta + \sin\delta\ \sin\alpha \leq 2$, where $\alpha > 0,\ \beta > 0,\ \gamma > 0,\ \delta > 0$, and $\alpha+\beta+\gamma+\delta = \pi$.
9. Let $x_1, x_2, \cdots, x_n$ be arbitrary real numbers.

   (a) Prove that, if $\lambda \geq 2$ or $\lambda < 0$, then it holds true the following inequality

$$\sum (\pm x_1 \pm x_2 \pm \cdots \pm x_n)^{\lambda} \geq 2^n \left(\sum_{i=1}^{n} x_i^2\right)^{\lambda/2}.$$

   (b) Prove that if $0 < \lambda < 2$, then one has the following inequality:

   $\sum |\pm x_1 \pm x_2 \pm \cdots \pm x_n|^{\lambda} \leq 2^n \left(\sum\limits_{i=1}^{n} x_i^2\right)^{\frac{\lambda}{2}}$ (on the left-hand side, the summation is over all combinations of plus and minus signs).
10. Prove that $\left(\frac{a^2-b^2}{2}\right)^2 \geq \sqrt{\frac{a^2+b^2}{2}} - \frac{a+b}{2}$, where $a \geq \frac{1}{2},\ b \geq \frac{1}{2}$.
11. Prove that $\sum\limits_{i=1}^{n} \left(x_i + \frac{1}{x_i}\right)^a \geq \frac{(n^2+1)^a}{n^{a-1}}$, where $a > 0,\ x_1 > 0,\ \ldots,\ x_n > 0$, and $\sum\limits_{i=1}^{n} x_i = 1$.
12. (a) Prove that $a_1x_1 + \cdots + a_nx_n \geq x_1^{a_1} \cdot \ldots \cdot x_n^{a_n}$, where $a_1 + \cdots + a_n = 1$ and $x_i > 0,\ a_i \geq 0,\ i = 1, \ldots, n$,

    (b) Let $a_{ij} \geq 0$ and $\sum\limits_{i=1}^{n} a_{ij} = 1,\ j = 1, 2, \ldots, n,\ \sum\limits_{j=1}^{n} a_{ij} = 1, i = 1, 2, \ldots, n$.

    Prove that

$$(a_{11}x_1 + a_{12}x_2 + \cdots + a_{1n}x_n)(a_{21}x_1 + a_{22}x_2 + \cdots + a_{2n}x_n) \cdot \ldots \cdot (a_{n1}x_1 + a_{n2}x_2 + \cdots + a_{nn}x_n) \geq x_1x_2 \cdot \ldots \cdot x_n,$$

    where $x_1 \geq 0,\ \cdots,\ x_n \geq 0$.

    (c) Prove that $(ux + vy + wz)(vx + wy + uz)(wx + uy + vz) \geq (y + z - x)(z + x - y)(x + y - z)$, where $u + v + w = 1,\ u \geq 0,\ v \geq 0,\ w \geq 0,\ x \geq 0,\ y \geq 0,\ z \geq 0$.
13. Prove that $\left(\frac{(1-x)(1-y)}{z}\right)^{\frac{(1-x)(1-y)}{z}} \cdot \left(\frac{(1-y)(1-z)}{x}\right)^{\frac{(1-y)(1-z)}{x}} \cdot \left(\frac{(1-z)(1-x)}{y}\right)^{\frac{(1-z)(1-x)}{y}} \geq \frac{256}{81}$, where $x > 0,\ y > 0,\ z > 0$ and $x + y + z = 1$
14. Find the largest and smallest values of the expression

    (a) $\frac{a}{a+b} + \frac{b}{b+c} - \frac{a}{a+c}$, where $a > 0,\ b > 0,\ c > 0$,

    (b) $\sum\limits_{j=1}^{n-1} \frac{x_j}{x_j+x_{j+1}} - \frac{x_1}{x_1+x_n}$, where $n \geq 3,\ \ x_1 > 0, \ldots, x_n > 0$.

15. Prove that $\frac{x_1}{\sqrt{1-x_1}} + \cdots + \frac{x_n}{\sqrt{1-x_n}} \geq \frac{\sqrt{x_1}+\cdots+\sqrt{x_n}}{\sqrt{n-1}}$, where $n \geq 2$, $x_i > 0$, $i = 1, \ldots, n$, and $x_1 + \cdots + x_n = 1$.
    Hint. Prove that $\frac{x_1}{\sqrt{1-x_1}} + \cdots + \frac{x_n}{\sqrt{1-x_n}} \geq n \cdot \frac{\frac{x_1+\cdots+x_n}{n}}{\sqrt{1-\frac{x_1+\cdots+x_n}{n}}}$.
16. Prove that $\frac{x_1}{1-x_1^2} + \cdots + \frac{x_n}{1-x_n^2} \geq n \cdot \frac{(x_1+\cdots+x_n)^3}{(x_1+\cdots+x_n)^2-(x_1^2+\cdots+x_n^2)^2}$, where $0 \leq x_i < 1$, $i = 1, \ldots, n$.
    Hint. Consider the function $f(x) = \frac{1}{1-x^2}$ in $[0, 1)$, taking $\alpha_i = \frac{x_i}{x_1+\cdots+x_n}$, $i = 1, \ldots, n$.
17. Prove that $(a_1b_1 + a_2b_2 + \cdots + a_nb_n)^k + (a_1b_2 + a_2b_3 + \cdots + a_nb_1)^k + \cdots + (a_1b_n + a_2b_1 + \cdots + a_nb_{n-1})^k \leq (b_1 + b_2 + \cdots + b_n)^k(a_1^k + a_2^k + \cdots + a_n^k)$, where $n \geq 2$, $a_i > 0$, $b_i > 0$, $i = 1, \cdots, n$, $k > 1$.
18. For which values of λ does the inequality $\sqrt{\frac{a_1^n}{a_1^n+\lambda a_1\cdot\ldots\cdot a_n}} + \cdots + \sqrt{\frac{a_n^n}{a_n^n+\lambda a_1\cdot\ldots\cdot a_n}} \geq \frac{n}{\sqrt{1+\lambda}}$ hold for all positive numbers $a_1, \ldots, a_n$?

Chapter 12
Inequalities of Sequences

In this section we consider inequalities related to sequences, in particular sequences given by recurrence relations, which are proved in various ways. While most inequalities can be proved by well-known proof techniques, there are not many tools for inequalities involving sequences. Below we provide some examples of problems on inequalities of sequences and demonstrate the proof techniques with which such problems can be attacked.

Problems

12.1. Prove that if $x_{n+1} = x_n + \frac{x_n^2}{n^2}$, $n = 1, 2, \ldots$, and $0 < x_1 < 1$, then the sequence (x_n) is bounded.

12.2. Prove that if $a_1 = 1$, $a_n = \frac{a_{n-1}}{2} + \frac{1}{a_{n-1}}$, $n = 2, 3, \ldots, 10$, then $0 < a_{10} - \sqrt{2} < 10^{-370}$.

12.3. Consider the sequence (u_n) such that $u_1 = 1$, $u_{n+1} = u_n + \frac{1}{u_n}$, $n = 1, 2, \ldots$. Prove that $14.2 < u_{100} < 14.22$.

12.4. Consider the sequence (u_n), such that $u_1 = 1$, $u_{n+1} = u_n + \frac{1}{u_n^2}$, $n = 1, 2, \ldots$. Prove that $30 < u_{9000} < 30.01$.

12.5. Prove that if $u_0 = 0.001$, $u_{n+1} = u_n(1 - u_n)$, $n = 0, 1, \ldots$, then $u_{1000} < \frac{1}{2000}$.

12.6. Consider the sequence $a_1 = 1$, $a_{n+1} = \frac{a_n}{2} + \sqrt{\frac{a_n^2}{4} + \frac{1}{a_n}}$, $n = 1, 2, \ldots$. Prove that this sequence is unbounded.

12.7. Given that $a_0 = a_n = 0$, $a_i > 0$, $i = 1, \ldots, n-1$, $n \geq 2$, and $\frac{a_{s-1}+a_{s+1}}{2} > a_s \cos\frac{\pi}{k}$, $k \in \mathbb{N}$, prove that $n \geq k$.

H. Sedrakyan and N. Sedrakyan, *Algebraic Inequalities*, Problem Books in Mathematics, https://doi.org/10.1007/978-3-319-77836-5_12

12.8. Let $x_1 \in [0, 1)$, and for $n = 1, 2, \ldots$, suppose that

$$x_{n+1} = \begin{cases} 0, & \text{if } x_n = 0, \\ \frac{1}{x_n} - \left[\frac{1}{x_n}\right] & \text{if } x_n \neq 0. \end{cases}$$

Prove that $x_1 + x_2 + \cdots + x_n < \frac{F_1}{F_2} + \frac{F_2}{F_3} + \cdots + \frac{F_n}{F_{n+1}}$, where $F_1 = F_2 = 1$ and $F_{n+2} = F_{n+1} + F_n$, $n = 1, 2, \ldots$.

12.9. Let $n \geq 2$ be a positive integer. Find the smallest possible value of the sum $a_0 + a_1 + \cdots + a_n$ if $a_0, a_1, \ldots, a_n$ are nonnegative numbers such that $a_0 = 1$ and $a_i \leq a_{i+1} + a_{i+2}$, $i = 0, 1, \ldots, n-2$.

12.10. Given a sequence of positive numbers $x_1, x_2, \ldots, x_n, \ldots$ such that $x_n^n = \sum\limits_{j=0}^{n-1} x_n^j$, $n = 1, 2, \ldots$, prove that $2 - \frac{1}{2^{n-1}} \leq x_n < 2 - \frac{1}{2^n}$.

12.11. Let $a_0, a_1, a_2, \ldots, a_n, \ldots$ be an infinite sequence of positive numbers. Prove that the inequality $1 + a_n > \sqrt[n]{2.7} \times a_{n-1}$ holds for infinitely many numbers n.

12.12. Consider the following sequence $a_n = \max\limits_{[0;1]}(x^n(1-x) + (1-x)^n x)$, $n = 1, 2, \ldots$. Prove that $a_{n+1} \geq \frac{1}{2} a_n$, $n = 1, 2, \ldots$.

12.13. Given that $a_0 = a_{n+1} = 0$ and $|a_{i-1} - 2a_i + a_{i+1}| \leq 1$, $i = 1, \ldots, n$, prove that $a_k \leq \frac{k(n+1-k)}{2}$, $k = 1, \ldots, n$.

12.14. Find the smallest value of the number C such that there exists a sequence (a_n) of positive numbers such that the inequality $a_1 + \cdots + a_{m+1} \leq C a_m$ holds for all positive integers m.

Proofs

12.1. Let us prove by induction that

$$x_n \leq \frac{n x_1}{(1 - \sqrt{x_1})n + \sqrt{x_1}}, \quad n = 1, 2, 3, \ldots.$$

If $n = 1$, then we have $x_1 \leq \frac{x_1}{(1-\sqrt{x_1}) + \sqrt{x_1}} = x_1$.
Assume that the given inequality holds for $n = k$, that is, $x_k \leq \frac{k x_1}{(1-\sqrt{x_1})k + \sqrt{x_1}}$.
Let us prove that the given inequality holds for $n = k+1$, that is, $x_{k+1} \leq \frac{(k+1)x_1}{(1-\sqrt{x_1})(k+1) + \sqrt{x_1}}$. Since $x_{k+1} = x_k + \frac{x_k^2}{k^2}$ and $x_k \leq \frac{k x_1}{(1-\sqrt{x_1})k + \sqrt{x_1}}$, we have

$$x_{k+1} \le \frac{kx_1}{k(1-\sqrt{x_1})+\sqrt{x_1}} + \frac{x_1^2}{(k(1-\sqrt{x_1})+\sqrt{x_1})^2} =$$

$$= x_1 \frac{k^2(1-\sqrt{x}_1)+k\sqrt{x_1}+x_1}{(k(1-\sqrt{x_1})+\sqrt{x_1})^2} \le \frac{(k+1)x_1}{(1-\sqrt{x_1})(k+1)+\sqrt{x_1}}, \text{ as}$$

$$\frac{k+1}{k(1-\sqrt{x_1})+1} - \frac{k^2(1-\sqrt{x_1})+k\sqrt{x_1}+x_1}{k^2(1-\sqrt{x_1})^2+2k(1-\sqrt{x_1})\sqrt{x_1}+x_1} =$$

$$= \frac{\sqrt{x_1}(1-\sqrt{x_1})^2}{(k(1-\sqrt{x_1})+1)(k^2(1-\sqrt{x_1})^2+2k(1-\sqrt{x_1})\sqrt{x_1}+x_1)} > 0.$$

Therefore, for every positive integer n we have $0 < x_n \le \frac{nx_1}{(1-\sqrt{x_1})n+\sqrt{x_1}} \le \frac{nx_1}{n(1-\sqrt{x_1})} = \frac{x_1}{1-\sqrt{x_1}}$, and hence the sequence (x_n) is bounded.

12.2. Prove that $a_n - \sqrt{2} > 0$, where $n = 2, 3, \ldots$.
We have $a_n = 0.5\left(a_{n-1} + \frac{2}{a_{n-1}}\right) \ge \sqrt{a_{n-1} \cdot \frac{2}{a_{n-1}}} = \sqrt{2}$, and hence $a_n - \sqrt{2} \ge 0$. There is never equality, since all terms of the sequence are rational numbers; hence $a_n - \sqrt{2} > 0$.
Now let us consider the sequence $b_n = a_n - \sqrt{2}$.
We have $b_n = a_n - \sqrt{2} = 0,5\left(a_{n-1} + \frac{2}{a_{n-1}}\right) - \sqrt{2} = \frac{a_{n-1}^2 - 2\sqrt{2}a_{n-1}+2}{2a_{n-1}} = \frac{(a_{n-1}-\sqrt{2})^2}{2a_{n-1}} = \frac{b_{n-1}^2}{2a_{n-1}}$, $n = 2, 3, \ldots$, and thus $b_n < \frac{b_{n-1}^2}{2\sqrt{2}}$.
Since $b_2 = a_2 - \sqrt{2} = \frac{3}{2} - \sqrt{2} < \frac{1}{10}$ and $b_n < \frac{b_{n-1}^2}{2\sqrt{2}}$, we have $b_3 < \frac{b_2^2}{2\sqrt{2}} < \frac{1}{10^2 2\sqrt{2}}$, $b_4 < \frac{1}{10^4(2\sqrt{2})^2 2\sqrt{2}}$, and continuing in a similar way, we deduce that

$$b_{10} < \frac{1}{10^{2^8}} \cdot \frac{1}{(2\sqrt{2})^{2^7}} \cdot \frac{1}{(2\sqrt{2})^{2^6}} \cdots \frac{1}{2\sqrt{2}} = \frac{1}{10^{256}} \cdot \frac{1}{(2\sqrt{2})^{255}}.$$

Now let us prove that

$$\frac{1}{10^{256}} \cdot \frac{1}{(2\sqrt{2})^{255}} < \frac{1}{10^{370}} \text{ or } 10^{114} < (2\sqrt{2})^{255} = 2^{382}\sqrt{2}.$$

Since $2^{10} > 10^3$, it follows that $10^{114} < 2^{380} < 2^{382}\sqrt{2}$.
Hence $b_{10} < \frac{1}{10^{256}} \cdot \frac{1}{(2\sqrt{2})^{255}} < \frac{1}{10^{370}}$, and thus it follows that $a_{10} - \sqrt{2} < 10^{-370}$.

12.3. We have $u_{n+1}^2 = u_n^2 + 2 + \frac{1}{u_n^2}$. Thus, it follows that

$$u_2^2 = u_1^2 + 2 + \frac{1}{u_1^2}, \quad u_3^2 = u_2^2 + 2 + \frac{1}{u_2^2}, \quad \ldots, \quad u_n^2 = u_{n-1}^2 + 2 + \frac{1}{u_{n-1}^2}.$$

Summing these equalities, we obtain

$$u_n^2 = 2n + \frac{1}{u_2^2} + \frac{1}{u_3^2} + \cdots + \frac{1}{u_{n-1}^2} \ (n \ge 3). \qquad (12.1)$$

Since $u_{n-1} > u_{n-2} > \cdots > u_2 = 2$, we have $\frac{1}{u_2^2} + \frac{1}{u_3^2} + \cdots + \frac{1}{u_{n-1}^2} < \frac{n}{4}$.
Therefore,

$$2n \le u_n^2 < \frac{9}{4}n \ (n \ge 2). \tag{12.2}$$

From the conditions (12.1) and (12.2), it follows that for $n = 100$ we have $200 + \frac{4}{9}\left(\frac{1}{2} + \frac{1}{3} + \cdots + \frac{1}{99}\right) < u_{100}^2 < 200 + \frac{1}{2}\left(\frac{1}{2} + \frac{1}{3} + \cdots + \frac{1}{99}\right)$.
Let us estimate from below the following sum; $\frac{1}{2} + \frac{1}{3} + \cdots + \frac{1}{99}$.
We have

$$\frac{1}{2}+\frac{1}{3}+\cdots+\frac{1}{99}=\left(\frac{1}{2}+\frac{1}{3}+\frac{1}{4}+\frac{1}{5}+\frac{1}{6}+\frac{1}{8}+\frac{1}{10}\right)+\left(\frac{1}{7}+\frac{1}{9}+\frac{1}{11}\right)+$$
$$+\left(\frac{1}{12}+\frac{1}{13}+\cdots+\frac{1}{33}\right)+\left(\frac{1}{34}+\frac{1}{35}+\cdots+\frac{1}{99}\right)>1.675+0.346+\frac{23^2}{23\cdot 23}+\frac{65^2}{65\cdot 67}>3.989.$$

In this proof, we have used that $\frac{a_1^2}{b_1} + \cdots + \frac{a_n^2}{b_n} \ge \frac{(a_1+\cdots+a_n)^2}{b_1+\cdots+b_n}$, where $b_1 > 0, \ldots, b_n > 0$.
Now let us estimate from above the following sum: $\frac{1}{2} + \frac{1}{3} + \cdots + \frac{1}{99}$.
According to Problem 7.1b we have

$$\frac{1}{12}+\frac{1}{13}+\cdots+\frac{1}{99}=\frac{1}{12}+\left(\frac{1}{13}+\cdots+\frac{1}{24}\right)+\left(\frac{1}{25}+\cdots+\frac{1}{48}\right)+\frac{1}{49}+\left(\frac{1}{50}+\cdots+\frac{1}{98}\right)+\frac{1}{99}$$
$$<\frac{1}{12}+\left(\frac{1}{13}+\cdots+\frac{1}{24}+\frac{1}{49}\right)+\left(\frac{1}{25}+\cdots+\frac{1}{48}+\frac{1}{97}\right)+\left(\frac{1}{50}+\cdots+\frac{1}{98}\right)$$
$$<\frac{1}{12}+\frac{25}{36}+\frac{25}{36}+\frac{25}{36}=\frac{13}{6}<2.2.$$

It follows that

$$\frac{1}{2}+\frac{1}{3}+\cdots+\frac{1}{99}=\left(\frac{1}{2}+\frac{1}{3}+\frac{1}{4}+\frac{1}{5}+\frac{1}{6}+\frac{1}{8}+\frac{1}{10}\right)+\left(\frac{1}{7}+\frac{1}{9}+\frac{1}{11}\right)+\left(\frac{1}{12}+\frac{1}{13}+\cdots+\frac{1}{99}\right)<$$
$$<2.02+\left(\frac{1}{12}+\frac{1}{13}+\cdots+\frac{1}{99}\right)<4.22.$$

Since $200 + \frac{4}{9}\left(\frac{1}{2} + \frac{1}{3} + \cdots + \frac{1}{99}\right) < u_{100}^2 < 200 + \frac{1}{2}\left(\frac{1}{2} + \frac{1}{3} + \cdots + \frac{1}{99}\right)$, we have $14.2^2 < 200 + \frac{4}{9} \cdot 3.989 < u_{100}^2 < 200 + \frac{1}{2} \cdot 4.22 < 14.22^2$, hence $14.2 < u_{100} < 14.22$.

12.4. We have $u_{n+1}^3 = u_n^3 + 3 + \frac{3}{u_n^3} + \frac{1}{u_n^6}$, where $n = 1, 2, \ldots$.
Therefore, $u_2^3 = u_1^3 + 3 + \frac{3}{u_1^3} + \frac{1}{u_1^6}$, $u_3^3 = u_2^3 + 3 + \frac{3}{u_2^3} + \frac{1}{u_2^6}$, $\ldots$, $u_{n+1}^3 = u_n^3 + 3 + \frac{3}{u_n^3} + \frac{1}{u_n^6}$.
Summing these equalities, we obtain
$u_{n+1}^3 = 3n + 3\left(\frac{1}{u_1^3} + \cdots + \frac{1}{u_n^3}\right) + \left(\frac{1}{u_1^6} + \cdots + \frac{1}{u_n^6}\right)$; hence
$u_{n+1}^3 > 3n + \frac{3}{u_1^3} = 3(n+1)$, or $u_{n+1} > \sqrt[3]{3(n+1)}$, thus $u_{9000} > \sqrt[3]{3 \cdot 9000} = 30$.

On the other hand,

$$u_{n+1}^3 = 3n + 3\left(\frac{1}{u_1^3} + \cdots + \frac{1}{u_n^3}\right) + \left(\frac{1}{u_1^6} + \cdots + \frac{1}{u_n^6}\right) <$$
$$< 3(n+1) + 1 + 3\left(\frac{1}{3 \cdot 2} + \cdots + \frac{1}{3 \cdot n}\right) + \left(\frac{1}{3^2 \cdot 2^2} + \cdots + \frac{1}{3^2 \cdot n^2}\right) =$$
$$= 3(n+1) + 1 + \left(\frac{1}{2} + \cdots + \frac{1}{n}\right) + \frac{1}{9}\left(\frac{1}{2^2} + \cdots + \frac{1}{n^2}\right).$$

Since $\frac{1}{2} + \frac{1}{3} + \cdots + \frac{1}{n} < \ln n$ (see Problem 12.3) and

$$\frac{1}{2^2} + \frac{1}{3^2} + \cdots + \frac{1}{n^2} < \frac{1}{1 \cdot 2} + \frac{1}{2 \cdot 3} + \cdots + \frac{1}{(n-1)n}$$
$$= \left(1 - \frac{1}{2}\right) + \left(\frac{1}{2} - \frac{1}{3}\right) + \cdots + \left(\frac{1}{n-1} - \frac{1}{n}\right) = 1 - \frac{1}{n} < 1,$$

we must have $u_{n+1}^3 < 3(n+1) + 1 + \ln n + \frac{1}{9}$; hence $u_{9000}^3 < 27002 + \ln 9000 < 27002 + \ln 2^{25} < 27027 < \left(30 + \frac{1}{100}\right)^3$.
Therefore, $u_{9000} < 30.01$

12.5. Since $u_{n+1} = u_n(1 - u_n)$, we have $\frac{1}{u_{n+1}} = \frac{1}{u_n(1-u_n)} = \frac{1}{u_n} + \frac{1}{1-u_n}$.
Therefore, $\frac{1}{u_1} = \frac{1}{u_0} + \frac{1}{1-u_0}$, $\frac{1}{u_2} = \frac{1}{u_1} + \frac{1}{1-u_1}, \ldots, \frac{1}{u_{1000}} = \frac{1}{u_{999}} + \frac{1}{1-u_{999}}$.
Summing these equalities, we obtain

$$\frac{1}{u_{1000}} = \frac{1}{u_0} + \frac{1}{1-u_0} + \cdots + \frac{1}{1-u_{999}}$$
$$= 10^3 + \frac{1}{1-u_0} + \cdots + \frac{1}{1-u_{999}}$$
$$> 10^3 + 1 + \cdots + 1 = 2000;$$

hence $u_{1000} < \frac{1}{2000}$.
In this proof, we have used the inequalities $0 < u_n < 1$, which can be proved in the following way:

$$u_{n+1} = u_n(1 - u_n) \le \left(\frac{u_n + (1 - u_n)}{2}\right)^2 = \frac{1}{4} < 1 \; (n = 0, 1, 2, \ldots).$$

If $u_n \le 0$, then from the equality $u_n = u_{n-1}(1 - u_{n-1})$ it follows that $u_{n-1} \le 0$. Continuing in a similar way, we deduce that $u_0 \le 0$, which leads to a contradiction; hence $u_n > 0$.

12.6. Assume that the sequence (a_n) is bounded, that is, $m < a_n < M$ $(n = 1, 2, 3, \ldots)$,
where $M > 0$, since $a_n > 0$.
We have $a_{n+1}^2 - a_{n+1}a_n = \frac{1}{a_n}$ or $\frac{1}{a_n} = \frac{1}{a_{n+1}} + \frac{1}{a_n^2 a_{n+1}^2}$.
Hence $\frac{1}{a_1} = \frac{1}{a_2} + \frac{1}{a_1^2 a_2^2}$, $\frac{1}{a_2} = \frac{1}{a_3} + \frac{1}{a_2^2 a_3^2}, \ldots, \frac{1}{a_n} = \frac{1}{a_{n+1}} + \frac{1}{a_n^2 a_{n+1}^2}$.

Summing these equalities, we deduce that

$$1 = \frac{1}{a_{n+1}} + \frac{1}{a_1^2 a_2^2} + \frac{1}{a_2^2 a_3^2} + \cdots + \frac{1}{a_n^2 a_{n+1}^2} > \underbrace{\frac{1}{M^4} + \frac{1}{M^4} + \cdots + \frac{1}{M^4}}_{n} = \frac{n}{M^4}, \text{ or } M^4 > n,$$

which leads to a contradiction, and hence the sequence (a_n) is bounded.

12.7. Assume that $n < k$. Thus, $\frac{\pi}{k} < \frac{\pi}{n}$, and it follows that $\cos \frac{\pi}{k} > \cos \frac{\pi}{n}$.
Hence we obtain $a_2 > 2a_1 \cos \frac{\pi}{n}, a_1 + a_3 > 2a_2 \cos \frac{\pi}{n}, \ldots, a_{n-2} > 2a_{n-1} \cos \frac{\pi}{n}$.
Let us multiply both sides of the first inequality by $\sin \frac{\pi}{n}$, and both sides of the second inequality by $\sin \frac{2\pi}{n}$, and so on, ending by multiplying both sides of the last inequality by $\sin \frac{(n-1)\pi}{n}$.
Summing all these inequalities, we obtain $0 > 0$, which is a contradiction. Therefore, our assumption is incorrect, since $n \geq k$.

12.8. Without loss of generality, one can assume that $x_n \neq 0$.
We have
$x_k = \left\{ \frac{1}{x_{k-1}} \right\} k = 2, \ldots, n$, whence $x_k \in (0, 1), k = 1, 2, \ldots, n$, and
$x_k = \frac{1}{x_{k+1} + \left[\frac{1}{x_k} \right]} \leq \frac{1}{x_{k+1}+1} = f(x_{k+1}) k = 1, 2, \ldots, n-1$ (∗), where $f(x) = \frac{1}{x+1}$.

Lemma *If the function $f(x)$ is decreasing on an interval I, and the function $x + f(x)$ is increasing there, then the function $g(x) = x + f(x) + f(f(x)) + \cdots + \underbrace{f(f(\ldots f}_{m}(x)\ldots))$ is increasing on I, where $m \in \mathbb{N}$.*

Note that the function $\underbrace{f(f(\ldots f}_{2k}(x)\ldots)), k \in \mathbb{N}$ is increasing on I, and hence the function $\underbrace{f(f(\ldots f}_{2k}(x)\ldots)) + \underbrace{f(f(\ldots f}_{2k+1}(x)\ldots))$ is increasing on I, and the function $g(x) = x + f(x) + f(f(x)) + f(f(f(x))) + \ldots$ is the sum of increasing functions. Thus, it follows that this function is increasing on I.
One can easily verify that the function $f(x) = \frac{1}{x+1}$ on $I = [0, 1]$ satisfies the assumptions of the lemma.
Therefore, by the lemma and (∗), we have

$$x_1 + x_2 + \cdots + x_n \leq f(x_2) + x_2 + x_3 + \cdots + x_n \leq x_3 + f(x_3) + f(f(x_3)) + $$
$$+ x_4 + \cdots + x_n \leq \ldots \leq x_n + f(x_n) + f(f(x_n)) + \cdots + \underbrace{f(f(\ldots f}_{n-1}(x_n)\ldots)) \leq$$
$$\leq 1 + f(1) + \underbrace{f(f(\ldots f}_{n-1}(1)\ldots)) = \frac{F_1}{F_2} + \frac{F_2}{F_3} + \cdots + \frac{F_n}{F_{n+1}},$$

as $1 = \frac{F_1}{F_2}$ *and* $f\left(\frac{F_i}{F_{i+1}}\right) = \frac{F_{i+1}}{F_{i+2}}, i = 1, 2, \ldots, n-1$.

12.9. Let us prove the following lemma by induction.

Lemma *If $n \geq 2$, $c_0 = 0$, $c_n \geq 0$, and $c_i \leq c_{i+1} + c_{i+2}$, $i = 1, 2, \ldots, n-2$, then $c_0 + c_1 + \cdots + c_n \geq 0$.*
Indeed, for $n = 2$ and $n = 3$, we have
$c_0 + c_1 + c_2 \geq 2c_0 = 0$ and $c_0 + c_1 + c_2 + c_3 \geq 2c_0 + c_3 \geq 0$.
Assume that for $n \leq k$ the lemma holds. Let us prove that the lemma holds for $n = k + 1$ ($k \geq 3$).
Consider the following two cases.

(a) *$c_1 \geq 0$. Since $0 \leq c_1 \leq c_2 + c_3$, for the numbers $0, c_2, c_3, \ldots, c_{k+1}$ the assumptions of the lemma hold, and therefore $0 + c_2 + c_3 + \cdots + c_{k+1} \geq 0$, whence $c_0 + c_1 + c_2 + \cdots + c_{k+1} \geq c_2 + c_3 + \cdots + c_{k+1} \geq 0$.*
(b) *$c_1 < 0$. We then have $c_2 \geq -c_1 > 0$, and for the numbers $0, c_3, c_4, \ldots, c_{k+1}$ the assumptions of the lemma hold. Hence $0 + c_3 + c_4 + \cdots + c_{k+1} \geq 0$. On the other hand,*

$$c_0 + c_1 + c_2 + c_3 + \cdots + c_{k+1} \geq 2c_0 + c_3 + \cdots + c_{k+1} \geq 0.$$

This ends the proof of the lemma.

Note that the numbers $c_i = a_i - \frac{F_{n-i}}{F_n}$ $i = 0, 1, \ldots, n$, where $F_0 = 0$, $F_1 = 1$, and $F_{i+2} = F_{i+1} + F_i$, $i = 0, 1, \ldots, n-2$, satisfy the assumptions of the lemma. Therefore, $c_0 + c_1 + \cdots + c_n \geq 0$, or $a_0 + a_1 + \cdots + a_n \geq \frac{F_n}{F_n} + \frac{F_{n-1}}{F_n} + \cdots + \frac{F_0}{F_n}$. On the other hand, the numbers $a_i = \frac{F_{n-i}}{F_n}$ satisfy the assumptions of the problem, and therefore, the smallest value of the sum $a_0 + a_1 + \cdots + a_n$ is equal to $\frac{F_n + F_{n-1} + \cdots + F_0}{F_n}$. One can prove by induction that $F_0 + F_1 + \ldots + F_n = F_{n+2} - 1$.

12.10. Note that $x_1 = 1$ and $x_n > 1$ for $n = 2, 3, \ldots$. Indeed, for $n > 1$ we have $x_n^n = \frac{x_n^n - 1}{x_n - 1}$, or $x_n^n(2 - x_n) = 1$.
Let $x_n = 2 - y_n$. Then $0 < y_n < 1$ for $n = 2, 3, \ldots$ and $y_n(2 - y_n)^n = 1$. We need to prove that $\frac{1}{2^n} < y_n \leq \frac{1}{2^{n-1}}$ for $n = 2, 3, \ldots$. We have $y_n = \frac{1}{(2-y_n)^n} > \frac{1}{2^n}$ (for $n = 2, 3, \ldots$).
By Bernoulli's inequality, we have
$1 = y_n(2 - y_n)^n = y_n(1 + (1 - y_n))^n \geq y_n(1 + n(1 - y_n))$, whence $y_n \leq \frac{1}{n}$, and therefore $y_n = \frac{1}{2^n(1-\frac{y_n}{2})^n} \leq \frac{1}{2^n(1-\frac{ny_n}{2})} \leq \frac{1}{2^{n-1}}$, for $n = 2, 3, \ldots$.

12.11. We argue by contradiction. Assume that the inequality $1 + a_n > \sqrt[n]{2.7} \times a_{n-1}$ holds for finitely many numbers n. Thus, there exists a positive integer n_0 such that for all $n \geq n_0$, one has $1 + a_n > \sqrt[n]{2.7} \times a_{n-1}$.
We have $\lim\limits_{n\to\infty} \left(1 + \frac{1}{n}\right)^n = e$; thus there exists a positive integer n_1 such that for $n \geq n_1$, $\left(1 + \frac{1}{n}\right)^n > 2.7$ ($e = 2.718281828\ldots$). Hence, for $n \geq \max(n_0, n_1)$, we have $1 + a_n \leq a_{n-1} \cdot \sqrt[n]{2,7} < \frac{n+1}{n} a_{n-1}$, and therefore $\frac{1}{n+1} + \frac{a_n}{n+1} < \frac{a_{n-1}}{n}$. It follows that $\frac{1}{m+1} < \frac{a_{m-1}}{m} - \frac{a_m}{m+1}, \frac{1}{m+2} < \frac{a_m}{m+1} - \frac{a_{m+1}}{m+2}, \ldots, \frac{1}{m+k} < \frac{a_{m+k-2}}{m+k-1} - \frac{a_{m+k-1}}{m+k}$. Summing these inequalities, we obtain $\frac{1}{m+1} + \frac{1}{m+2} + \cdots + \frac{1}{m+k} < \frac{a_{m-1}}{m} - \frac{a_{m+k-1}}{m+k} < \frac{a_{m-1}}{m}$, and hence $\frac{1}{m+1}\left(1 + \frac{1}{2} + \cdots + \frac{1}{k}\right) < \frac{a_{m-1}}{m}$, and it follows that the

sequence $x_k = 1 + \frac{1}{2} + \cdots + \frac{1}{k}$ is bounded, which leads to a contradiction, since $x_{2n} - x_n > \frac{1}{2}$, where $n \in \mathbb{N}$.
Indeed, $x_{2n} = x_n + \frac{1}{n+1} + \cdots + \frac{1}{2n} > x_n + n \cdot \frac{1}{2n} = x_n + \frac{1}{2}$.

12.12. We proceed by induction.
We have $a_1 = \frac{1}{2}$, $a_2 = \frac{1}{4}$, and therefore $a_2 \geq \frac{1}{2}a_1$.
Let $a_{m+1} \geq \frac{1}{2}a_m$, where $m \in \mathbb{N}$, and let us prove that $a_{m+2} \geq \frac{1}{2}a_{m+1}$.
Let $x_0 \in [0, 1]$ be such that $a_{m+1} = x_0^{m+1}(1 - x_0) + (1 - x_0)^{m+1}x_0$. Then

$$a_{m+2} \geq x_0^{m+2}(1 - x_0) + (1 - x_0)^{m+2}x_0 = (x_0^{m+1}(1 - x_0) + (1 - x_0)^{m+1}x_0)(x_0 + (1 - x_0))$$
$$- x_0(1 - x_0)(x_0^m(1 - x_0) + (1 - x_0)^m x_0) \geq a_{m+1} - x_0(1 - x_0)a_m = \frac{a_{m+1}}{2} + \frac{a_{m+1}}{2} - x_0(1 - x_0)a_m$$
$$\geq \frac{a_{m+1}}{2} + \frac{a_m}{4} - x_0(1 - x_0)a_m = \frac{a_{m+1}}{2} + a_m\left(\frac{1}{2} - x_0\right)^2 \geq \frac{a_{m+1}}{2},$$

and therefore, $a_{m+2} \geq \frac{1}{2}a_{m+1}$.
Hence, $a_{n+1} \geq \frac{1}{2}a_n$, $n = 1, 2, \ldots$.

12.13. Let us set $a_{i+1} - a_i = b_i$, $i = 0, 1, \ldots, n$, and $b_{i+1} - b_i = c_i$, $i = 0, 1, \ldots, n-1$. Then $|c_i| \leq 1$, $i = 0, 1, \ldots, n-1$. We have $b_{i+1} = c_0 + \cdots + c_i + b_0$, $i = 0, 1, \ldots, n-1$, and $a_{i+1} = b_0 + \cdots + b_i = (i+1)b_0 + ic_0 + (i-1)c_1 + \cdots + c_{i-1}$. Therefore, $0 = a_{n+1} = (n+1)b_0 + nc_0 + \cdots + c_{n-1}$, whence, $b_0 = -\frac{n}{n+1}c_0 - \cdots - \frac{1}{n+1}c_{n-1}$ and

$$a_{i+1} = \left(i - \frac{(i+1)n}{n+1}\right)c_0 + \cdots + \left(1 - \frac{(i+1)(n-i+1)}{n+1}\right)c_{i-1}$$
$$- \frac{n-i}{n+1}c_i - \cdots - \frac{1}{n+1}c_{n-1},$$

where we note that $i - k - \frac{(i+1)(n-k)}{n+1} \leq 0$, for $k = 0, \ldots, i-1$. Thus, a_{i+1} is maximal if $c_0 = \cdots = c_{n-1} = -1$, in which case

$$a_{i+1} = (i+1)b_0 - i - (i-1) - \cdots - 1$$
$$= (i+1) \cdot \frac{1 + \cdots + n}{n+1} - \frac{i(i+1)}{2} = \frac{(i+1)(n-i)}{2}.$$

Therefore, $a_k \leq \frac{k(n+1-k)}{2}$, $k = 1, \ldots, n$.

12.14. Let the sequence of the positive numbers (a_n) be such that
$a_1 + \cdots + a_{m+1} \leq Ca_m$, (1) $m = 1, 2, \ldots$, and let us prove that $C \geq 4$.
Indeed, for $m = 2, 3, \ldots$, we have $S_{m+1} \leq C(S_m - S_{m-1})$, where $S_n = a_1 + \cdots + a_n$ $(n \in \mathbb{N})$.
Therefore, $CS_m \geq S_{m+1} + CS_{m-1} \geq 2\sqrt{S_{m+1} \cdot CS_{m-1}}$, whence $\sqrt{C} \geq \frac{2\sqrt{S_{m+1} \cdot S_{m-1}}}{S_m}$.

Therefore, $\left(\sqrt{C}\right)^n \geq \frac{2\sqrt{S_{n+2}\cdot S_n}}{S_{n+1}} \cdot \ldots \cdot \frac{2\sqrt{S_3\cdot S_1}}{S_2} = 2^n \cdot \sqrt{\frac{S_{n+2}\cdot S_1}{S_{n+1}\cdot S_2}} > 2^n \cdot \sqrt{\frac{S_1}{S_2}}$, and hence $\sqrt{C} \geq 2 \cdot \sqrt[n]{q}$, where $q = \sqrt{\frac{S_1}{S_2}}$. Thus letting $n \to +\infty$, it follows that $\sqrt{C} \geq 2$, and we obtain that $C \geq 4$.

If we prove that for $C = 4$ there exists a sequence of positive numbers (a_n) such that inequality (1) holds, then we obtain that the smallest possible value of the number C is 4.

For the sequence $a_n = 2^{n-1}$, we have $a_1 + \cdots + a_{m+1} = 1 + \cdots + 2^m = 2^{m+1} - 1 < 4 \cdot 2^{m-1} = 4a_m$, and hence $a_1 + \cdots + a_{m+1} < 4a_m$.

This ends the proof.

Problems for Independent Study

1. Consider the sequence (u_n) such that $u_1 = 10^9$, $u_{n+1} = \frac{u_n^2+2}{2u_n}$, $n = 1, 2, \ldots$.
   Prove that $0 < u_{36} - \sqrt{2} < 10^{-13}$.
2. Let $a_1, a_2, \ldots, a_n$ be real numbers such that $a_1 = 0, |a_2| = |a_1+1|, |a_3| = |a_2+1|$, ..., $|a_n| = |a_{n-1} + 1|$. Prove that $\frac{a_1+a_2+\cdots+a_n}{n} \geq -\frac{1}{2}$.
3. Consider the sequence $a_1, a_2, \ldots, a_n, \ldots$ such that $a_1 = 1$, $a_{n+1} = a_n + \frac{1}{a_n^2}$, $n = 1, 2, \ldots$.
   Is this sequence bounded?
4. Suppose the sequence (a_n) is nondecreasing and $a_0 = 0$. Given that the sequence $b_n = a_n - a_{n-1}, n = 1, 2, \ldots$, is not increasing, prove that the sequence $c_n = \frac{a_n}{n}$, $n = 1, 2, \ldots$, is not increasing.
5. Consider the sequence (x_n) such that $x_1 = 2$, $x_{n+1} = \frac{x_n^4+9}{10x_n}$, $n = 1, 2, \ldots$. Prove that $\frac{4}{5} < x_n \leq \frac{5}{4}$, for all $n > 1$.
6. Consider the sequence $a_1 = 1$, $a_{n+1} = \frac{a_n}{2} + \sqrt{\frac{a_n^2}{4} + \frac{1}{a_n}}$, $n = 1, 2, \ldots$. Prove that $a_{250} < 10$.
7. Consider the sequence $x_1 = 2$, $x_{n+1} = x_n^2 - x_n + 1$, $n = 1, 2, \ldots$.
   Prove that $\frac{1}{x_1} + \frac{1}{x_2} + \cdots + \frac{1}{x_n} < 1$.
8. Consider the sequences (a_n) and (b_n) such that $a_1 = 1$, $a_{n+1} = a_n + \sqrt{a_n^2 + 1}$, $b_n = \frac{a_n}{2^n}$, $n = 1, 2, \ldots$. Prove that the sequence (b_n) is bounded.
9. Given that $R_1 = 1$, $R_{n+1} = 1 + \frac{n}{R_n}$, $n = 1, 2, \ldots$, prove that $\sqrt{n} \leq R_n \leq \sqrt{n}+1$.
10. Given that $a_{n+1} = a_n + \frac{a_n^2}{n}$, prove that there exists a number A such that $0 < n(A - a_n) < A^3$.
11. Consider the following sequence: $a_0 = 2$, $a_{n+1} = a_n + \frac{1}{a_n}$, $n = 0, 1, \ldots$.
    Prove that $12 < a_{70} < 12.25$.
12. Consider the following sequences: (a_n) $(a_n > 0, n = 1, 2, \ldots)$ and $x_1 = 1$, $x_2 = 2$, $x_{n+2} + a_n x_{n+1} + x_n = 0$, $n = 1, 2, \ldots$. Prove that the sequence (x_n) contains an infinite number of positive terms and an infinite number of negative terms.

13. Consider the following sequence: $a_1 = 1$, $a_n = a_{n-1} + \frac{1}{a_{n-1}}$, $n = 2, 3, \ldots$.
Prove that $\sqrt{2n-1} \leq a_n \leq \sqrt{3n-2}$.
14. Consider the following sequence: $x_1 = 2$, $x_{n+1} = \frac{x_n^4+9}{10x_n}$, $n = 1, 2, \ldots$.
Prove that $1 - 10^{-10} < x_{100} \leq 1 + 10^{-10}$.
15. (a) Consider the following sequence: $a_1 \geq 2$, $a_{n+1} = a_n^2 - 2$, $n = 1, 2, \ldots$.
Prove that $\frac{1}{a_1} + \frac{1}{a_1 a_2} + \frac{1}{a_1 a_2 a_3} + \cdots + \frac{1}{a_1 a_2 \ldots a_n} < \frac{a_1}{2} - \sqrt{\frac{a_1^2}{4} - 1}$.
(b) Let $a > 2$ and $a_0 = 1$, $a_1 = a$, $a_{n+1} = \left(\frac{a_n^2}{a_{n-1}^2} - 2\right) a_n$, $n = 1, 2, \ldots$.
Prove that $\frac{1}{a_0} + \frac{1}{a_1} + \cdots + \frac{1}{a_k} < \frac{1}{2}\left(2 + a - \sqrt{a^2 - 4}\right)$.
16. For which values of x_1 are all terms of the sequence $x_{n+1} = 2x_n^2 - 1$, $n = 1, 2, \ldots$, negative?
17. Consider the following sequence: $x_n = \sum\limits_{i=2}^{n+1} \frac{1}{i}$. Prove that $x_{n^k - 1} > k x_{n-1}$, where $n, k \in \mathbb{N}$ and $n > 1, k > 1$.
18. Consider the following sequence: $a_1 = 1$, $a_{n+1} = \sqrt{(a_1 + \cdots + a_n)^2 + 1}$, $n = 1, 2, \ldots$.
Prove that $\frac{1}{a_1} + \frac{1}{a_2} + \cdots + \frac{1}{a_n} < 2.5$.
19. Consider the following sequence: $u_1 = 1$, $u_n = n! + \frac{n-1}{n} u_{n-1}$, $n = 2, 3, \ldots$.
Prove that $\frac{1}{u_1} + \frac{1}{u_2} + \cdots + \frac{1}{u_n} < 1$.
20. Consider a monotonic sequence (x_n) of positive numbers. Prove that $\frac{x_1}{x_2} + \cdots + \frac{x_{n-1}}{x_n} + \frac{x_n}{x_1} \geq n + \frac{(x_2 - x_1)^2}{x_1 x_2} + \cdots + \frac{(x_n - x_{n-1})^2}{x_{n-1} x_n}$, where $n \geq 2$.
Hint. Let $\frac{x_2}{x_1} = 1 + \alpha_1, \ldots, \frac{x_n}{x_{n-1}} = 1 + \alpha_{n-1}$. Then $\frac{x_n}{x_1} = (1 + \alpha_1) \ldots (1 + \alpha_{n-1})$.
21. Let $0 \leq a_i \leq a$, $i = 1, 2, \ldots$, and $|a_i - a_j| \geq \frac{1}{i+j}$, for all $i < j$. Prove that $a \geq 1$.
Hint. Let $a_{i_1} \leq \ldots \leq a_{i_n}$, where $\{i_1, \ldots, i_n\} = \{1, \ldots, n\}$. Then by inequality (8.4) (see Chapter 8) and the assumptions of the problem, it follows that $a \geq a_{i_n} - a_{i_1} = |a_{i_n} - a_{i_{n-1}}| + \cdots + |a_{i_2} - a_{i_1}| \geq \frac{(n-1)^2}{n(n+1)}$.
22. Let the sequence (u_n) be such that $|u_{m+n} - u_m - u_n| \leq \frac{m}{n}$, for all $m > n$, $m, n \in \mathbb{N}$.
Prove that $u_n = n u_1$, $n = 1, 2, \ldots$.
Hint. Prove that $\lim\limits_{n \to \infty} (u_{m+n} - u_n) = u_m$, then $u_{m+1} = \lim\limits_{n \to \infty} (u_{m+1+n} - u_n) = \lim\limits_{n \to \infty} (u_{m+1+n} - u_{n+1}) + \lim\limits_{n \to \infty} (u_{n+1} - u_n) = u_m + u_1$.

Chapter 13
Algebraic Inequalities in Number Theory

Inequalities arise not only in algebra, but very often in number theory as well.

In this chapter we consider inequalities in number theory and prove them using *algebraic inequalities*. In order to explain how a large number of problems may be attacked, we provide a list of problems and their proof techniques.

Problems

Prove the following inequalities (13.1–13.3).

13.1. $\frac{m}{n} < \sqrt{2}\left(1 - \frac{1}{4n^2}\right)$, where $\frac{m}{n} < \sqrt{2}, \quad m, n \in \mathbb{N}$.

13.2. $\left(n\sqrt{d} + 1\right)\left|\sin\left(\pi\sqrt{d}n\right)\right| \geq 1$, where $n, d \in \mathbb{N}$ and $\sqrt{d} \notin \mathbb{N}$.

13.3. Let $S, p, q \in \mathbb{N}$, and suppose that S is divisible by q and leaves a remainder 1 on division by p. Prove that $S \leq pq - \frac{q(p-1)}{q-p}$ if $q > p$, $S < pq$.

13.4. Let $m, k \in \mathbb{N}$ and suppose that all prime divisors of the number m are less than or equal to an integer n such that $m \leq n^{\frac{k+1}{2}}$. Prove that m can be represented as the product of k positive integers each of which is less than or equal to n.

13.5. Let $n \in \mathbb{N}$ and let $\sigma(n)$ denote the sum of all divisors of the number n (note that $\sigma(3) = 4, \sigma(6) = 12, \sigma(12) = 28$). Prove that if $\sigma(a) > 2a$ and $b \in \mathbb{N}$, then $\sigma(ab) > 2ab$.

13.6. Find all numbers α such that for every positive integer n there exists a positive integer m such as $\left|\alpha - \frac{m}{n}\right| < \frac{1}{3n}$.

13.7. A number is called an *interesting number* if it is equal to the product of two primes. What is the greatest number of consecutive interesting numbers?

13.8. Prove that there is an infinite number of positive integers n such that $S(2^n) > S(2^{n+1})$, where $S(a)$ is the sum of the digits of the positive integer a.

13.9. Let the positive integer n be represented as a sum of natural numbers. We denote the number of all such representations by $P(n)$ (for example, $P(4) =$

H. Sedrakyan and N. Sedrakyan, *Algebraic Inequalities*, Problem Books in Mathematics, https://doi.org/10.1007/978-3-319-77836-5_13

5, since $4 = 4$, $4 = 3+1$, $4 = 2+2$, $4 = 2+1+1$, $4 = 1+1+1+1$). Prove that $P(n+1) + P(n-1) \geq 2P(n)$, $n = 2, 3, \ldots$.

13.10. Let the positive integer a be divided (long division) by a positive integer b yielding the number $c_0.c_1c_2c_3\ldots$. Given that $c_{n+1} = c_{n+2} = \cdots = c_{n+k} = 9$ for some $n \in \{0, 1, \ldots\}$, $k \in \mathbb{N}$, prove that, (a) $k \leq \left[\frac{b}{10}\right]$, (b) $k \leq \left[\log b\right]$.

13.11. Find for every positive integer n the greatest number $f(n)$ such that from the numbers $1, 2, \ldots n$ one can choose $f(n)$ numbers such that no two of them have ratio equal to 2.

13.12. Let n be a positive integer. Prove that n can be represented as the sum of the squares of m positive integers, where $3\left[\log_4 n\right] + 5 \leq m \leq n - 14$.

13.13. Let a and b be distinct positive integers greater than or equal to 3. Prove that $\left|a^b - b^a\right| > 0.62b^{a-1}$.

Proofs

13.1. We have $\sqrt{2} - \frac{m}{n} = \frac{\sqrt{2}n - m}{n} = \frac{2n^2 - m^2}{n(\sqrt{2}n+m)} \geq \frac{1}{n(\sqrt{2}n+m)}$, since $2n^2 - m^2$ is a positive integer. On the other hand, $m < \sqrt{2}n$, and therefore $\sqrt{2} - \frac{m}{n} \geq \frac{1}{n(\sqrt{2}n+m)} > \frac{1}{n(\sqrt{2}n+\sqrt{2}n)} = \frac{1}{2\sqrt{2}n^2}$. Hence, we have obtained that $\sqrt{2} - \frac{m}{n} > \frac{1}{2\sqrt{2}n^2}$, or $\frac{m}{n} < \sqrt{2}(1 - \frac{1}{4n^2})$.

13.2. Let $\left[\sqrt{dn}\right] = k$, whence $k < \sqrt{dn} < k + 1$.

Consider the following two cases.

(a) If $k < \sqrt{dn} < k + \frac{1}{2}$, then

$$\left(n\sqrt{d}+1\right)\left|\sin(\pi\sqrt{dn})\right| = \left(n\sqrt{d}+1\right)\left|\sin\pi\left(\sqrt{dn}-k\right)\right| \geq$$
$$\geq \left(n\sqrt{d}+1\right)2\left(\sqrt{dn}-k\right) = \left(2n\sqrt{d}+2\right)\frac{dn^2-k^2}{\sqrt{dn}+k} \geq \frac{2n\sqrt{d}+2}{\sqrt{dn}+k} > 1.$$

Here we have used that $\sin x \geq \frac{2}{\pi}x$, where $0 \leq x \leq \frac{\pi}{2}$ (this can be proved using derivatives).

(b) If $k + \frac{1}{2} < \sqrt{dn} < k + 1$, then $\left(n\sqrt{d}+1\right)\left|\sin\pi\left(k+1-\sqrt{dn}\right)\right| \geq \frac{2n\sqrt{d}+2}{k+1+\sqrt{dn}} > 1$.

13.3. By to the assumption of the problem, we have that S can be represented in the following way: $S = p(q-i) + 1 = pq - (pi - 1) = pq - qj$, where $i, j \in \mathbb{N}$.

We have $pi - 1 = qj$ and $q > p$, and hence $i > j$, where $i \geq j + 1$ and $qj \geq p(j+1) - 1$, or $(q-p)j \geq p - 1$. Therefore, since $q > p$, we have $S = pq - qj \leq pq - \frac{q(p-1)}{q-p}$.

13.4. Consider all possible representations of the number $m = a_1 \cdot \cdots \cdot a_k$, where $a_1 \leq a_2 \leq \ldots \leq a_k$ and $a_1, a_2, \ldots, a_k \in \mathbb{N}$. Note that the number of such representations is finite. Let $m = a_1 a_2 \cdots a_k$ be the representation among these representations such that a_k is the smallest possible.

If $a_k \le n$, then this ends the proof. Assume that $a_k > n$. Then a_k is not a prime number, since $m | a_k$. Let p be the smallest prime divisor of a_k. Therefore, $p \le \sqrt{a_k}$.

We have $m = (pa_1) a_2 \cdot \cdots \cdot a_{k-1} \cdot \frac{a_k}{p}$, and hence $pa_1 \ge a_k$, where $a_1 \ge \sqrt{a_k} > \sqrt{n}$.

Therefore, $m = a_1 a_2 \cdot \cdots \cdot a_{k-1} a_k > \left(\sqrt{n}\right)^{k-1} \cdot n = n^{\frac{k+1}{2}}$, which leads to a contradiction.

13.5. Let the numbers $1 = a_1 < \cdots < a_k = a$ be all the divisors of a. Then the numbers $ba_1 < \cdots < ba_k$ are the divisors of ab, whence $\sigma(ab) \ge ba_1 + \cdots + ba_k = b\sigma(a) > 2ab$.

13.6. Without loss of generality one can assume that $\alpha \in [0, 1)$, since $\left|\alpha - \frac{m}{n}\right| = \left|\{\alpha\} - \frac{m - [\alpha] n}{n}\right|$.

Note that $\alpha = 0$ satisfies the condition of the problem, and it is sufficient to take $m = 0$. Let us prove that if $0 < \alpha < 1$, then there exists a positive integer n such that for every integer m we have $\left|\alpha - \frac{m}{n}\right| \ge \frac{1}{3n}$.

If $\alpha \in \mathbb{Q}$, then $\alpha = \frac{p}{q}$, where $0 < p < q$, $p, q \in \mathbb{Z}$. Assume that α satisfies the condition of the problem. If $n = q, q+1, \ldots, 2q$, then the numbers $q\alpha = p, (q+1)\alpha, \ldots, 2q\alpha = 2p$ belong to the union of the sets $\left(p - \frac{1}{3}, p + \frac{1}{3}\right), \left(p + 1 - \frac{1}{3}, p + 1 + \frac{1}{3}\right), \ldots, \left(2p - \frac{1}{3}, 2p + \frac{1}{3}\right)$. Since the number of these numbers is equal to $q + 1$, and the number of intervals is equal to $p + 1$, we have according to the Dirichlet's principle that there exist $i, j \in \{q, q+1, \ldots, 2q\}$ $i \ne j$, such that the numbers α_i, α_j belong to the same interval. Hence we obtain that $\alpha < \frac{2}{3}$. On the other hand, there exists a number $l \in \{q, q+1, \ldots, 2q\}$ such that the numbers αl and $\alpha(l+1)$ belong to different intervals. Therefore $\alpha > \frac{1}{3}$. Hence if $\frac{1}{3} < \alpha < \frac{2}{3}$, then for $n = 1$ there does not exist an integer m such that $|\alpha - m| < \frac{1}{3}$. This leads to a contradiction.

If α is an irrational number, then there exists $n_0 \in \mathbb{N}$ such that $l + \frac{2}{5} < \alpha n_0 < l + \frac{3}{5}$, where $l \in \mathbb{Z}$ (see Problem 14.25).

Assume that $m_0 \in \mathbb{Z}$ and $\left|\alpha - \frac{m_0}{n_0}\right| < \frac{1}{3}$.

If $l = m_0$, then $|\alpha n_0 - m_0| = |\alpha n_0 - l + l - m_0| \ge |l - m_0| - |\alpha n_0 - l| > 1 - \frac{3}{5} = \frac{2}{5} > \frac{1}{3}$.

If $l = m_0$, then $|\alpha n_0 - m_0| = |\alpha n_0 - l| > \frac{2}{5} > \frac{1}{3}$.

Thus, only integers satisfy the condition of the problem.

Alternative solution. Let $n \in \mathbb{N}$, $k, m \in \mathbb{Z}$ and $\left|\alpha - \frac{m}{n}\right| < \frac{1}{3n}$ and $\left|\alpha - \frac{k}{2n}\right| < \frac{1}{6n}$.

If $k \ne 2m$, then $|k - 2m| \ge 1$; thus $\frac{1}{2n} \le \left|\frac{m}{n} - \frac{k}{2n}\right| \le \left|\alpha - \frac{k}{2n}\right| + \left|\frac{m}{n} - \alpha\right| < \frac{1}{3n} + \frac{1}{6n} = \frac{1}{2n}$, which leads to a contradiction. Let $k = 0, 1, 2, \ldots$, $\left|\alpha - \frac{m_k}{2^k}\right| < \frac{1}{3 \cdot 2^k}$, and $m_k \in \mathbb{Z}$. Then $m_0 = \frac{m_1}{2} = \cdots = \frac{m_k}{2^k} = \cdots$, and hence $|\alpha - m_0| < \frac{1}{3 \cdot 2^k}$. Therefore $\alpha = m_0 \in \mathbb{Z}$.

If $\alpha \in \mathbb{Z}$, then it is obvious that for $m = \alpha n$ we have $m \in \mathbb{Z}$ and $\left|\alpha - \frac{m}{n}\right| = 0 < \frac{1}{3n}$.

13.7. Let us provide an example of three consecutive interesting numbers. $33 = 3 \cdot 11, 34 = 2 \cdot 17, 35 = 5 \cdot 7$.
Let us prove that there do not exist four consecutive interesting numbers. We argue by contradiction. Assume that there exist four consecutive interesting numbers, Then one of these numbers is divisible by 4. Therefore, that number is equal to 4. Thus, it follows that one of the numbers is equal to either 5 or 3, neither of which is an interesting number. This is the desired contradiction.

13.8. Assume that n satisfies the inequality $S(2^n) > S(2^{n+1})$. Therefore, there exists a positive integer n_0 such that for all $n \geq n_0$, $n \in \mathbb{N}$,

$$S(2^n) \leq S(2^{n+1}) \tag{1}$$

Note that $S(2^n) = S(2^{n+1})$ is impossible, for otherwise, we would have that $2^n = 2^{n+1} - 2^n$ is divisible by 9, which leads to a contradiction.
Let us prove the following lemmas.

Lemma 1 *If $n \geq n_0$, then $S(2^{n+6}) \geq S(2^n) + 27$.*
Indeed, on dividing the numbers $2^n, 2^{n+1}, 2^{n+2}, \ldots, 2^{n+6}$ by 9, we obtain as remainders seven consecutive terms of the following periodic sequence: $1, 2, 4, 8, 7, 5, 1, 2, 4, 8, 7, 5, 1, 2, \ldots$. Hence from (1), it follows that

$$S(2^{n+6}) - S(2^n) = \left(S(2^{n+6}) - S(2^{n+5})\right) + \left(S(2^{n+5}) - S(2^{n+4})\right) + \ldots + \left(S(2^{n+1}) - S(2^n)\right) \geq$$
$$\geq 1 + 2 + 4 + 8 + 7 + 5 = 27.$$

Using Lemma 1 and induction, one can easily prove the following Lemma 2.

Lemma 2 *$S(2^{n_0+6k}) \geq S(2^{n_0}) + 27k$, where $k \in \mathbb{N}$.*

Lemma 3 *$S(2^{n_0+6k}) < 9m + 18k$, where $k \in \mathbb{N}$, and m is the number of digits of 2^{n_0}.*

Indeed, we have $2^{n_0} < 10^m$, where $2^{n_0+6k} < 10^{m+2k}$; hence the number of the digits of 2^{n_0+6k} is not greater than $(m + 2k)$, and therefore, $S(2^{n_0+6k}) < 9(m + 2k)$.
Note that Lemmas 2 and 3 contradict each other.
This leads to a contradiction.

13.9. Let $x_1, x_2, \ldots, x_k \in \mathbb{N}$, $x_1 \geq x_2 \geq \cdots \geq x_k$, and $n = x_1 + x_2 + \cdots + x_k$. Therefore, $n + 1 = x_1 + x_2 + \cdots + x_k + 1$; hence one can create a one-to-one correspondence between partitions of the number n and the partitions of the number $n + 1$ that contain 1. It follows that

$$(x_1, x_2, \ldots, x_k) \longleftrightarrow (x_1, x_2, \ldots, x_k, 1).$$

Hence, the number $p(n+1) - p(n)$ is equal to the number of the partitions of $n+1$ not containing 1.
If $n = y_1 + y_2 + \cdots + y_l$, where $y_1 \geq y_2 \geq \cdots \geq y_l \geq 2$, then $n+1 = (y_1+1) + y_2 + \cdots + y_l$, and the number of partitions of the number $n+1$ not containing 1 is greater than or equal to the number of similar partitions of the number n.
Therefore, $p(n+1) - p(n) \geq p(n) - p(n-1)$, or $p(n+1) + p(n-1) \geq 2p(n)$, $n = 2, 3, \ldots$.

13.10. Let

$$\begin{array}{l|l} a & \underline{\quad b \quad} \\ \ \ddots & \ldots c_{n+1}c_{n+2}\ldots \\ \overline{\quad 10a_{n+1}} & \\ -\ \underline{c_{n+1}b} & \\ \quad 10a_{n+2} & \\ -\ \underline{c_{n+2}b} & \\ \qquad \ddots & \end{array}$$

We have that $10a_{n+1} - 9b = a_{n+2}$, $10a_{n+2} - 9b = a_{n+3}, \ldots, 10a_{n+k} - 9b = a_{n+k+1}$,
where $0 < a_{n+i} < b, i = 1, 2, \ldots, k, a_{n+k+1} \geq 0$, whence $\frac{9b}{10} \leq a_{n+1}, a_{n+2}, \ldots, a_{n+k} < b$.

(a) Assume that $k > \left[\frac{b}{10}\right]$. Then among the numbers $a_{n+1}, a_{n+2}, \ldots, a_{n+k}$ there are two equal numbers. Therefore,
$\mathrm{A} = \max\limits_{1\leq i\leq k} a_{n+i} - \min\limits_{1\leq i\leq k} a_{n+i} \leq b - 1 - \frac{9b}{10} = \frac{b}{10} - 1$, hence $\mathrm{A} \leq \left[\frac{b}{10}\right] - 1 < k-1$. It follows that there exist numbers $p < q$ such that $a_{n+p} = a_{n+q}$.
Hence, we obtain
$10a_{n+p} - 9b = a_{n+p+1}$, $10a_{n+p+1} - 9b = a_{n+p+2}, \ldots, 10a_{n+q-1} - 9b = a_{n+q} = a_{n+p}$, and summing these equations, we deduce that $a_{n+p} + \ldots + a_{n+q-1} = (q-p)b$, which leads to a contradiction. Hence, we obtain that $k \leq \left[\frac{b}{10}\right]$.

(b) Define $a_{n+i} = b - b_{n+i}$ $i = 1, 2, \ldots, k+1$. It then follows that $1 \leq b_{n+1}, b_{n+2}, \ldots, b_{n+k} < b$ and $1 \leq b_{n+k+1} \leq b$.
Note that $b_{n+2} = 10b_{n+1}, b_{n+3} = 10b_{n+2}, \ldots, b_{n+k+1} = 10b_{n+k}$, and therefore, $b_{n+k+1} = 10^k b_{n+1}$, where $10^k \leq 10^k b_{n+1} \leq b$, or $k \leq \left[\log b\right]$.

Remark If $b = 10^k$ and $a = \underbrace{99\ldots9}_{k}$, then $\frac{a}{b} = 0,\underbrace{99\ldots9}_{\log b}$.

13.11. Consider the set $X = \{4^\alpha q \mid 4^\alpha q \leq n, \alpha \in \mathbb{Z}_0, q \text{ odd}\}$. It is clear that the ratio of two elements of the set X is never equal to 2, and therefore $f(n) \geq |X|$, where we denote by $|X|$ the cardinality of the set X.
Let $\Upsilon \subset \{1, 2, \ldots, n\}$ and $|\Upsilon| > |X|$, and let us put every number a from X into correspondence with the pair $(a, 2a)$. We have that $2a \notin X$ and $|\{(a, 2a)\}| = |X|$. Therefore, for some $a \in \Upsilon$, we have $2a \in \Upsilon$. It follows that $f(n) \leq |X|$; hence $f(n) = |X|$.

Let $n = 2^{k_1}+2^{k_2}+\ldots+2^{k_m}$, where $k_1 > k_2 > \ldots > k_m \geq 0$ and $k_1, k_2, \ldots, k_m$ are integers.
Consider the following sets:

$$X_0 = \{2^{k_1} + 2^{k_2} + \cdots + 2^{k_i} | i \in \{1, 2, \ldots, m\}, k_i \text{ even}\},$$
$$X_i = \{2^{r_1} + 2^{r_2} + \cdots + 2^{r_s} | s \geq i,\ r_1 > r_2 > \ldots > r_s \geq 0,\ r_s \text{ even and}$$
$$r_1 = k_1, r_2 = k_2, \ldots, r_{i-1} = k_{i-1},\quad r_i < k_i,\}\ i = 1, 2, \cdots, m.$$

Obviously, $X_0, X_1, \ldots, X_m$ are mutually disjoint sets and $X = X_0 \cup X_1 \cup \cdots \cup X_m$. Therefore, $|X| = |X_0| + |X_1| + \cdots + |X_m|$.
Let us denote by λ the number of even numbers among the numbers $k_1, k_2, \ldots, k_m$
and the number of odd numbers by μ.
We have $|X_0| = \lambda$. Let us prove that

$$|X_i| = \begin{cases} \dfrac{2^{k_i+1}-1}{3} \text{ if } k_i \text{ is odd,} \\ \dfrac{2^{k_i+1}-2}{3} \text{ if } k_i \text{ is even.} \end{cases}$$

Indeed, let k_i be an even number. We have that $2^{r_i} + \ldots + 2^{r_s} \leq 2^{k_i-1} + \ldots + 2^0 = 2^{k_i} - 1$. Note that $|X_i|$ is one less than the number of numbers from $1, 2, 3, \cdots, 2^{k_i}$ whose canonical decomposition contains an even power of 2, that is,

$$2^{k_i} - 2^{k_i-1} + 2^{k_i-2} - 2^{k_i-3} + \cdots + 2^2 - 2 + 1 - 1$$
$$= \frac{(-2)^{k_i+1} - 1}{-2-1} - 1 = \frac{2^{k_i+1} - 2}{3}.$$

If k_i is odd, then the proof can be obtained in a similar way. Thus, it follows that $f(n) = \lambda + \frac{2}{3}n - \frac{2}{3}\lambda - \frac{\mu}{3} = \frac{2}{3}n + \frac{\lambda-\mu}{3} = \frac{2}{3}n + \frac{(-1)^{k_1}+\cdots+(-1)^{k_m}}{3}$.

13.12. We have that $n = \underbrace{1^2 + \ldots + 1^2}_{n}$, $n = 3^2 + 3^2 + \underbrace{1^2 + \ldots + 1^2}_{n-18}$ and $n = 2^2 + 2^2 + 3^2 + \underbrace{1^2 + \ldots + 1^2}_{n-17}$. Therefore, n can be represented as the sum of the squares of m positive integers, where $m \in \{n - 16, n - 14, n\}$.
On the other hand, if $n = 4^k + 4^k + 4^k + 4^k + x_1^2 + \ldots + x_{m-4}^2$, then $n = 4^{k+1} + x_1^2 + \ldots + x_{m-4}^2$, where $k = 0, 1, \ldots$; hence n can be represented as the sum of the squares of $m - 3$ positive integers.
Thus, taking into consideration that $n = \underbrace{1^2 + \cdots + 1^2}_{n}$, it follows that n can be represented as the sum of the squares of m positive integers, where $m \in \{n, n - 3, n - 6, \ldots\}$ and $m \geq c_0 + c_1 + \cdots + c_k$, where $\overline{c_k c_{k-1} \ldots c_0}$ represents n written in the base-4 number system.

Therefore, $c_0+c_1+\cdots+c_k \leq 3(k+1)$ and $4^k \leq n < 4^{k+1}$; hence $k = \left[\log_4 n\right]$. In order to complete the proof, it is sufficient to note that

$$\{n, n-3, n-6, \ldots\} \cup \{n-14, n-17, n-20, \ldots\} \cup \{n-16, n-19, n-22, \ldots\} \supset$$
$$\supset \left\{n-14, n-15, n-16, \ldots, 3\left[\log_4 n\right]+5\right\}.$$

13.13. Let us prove the following lemmas.

Lemma 1 *If $n \in \mathbb{N}$, then $\left(1+\frac{1}{n}\right)^n < 3$.*
By mathematical induction, one can easily prove that if $k \in \mathbb{N}$ and $n = 1, 2, \ldots, k$, then

$$\left(1+\frac{1}{k}\right)^n < 1+\frac{n}{k}+\frac{n^2}{k^2} \tag{1}$$

If $n = 1$, then we have $\left(1+\frac{1}{k}\right)^n = 1+\frac{1}{k} < 1+\frac{1}{k}+\frac{1}{k^2}$.
Assume that for $n = m \leq k-1$, inequality (1) holds, and let us prove that it holds for $n = m+1$.

Note that $\left(1+\frac{1}{k}\right)^{m+1} = \left(1+\frac{1}{k}\right)\cdot\left(1+\frac{1}{k}\right)^m < \left(1+\frac{1}{k}\right)\cdot\left(1+\frac{m}{k}+\frac{m^2}{k^2}\right) < 1+\frac{m+1}{k}+\frac{(m+1)^2}{k^2}$, since $m^2 < (m+1)k$.
This ends the proof of inequality (1).
For $n = k$ we obtain $\left(1+\frac{1}{k}\right)^k < 3$.

Lemma 2 *If $a, b \in \mathbb{N}$ and $a > b \geq 3$. Then $b^{a-1} > a^{b-1}$.*
For $n \in \mathbb{N}$ and $n \geq 3$, using Lemma 1, we deduce that $n \geq 3 > \left(1+\frac{1}{n}\right)^n$. Therefore,

$$n^{\frac{1}{n}} > (n+1)^{\frac{1}{n+1}}. \tag{2}$$

Thus, from (2) it follows that $b^{\frac{1}{b}} > a^{\frac{1}{a}}$, and therefore $b^a > a^b > b \cdot a^{b-1}$, whence $b^{a-1} > a^{b-1}$.
From Lemma 2 it follows that without loss of generality one can assume that $a > b$. Let $a = b+c$, where $c \in \mathbb{N}$.
Consider the following two cases:

(a) $b \geq 4$.
We obtain

$$\left|a^b - b^a\right| = \left|(b+c)^b - b^{b+c}\right| = b^b\left|\left(1+\frac{c}{b}\right)^b - b^c\right| =$$
$$= b^b\left|b-\left(1+\frac{c}{b}\right)^{\frac{b}{c}}\right|\left(b^{c-1}+b^{c-2}\left(1+\frac{c}{b}\right)^{\frac{b}{c}}+\ldots+\left(\left(1+\frac{c}{b}\right)^{\frac{b}{c}}\right)^{c-1}\right) \geq b^b|b-(1+\frac{c}{b})^{\frac{b}{c}}|b^{c-1} =$$
$$= b^{a-1}\left|b-\left(1+\frac{c}{b}\right)^{\frac{b}{c}}\right|, \text{ therefore, } \left|a^b-b^a\right| > b^{a-1}\left|b-\left(1+\frac{c}{b}\right)^{\frac{b}{c}}\right|. \tag{3}$$

For $c > 1$, *using Problem 2.1 for* $\underbrace{(1+c/b), \ldots, (1+c/b)}_{c-1}, 1$, *we deduce that* $\frac{(1+\frac{c}{b})+\ldots+(1+\frac{c}{b})+1}{c} > \sqrt[c]{\left(1+\frac{c}{b}\right)^{c-1}}$, *or* $\left(1+\frac{c-1}{b}\right)^{\frac{b}{c-1}} > \left(1+\frac{c}{b}\right)^{\frac{b}{c}}$.
Hence, we have $\left(1+\frac{c}{b}\right)^{\frac{b}{c}} \leq \left(1+\frac{1}{b}\right)^{b} < 3$, *and therefore,* $\left|b - \left(1+\frac{c}{b}\right)^{\frac{b}{c}}\right| > 1$.
From (3), it follows that $\left|a^b - b^a\right| > b^{a-1}$.

(*b*) *If* $b = 3$, *then by induction one can easily prove that* $2.38 \cdot 3^{a-1} > a^3$, *where* $a = 4, 5, \ldots$.

This ends the proof.

Problems for Independent Study

Prove the following inequalities (1–4).

1. $\left|\sqrt{a} - \frac{p+aq}{p+q}\right| \leq \frac{p}{q}$, where $a, p, q \in \mathbb{N}$, $\sqrt{a} \notin \mathbb{N}$ and $\left|\sqrt{a} - \frac{p}{q}\right| \leq 1$.
2. (a) $S(mn) \leq S(m)S(n)$, where $n, m \in \mathbb{N}$.
   (b) $S(m+n) \leq S(m) + S(n)$, where $n, m \in \mathbb{N}$
   (see Problem 13.8).
3. $S(1981^n) \geq 19$, where $n \in \mathbb{N}$.
4. $S(1998^n) > 10^6$ for all positive integers n starting from some number.
5. Prove that for every number M there exists a positive integer n such that $\frac{S(n)}{S(n^2)} > M$.
6. Prove that 5^{2l} can be represented as the sum of the squares of m numbers, where $l \in \mathbb{N}$ and $1 \leq m \leq 2^l$.
7. Find the smallest positive integer n such that for every partition of the positive integers $1, 2, \ldots, n$ into two groups, in one of them there are three numbers forming a geometric progression.
8. Let $a_1, a_2, \ldots, a_k$ be integers such that $1 < a_1 < a_2 < \cdots < a_k \leq n$ and $a_1 \cdot a_2 \cdots a_k$ is not divisible by a_i^2, $i = 1, 2, \ldots, k$. Prove that $k \leq \pi(n)$, where $\pi(n)$ is the number of primes less than or equal to n.
9. Let k be a given positive integer. Prove that the sequence $\frac{S(n)}{S(kn)}$ is bounded if and only if $k = 2^\alpha \cdot 5^\beta$, where $\alpha, \beta \in \{0, 1, \ldots\}$.
10. Let $x \geq 1$ and let $A(x)$ be the least common multiple of the numbers $1, 2, \ldots, [x]$. Prove that

    (a) $A\left(\frac{x}{2}\right) \cdot \frac{[x]!}{\left(\left[\frac{x}{2}\right]!\right)^2} \geq A(x)$, where $x \geq 2$,
    (b) $A(x) < 5^x$,
    (c) $A(2n) \cdot A\left(\frac{2n}{3}\right) \geq A(n) \cdot \frac{(2n)!}{(n!)^2}$, where $n \geq 2$, $n \in \mathbb{N}$,
    (d) $A(2n) > A(n)A\left(\sqrt{2n}\right)$, where $n \in \mathbb{N}$, $n \geq 100$,

(e) there exists a prime number belonging to $(n, 2n)$, where $n \in \mathbb{N}, n > 1$.

Remark If we assume that there is no prime number belonging to $(n, 2n)$, prove that $A(2n) | A(n)A\left(\sqrt{2n}\right)$ (i.e., $A(n)A\left(\sqrt{2n}\right)$ divides $A(2n)$).

11. Prove that $\frac{d(1)}{1^2} + \frac{d(2)}{2^2} + \cdots + \frac{d(n)}{n^2} < 2.78$, where $d(k)$ is the number of the positive integer divisors of k.

Remark $\left(\sum_{k=1}^{\infty} \frac{1}{k^2}\right)^2 = \left(\sum_{i=1}^{\infty} \frac{1}{i^2}\right)\left(\sum_{j=1}^{\infty} \frac{1}{j^2}\right) = \left(\sum_{k=1}^{\infty} \frac{d(k)}{k^2}\right)$.

12. Find all numbers α such that the following condition holds: for every positive integer n there exists an integer m such as $\left|\alpha - \frac{m}{n}\right| \le \frac{1}{3n}$.
13. Let $a_1 \le a_2 \le \cdots \le a_n$ be positive integers such that no sum of some of them is equal to the sum of some of the others. Prove that

(a) $a_1 + a_2 + \cdots + a_k \ge 2^k - 1, k = 1, 2, \ldots, n$,
(b) $\frac{1}{a_1} + \frac{1}{a_2} + \cdots + \frac{1}{a_n} \le 1 + \frac{1}{2} + \cdots + \frac{1}{2^{n-1}}$.

14. Let $a_1, \ldots, a_n$ are distinct positive integers. Prove that

$$a_1^3 + \cdots + a_n^3 \ge (a_1 + \cdots + a_n)^2.$$

Remark Let $a_1 < \cdots < a_n$ and $a_i = i + b_i$. Then $b_1 \le \cdots \le b_n$.

Chapter 14
Miscellaneous Inequalities

In this chapter we consider miscellaneous inequalities, and we mostly use various proof techniques not included in the previous chapters.

Problems

Prove the following inequalities (14.1–14.5).

14.1. $\sqrt{a_1^2+b_1^2}+\cdots+\sqrt{a_n^2+b_n^2}\geq\sqrt{(a_1+\cdots+a_n)^2+(b_1+\cdots+b_n)^2}$.

14.2. $\frac{(a-b)^2}{2(a+b)}\leq\sqrt{\frac{a^2+b^2}{2}}-\sqrt{ab}\leq\frac{(a-b)^2}{\sqrt{2}(a+b)}$, where $a,b>0$.

14.3. $x_2^{\frac{1}{n}}-x_1^{\frac{1}{n}}\leq(x_2-\alpha)^{\frac{1}{n}}-(x_1-\alpha)^{\frac{1}{n}}$, where $0\leq\alpha\leq x_1\leq x_2,\quad n\in\mathbb{N}$.

14.4. $\left|\sqrt[n]{a_1^n+\cdots+a_k^n}-\sqrt[n]{b_1^n+\cdots+b_k^n}\right|\leq|a_1-b_1|+\cdots+|a_k-b_k|$, where $n\geq 2,\quad n\in\mathbb{N}$, $a_1>0,\ldots,a_k>0,b_1>0,\ldots,b_k>0$.

14.5. $\frac{1}{1+\sqrt{3}}+\frac{1}{\sqrt{5}+\sqrt{7}}+\cdots+\frac{1}{\sqrt{9997}+\sqrt{9999}}>24$.

14.6. Given that $a^2-4a+b^2-2b+2\leq 0$, $c^2-4c+d^2-2d+2\leq 0$ and $e^2-4e+f^2-2f+2\leq 0$, find the greatest possible value of the expression $(a-c)(f-d)+(c-e)(b-d)$.

14.7. Prove that if an increasing sequence of positive numbers $a_1,a_2,\ldots,a_n,\ldots$ is unbounded, then there exists a sufficiently large index k such that the following inequalities hold:

(a) $\frac{a_1}{a_2}+\frac{a_2}{a_3}+\cdots+\frac{a_k}{a_{k+1}}<k-\frac{1}{2}$,
(b) $\frac{a_1}{a_2}+\frac{a_2}{a_3}+\cdots+\frac{a_k}{a_{k+1}}<k-1985$.

14.8. *Minkowski's inequality*: Prove that $\left(\sum\limits_{i=1}^{n}(a_i+b_i)^r\right)^{\frac{1}{r}}\leq\left(\sum\limits_{i=1}^{n}a_i^r\right)^{\frac{1}{r}}+\left(\sum\limits_{i=1}^{n}b_i^r\right)^{\frac{1}{r}}$, where $a_i\geq 0,\ b_i\geq 0,\quad i=1,\ldots,n$.

H. Sedrakyan and N. Sedrakyan, *Algebraic Inequalities*, Problem Books in Mathematics, https://doi.org/10.1007/978-3-319-77836-5_14

14.9. *Young's inequality*: Prove that if the function $f(x)$ is increasing and continuous on $[0, +\infty)$ and $f(0) = 0$, then $\int_0^a f(x)dx + \int_0^b f^{-1}(y)dy \geq ab$, where the function $f^{-1}(x)$ is the inverse function of the function $f(x)$ and $a > 0,\ b > 0$.

14.10. (a) Prove that $x_1y_1 + \cdots + x_ny_n = x_1(y_1 - y_2) + (x_1 + x_2)(y_2 - y_3) + \ldots + (x_1 + \cdots + x_{n-1})(y_{n-1} - y_n) + (x_1 + \ldots + x_n)y_n$.

(b) Prove that if $x_1 \geq x_2 \geq \cdots \geq x_n$ and $y_1 \geq y_2 \geq \cdots \geq y_n$, then $x_1y_n + x_2y_{n-1} + \cdots + x_ny_1 \leq x_1y_{i_1} + x_2y_{i_2} + \cdots + x_ny_{i_n} \leq x_1y_1 + \cdots + x_ny_n$, where $i_1, i_2, \ldots, i_n$ is some permutation of the numbers $1, 2, \ldots, n$.

14.11. Prove that

(a) $x_1 + \cdots + x_n \geq n$, where $n \geq 2,\ x_1 > 0, \ldots,\ x_n > 0$, and $x_1 \cdots x_n = 1$,

(b) $\frac{1}{a_1(a_2+1)} + \cdots + \frac{1}{a_{n-1}(a_n+1)} + \frac{1}{a_n(a_1+1)} \geq \frac{n}{1+a_1 \cdots a_n}$, where $n \geq 2,\ (n-3)(a_i - 1) \geq 0,\ a_i > 0,\quad i = 1, \ldots, n$,

(c) $3 + (A + M + S) + \left(\frac{1}{A} + \frac{1}{M} + \frac{1}{S}\right) + \left(\frac{A}{M} + \frac{M}{S} + \frac{S}{A}\right) \geq \frac{3(A+1)(M+1)(S+1)}{AMS+1}$, where $A > 0,\ M > 0,\ S > 0$,

(d) $\left(\frac{1}{a} + \frac{1}{b} + \frac{1}{c}\right)\left(\frac{1}{1+a} + \frac{1}{1+b} + \frac{1}{1+c}\right) \geq \frac{9}{1+abc}$, where $a > 0,\ b > 0,\ c > 0$,

(e) $\frac{a}{\sqrt{a+b}} + \frac{b}{\sqrt{b+c}} + \frac{c}{\sqrt{c+a}} \geq \frac{\sqrt{a}+\sqrt{b}+\sqrt{c}}{\sqrt{2}}$, where $a > 0,\ b > 0,\ c > 0$.

14.12. *Schur's inequality*: Prove that $\varphi(x_1) + \cdots + \varphi(x_n) \geq \varphi(y_1) + \cdots + \varphi(y_n)$, where $y_1 \geq y_2 \geq \cdots \geq y_n$, $x_1 + \cdots + x_k \geq y_1 + \cdots + y_k,\quad k = 1, \ldots, n-1$, $x_1 + \cdots + x_n = y_1 + \cdots + y_n$, and for every x in I, one has $\varphi''(x) > 0$ $(x_1, \ldots, x_n, y_1, \ldots, y_n \in I)$.

14.13. Prove that $b_1 + \cdots + b_n \geq a_1 + \cdots + a_n$, where $a_1 \geq \cdots \geq a_n > 0,\ b_1 \geq a_1$, $b_1b_2 \geq a_1a_2, \ldots, b_1 \cdot \ldots \cdot b_n \geq a_1 \cdot \ldots \cdot a_n$.

14.14. Let α, β, γ be angles of an acute triangle. Prove that $\sqrt{2} < \sin\frac{\alpha}{2} + \sin\frac{\beta}{2} + \sin\frac{\gamma}{2} \leq \frac{3}{2}$.

14.15. Let α, β, γ be angles of an obtuse triangle. Prove that $0 < \sqrt{\sin\alpha} + \sqrt{\sin\beta} + \sqrt{\sin\gamma} < 1 + \sqrt[4]{8}$.

14.16. Prove that $1 + \frac{1}{\sqrt{2}} + \frac{1}{\sqrt{3}} + \cdots + \frac{1}{\sqrt{n}} > 2\sqrt{n} - 2$, where $n \in \mathbb{N}$.

14.17. Prove that $a > 0,\ b > 0,\ c > 0,\ d > 0$, if $S_1 = a + b + c + d$, $S_2 = ab + bc + cd + ad + ac + bd > 0$, $S_3 = abc + bcd + cda + abd > 0$, $S_4 = abcd > 0$.

14.18. Prove that $n! < n\left(\frac{n}{e}\right)^n$, where $n \geq 8$ and $n \in \mathbb{N}$.

14.19. Prove that $\left(1 + \frac{1}{2^1}\right) \cdots \left(1 + \frac{1}{2^n}\right) < 3$, where $n \in \mathbb{N}$.

14.20. Prove that if g is a convex function on $[0, a_1]$ (see Chapter 11) and $a_1 \geq a_2 \geq \cdots \geq a_{2m} \geq a_{2m+1} \geq 0$, then $g(a_1) - g(a_2) + g(a_3) - \cdots + g(a_{2m+1}) \geq g(a_1 - a_2 + a_3 - \cdots + a_{2m+1})$.

14.21. Let the absolute value of the polynomial $p(x)$ on $[-1, 1]$ be less than or equal to 1. Prove that

(a) $|a| + |b| + |c| \leq 3$, where $p(x) = ax^2 + bx + c$,

(b) $a^2 + b^2 + c^2 \leq 5$, where $p(x) = ax^2 + bx + c$,

(c) $|a| \leq 4$, where $p(x) = ax^3 + bx^2 + cx + d$.

14.22. Let the polynomial $p(x) = x^n + a_{n-1}x^{n-1} + \cdots + a_0$ $(n > 1)$ have n negative roots. Prove that $a_1 p(1) \geq 2n^2 a_0$.

14.23. The complete graph on n vertices has each of its edges colored in one of two given colors. We denote by $t(n)$ the number of triangles whose sides are colored in the same color. Prove that

$$t(n) \geq \begin{cases} \frac{k(k-1)(k-2)}{3}, & \text{if } n = 2k, \\ \frac{2}{3}k(k-1)(4k+1), & \text{if } n = 4k+1, \\ \frac{2}{3}k(k+1)(4k-1), & \text{if } n = 4k+3. \end{cases}$$

14.24. Consider a graph having n $(n \geq 3)$ vertices. The number of its edges is greater than $\frac{n^2}{4}$. Prove that the number of triangles formed by these edges is not less than $\left[\frac{n}{2}\right]$.

14.25. Prove that if α, $(\alpha > 0)$ is an irrational number and $0 < a < b < 1$, then there exists a positive integer n such that $a < \{n\alpha\} < b$.

14.26. Let $x_1 + \cdots + x_n = 0$ and $x_i \in [m, M]$, $i = 1, \ldots, n$. Prove that

(a) $\sum\limits_{i=1}^{n} x_i^2 \leq -mMn$,

(b) $\sum\limits_{i=1}^{n} x_i^4 \leq -mMn \cdot \left(m^2 + M^2 + mM\right)$.

14.27. Prove that $\left(x^2 + y^2\right)^m \geq 2^m x^m y^m + (x^m - y^m)^2$, where $m \in \mathbb{N}$.

14.28. Prove that $cos(\alpha - \beta)cos(\beta - \gamma)cos(\gamma - \alpha) \geq 8\, cos\alpha\, cos\beta\, cos\gamma$, where α, β, γ are the angles of some triangle.

14.29. Prove that $\left|\sin x + \frac{\sin 2x}{2} + \cdots + \frac{\sin nx}{n}\right| < 3$.

14.30. Prove that $\frac{a_1+a_2}{2} \cdot \frac{a_2+a_3}{2} \cdots \frac{a_{n-1}+a_n}{2} \cdot \frac{a_n+a_1}{2} \leq \frac{a_1+a_2+a_3}{3} \cdot \frac{a_2+a_3+a_4}{3} \cdots \frac{a_n+a_1+a_2}{3}$, where $0 < a_1 \leq a_2 \leq \cdots \leq a_n$ and $n \geq 3$.

14.31. Prove that if $x, y, z \geq 0$ and $x^2 + y^2 + z^2 = 1$, then

(a) $1 \leq \frac{x}{1-yz} + \frac{y}{1-zx} + \frac{z}{1-xy} \leq \frac{3\sqrt{3}}{2}$,

(b) $1 \leq \frac{x}{1+yz} + \frac{y}{1+zx} + \frac{z}{1+xy} \leq \sqrt{2}$.

14.32. For which values of λ does the inequality

$$\frac{a}{\sqrt{a^2 + \lambda bc}} + \frac{b}{\sqrt{b^2 + \lambda ca}} + \frac{c}{\sqrt{c^2 + \lambda ab}} \geq \frac{3}{\sqrt{1 + \lambda}} \tag{14.1}$$

hold for arbitrary positive numbers a, b, c?

14.33. Prove that for every triangle with sides a, b, c, one has

$$\sqrt{a^2 + ab + b^2} + \sqrt{b^2 + bc + c^2} + \sqrt{c^2 + ca + a^2}$$
$$\leq \sqrt{5a^2 + 5b^2 + 5c^2 + 4ab + 4bc + 4ca}.$$

14.34. Prove that

(a) $\frac{x_1^3+\cdots+x_n^3}{x_1\cdots x_n} \geq \frac{(1-x_1)^3+\cdots+(1-x_n)^3}{(1-x_1)\cdots(1-x_n)}$, where $n \geq 3$, $0 \leq x_i \leq \frac{1}{2}$, $i = 1, \ldots, n$,
(b) $\frac{x_1^4+\cdots+x_n^4}{x_1\cdots x_n} \geq \frac{(1-x_1)^4+\cdots+(1-x_n)^4}{(1-x_1)\cdots(1-x_n)}$, where $n \geq 4$, $0 \leq x_i \leq \frac{1}{2}$, $i = 1, \ldots, n$.

14.35. Let $n \geq 2$ and let $a_1, \ldots, a_n$ be distinct positive integers. Prove that

$$\frac{a_1^{2015} + \cdots + a_n^{2015}}{a_1^{2000} + \cdots + a_n^{2000}} \geq \frac{1^{2015} + \cdots + n^{2015}}{1^{2000} + \cdots + n^{2000}}.$$

14.36. Prove that $\sqrt{1+x^2} + \sqrt{1+y^2} + \sqrt{(1-x)^2 + (1-y)^2} \geq \left(1+\sqrt{5}\right)(1-xy)$, where $0 \leq x \leq 1$, $0 \leq y \leq 1$.

14.37. Prove that if $x_i > 0$, $i = 1, \ldots, n$ and $x_1 \cdots x_n = 1$, then

(a) $\frac{1}{\sqrt{1+x_1}} + \cdots + \frac{1}{\sqrt{1+x_n}} \leq \frac{\sqrt{2}}{2} \cdot n$, where $n = 2, 3, \ldots$,
(b) $\frac{1}{\sqrt{1+x_1}} + \cdots + \frac{1}{\sqrt{1+x_n}} < n - 1$, where $n \geq 4$.

14.38. Prove that
$\sum_{i=1}^{n} \frac{x_i^\alpha - x_i}{x_1+\cdots+x_n+x_i^\alpha - x_i} \geq 0$, where $n \geq 2$, $x_i > 0$, $i = 1, \ldots, n$, $x_1 \cdots x_n \geq 1$, and $\alpha \geq 1$.

Proofs

14.1. Consider the points

$$A_1(0,0), A_2(a_1, b_1), A_3(a_1 + a_2, b_1 + b_2), \ldots, A_{n+1}(a_1 + \cdots + a_n, b_1 + \cdots + b_n).$$

Since $A_1A_2 + A_2A_3 + \cdots + A_nA_{n+1} \geq A_1A_{n+1}$, it follows that

$$\sqrt{a_1^2 + b_1^2} + \sqrt{a_2^2 + b_2^2} + \cdots + \sqrt{a_n^2 + b_n^2} \geq \sqrt{(a_1 + \cdots + a_n)^2 + (b_1 + \cdots + b_n)^2}.$$

14.2. Let us rewrite the expression $\sqrt{\frac{a^2+b^2}{2}} - \sqrt{ab}$ in the following way:

$$\sqrt{\frac{a^2+b^2}{2}} - \sqrt{ab} = \frac{\left(\sqrt{\frac{a^2+b^2}{2}} - \sqrt{ab}\right)\left(\sqrt{\frac{a^2+b^2}{2}} + \sqrt{ab}\right)}{\sqrt{\frac{a^2+b^2}{2}} + \sqrt{ab}} = \frac{(a-b)^2}{2\left(\sqrt{\frac{a^2+b^2}{2}} + \sqrt{ab}\right)}.$$

It is left to prove that $\frac{a+b}{\sqrt{2}} \leq \sqrt{\frac{a^2+b^2}{2}} + \sqrt{ab} \leq a + b$. The left-hand side of this inequality obviously holds, since $\left(\sqrt{\frac{a^2+b^2}{2}} + \sqrt{ab}\right)^2 > \frac{a^2+b^2}{2} + ab = \frac{(a+b)^2}{2}$.
In order to prove the right-hand side, let us prove the following inequality:
$\sqrt{\frac{a^2+b^2}{2}} - \frac{a+b}{2} \leq \frac{a+b}{2} - \sqrt{ab}$, or $\frac{(a-b)^2}{4\left(\sqrt{\frac{a^2+b^2}{2}} + \frac{a+b}{2}\right)} \leq \frac{(a-b)^2}{4\left(\frac{a+b}{2} + \sqrt{ab}\right)}$.

Note that this inequality holds, because $\sqrt{\frac{a^2+b^2}{2}} \geq \sqrt{ab}$.

14.3. We have

$$x_2^{\frac{1}{n}} - x_1^{\frac{1}{n}} = \frac{x_2 - x_1}{\left(x_2^{\frac{1}{n}}\right)^{n-1} + \left(x_2^{\frac{1}{n}}\right)^{n-2} \cdot x_1^{\frac{1}{n}} + \cdots + \left(x_1^{\frac{1}{n}}\right)^{n-1}}$$
$$\leq \frac{x_2 - x_1}{\left((x_2-\alpha)^{\frac{1}{n}}\right)^{n-1} + \left((x_2-\alpha)^{\frac{1}{n}}\right)^{n-2} \cdot (x_1-\alpha)^{\frac{1}{n}} + \cdots + \left((x_1-\alpha)^{\frac{1}{n}}\right)^{n-1}}$$
$$= (x_2-\alpha)^{\frac{1}{n}} - (x_1-\alpha)^{\frac{1}{n}}.$$

14.4. Let us rewrite the expression $\sqrt[n]{a_1^n + \cdots + a_k^n} - \sqrt[n]{b_1^n + \cdots + b_k^n}$ in the following way:

$$\left|\sqrt[n]{a_1^n + \cdots + a_k^n} - \sqrt[n]{b_1^n + \cdots + b_k^n}\right| = \frac{\left|a_1^n + \cdots + a_k^n - b_1^n - \cdots - b_k^n\right|}{A^{n-1} + A^{n-2}B + \cdots + B^{n-1}},$$

where $A = \sqrt[n]{a_1^n + \cdots + a_k^n}$, $\quad B = \sqrt[n]{b_1^n + \cdots + b_k^n}$.
Therefore,

$$|A-B| = \frac{\left|(a_1-b_1)\left(a_1^{n-1} + \cdots + b_1^{n-1}\right) + \cdots + (a_k-b_k)\left(a_k^{n-1} + a_k^{n-2}b_k + \cdots + b_k^{n-1}\right)\right|}{A^{n-1} + A^{n-2}B + \cdots + B^{n-1}}$$
$$\leq |a_1-b_1| \cdot \left|\frac{a_1^{n-1} + \cdots + b_1^{n-1}}{A^{n-1} + \cdots + B^{n-1}}\right| + \cdots + |a_k-b_k| \cdot \left|\frac{a_k^{n-1} + \cdots + b_k^{n-1}}{A^{n-1} + \cdots + B^{n-1}}\right|$$
$$\leq |a_1-b_1| + \cdots + |a_k-b_k|,$$

as $A > a_i > 0, \quad B > b_i > 0 \quad (i = 1, \ldots, n)$.

14.5. We have

$$2 \cdot \left(\frac{1}{1+\sqrt{3}} + \frac{1}{\sqrt{5}+\sqrt{7}} + \cdots + \frac{1}{\sqrt{9997}+\sqrt{9999}}\right)$$
$$= \frac{1}{1+\sqrt{3}} + \frac{1}{1+\sqrt{3}} + \frac{1}{\sqrt{5}+\sqrt{7}} + \frac{1}{\sqrt{5}+\sqrt{7}} + \cdots + \frac{1}{\sqrt{9997}+\sqrt{9999}} + \frac{1}{\sqrt{9997}+\sqrt{9999}}$$
$$> \frac{1}{1+\sqrt{3}} + + \frac{1}{\sqrt{3}+\sqrt{5}} + \frac{1}{\sqrt{5}+\sqrt{7}} + \frac{1}{\sqrt{7}+\sqrt{9}} + \cdots + \frac{1}{\sqrt{9997}+\sqrt{9999}} + \frac{1}{\sqrt{9999}+\sqrt{10001}}$$
$$= \frac{\sqrt{3}-1}{2} + \frac{\sqrt{5}-\sqrt{3}}{2} + \frac{\sqrt{7}-\sqrt{5}}{2} + \cdots + \frac{\sqrt{10001}-\sqrt{9999}}{2} = \frac{\sqrt{10001}-1}{2} > \frac{100-1}{2} > 48,$$

whence $\frac{1}{1+\sqrt{3}} + \frac{1}{\sqrt{5}+\sqrt{7}} + \cdots + \frac{1}{\sqrt{9997}+\sqrt{9999}} > 24$.

14.6. We have that the points $A(a, b)$, $\quad B(c, d)$, $\quad C(e, f)$ are inside the circle with center $O(2, 1)$ and radius $\sqrt{3}$.

Consider the vectors $\overrightarrow{BA}\{a-c, b-d\}, \overrightarrow{BC}\{e-c, f-d\}$ and $\overrightarrow{BD}\{f-d, c-e\}$. Note that $\overrightarrow{BC}\cdot\overrightarrow{BD} = 0$, and therefore, $\left(\overrightarrow{BA}, \overset{\wedge}{} \overrightarrow{BD}\right) =$

$|90^\circ - \beta|$, or $\left(\overrightarrow{BA}, \overset{\wedge}{} \overrightarrow{BD}\right) = |90^\circ + \beta|$, where $\left(\overrightarrow{BA}, \overset{\wedge}{} \overrightarrow{BC}\right) = \beta$.

It follows that $\left|\overrightarrow{BA}\cdot\overrightarrow{BD}\right| = BA \cdot BD \cdot |\cos(90^\circ \pm \beta)| = BA \cdot BD \sin\beta = BA \cdot BC \sin\beta = 2S_{\Delta ABC}$.

Hence, $|(a-c)(f-d) + (c-e)(b-d)| = 2S_{\Delta ABC} \leq 2\cdot\frac{9\sqrt{3}}{4} = \frac{9\sqrt{3}}{2}$, since among all triangles inscribed in a circle of radius $\sqrt{3}$, the one with the largest area is an equilateral triangle (see the proof of Problem 2.7).

It is left to note that the expression $(a-c)(f-d)+(c-e)(b-d)$ can assume the value $\frac{9\sqrt{3}}{2}$.

Thus, the greatest possible value of the given expression is $\frac{9\sqrt{3}}{2}$.

14.7. (a) If we define $\frac{a_1}{a_2} = 1+\alpha_1,\ \frac{a_2}{a_3} = 1+\alpha_2, \dots, \frac{a_k}{a_{k+1}} = 1+\alpha_k$, then we have $-1 < \alpha_1 < 0, \dots, -1 < \alpha_k < 0$.

We have $(1+\alpha_1)\cdots(1+\alpha_k) = \frac{a_1}{a_{k+1}}$.

Since the sequence (a_k) is unbounded, there exists a number k_0 such that $k \geq k_0\ \frac{a_1}{a_{k+1}} < \frac{1}{2}$, and therefore, $\frac{1}{2} > (1+\alpha_1)\cdots(1+\alpha_k) \geq 1+\alpha_1+\dots+\alpha_k$ (see Problem 10.6), and hence $\frac{a_1}{a_2}+\frac{a_2}{a_3}+\dots+\frac{a_k}{a_{k+1}} < k-\frac{1}{2}$.

(b) Let the numbers $k_1, k_2, \dots, k_{2\cdot 1985}$ be such that $\frac{a_1}{a_2}+\dots+\frac{a_{k_1}}{a_{k_1+1}} < k_1-\frac{1}{2}$, $\frac{a_{k_1+1}}{a_{k_1+2}}+\dots+\frac{a_{k_2}}{a_{k_2+1}} < (k_2-k_1)-\frac{1}{2}, \dots, \frac{a_{k_{s-1}+1}}{a_{k_{s-1}+2}}+\dots+\frac{a_{k_s}}{a_{k_s+1}} < (k_s-k_{s-1})-\frac{1}{2}$ $(S = 2\cdot 1985)$.

Summing these inequalities, we obtain

$\frac{a_1}{a_2}+\frac{a_2}{a_3}+\dots+\frac{a_{k_s}}{a_{k_s+1}} < k_s - S\cdot\frac{1}{2} = k_s - 1985$; hence for $k \geq k_S$, it follows that $\frac{a_1}{a_2}+\frac{a_2}{a_3}+\dots+\frac{a_k}{a_{k+1}} < k-1985$, since $\frac{a_m}{a_{m+1}} < 1$.

14.8. $\sum\limits_{i=1}^{n}(a_i+b_i)^r = \sum\limits_{i=1}^{n} a_i(a_i+b_i)^{r-1} + \sum\limits_{i=1}^{n} b_i(a_i+b_i)^{r-1}$.

By Problem 11.12, we have

$$\sum_{i=1}^{n} a_i(a_i+b_i)^{r-1} \leq \left(\sum_{i=1}^{n} a_i^r\right)^{1/r}\cdot\left(\sum_{i=1}^{n}\left((a_i+b_i)^{r-1}\right)^{r/(r-1)}\right)^{(r-1)/r}$$

and

$$\sum_{i=1}^{n} b_i(a_i+b_i)^{r-1} \leq \left(\sum_{i=1}^{n} b_i^r\right)^{1/r}\cdot\left(\sum_{i=1}^{n}\left((a_i+b_i)^{r-1}\right)^{r/(r-1)}\right)^{(r-1)/r}.$$

It follows that

$$\sum_{i=1}^{n}(a_i+b_i)^r \le \left(\sum_{i=1}^{n}(a_i+b_i)^r\right)^{(r-1)/r}\left(\left(\sum_{i=1}^{n}a_i^r\right)^{1/r}+\left(\sum_{i=1}^{n}b_i^r\right)^{1/r}\right),$$

or

$$\left(\sum_{i=1}^{n}(a_i+b_i)^r\right)^{1/r} \le \left(\sum_{i=1}^{n}a_i^r\right)^{1/r}+\left(\sum_{i=1}^{n}b_i^r\right)^{1/r}.$$

14.9. For $b < f(a)$ we have that $\int\limits_0^a f(x)dx = S_1 + S_2$, $\int\limits_0^b f^{-1}(y)dy = S_3$, where $ab = S_1 + S_3$, and therefore, $ab \le \int\limits_0^a f(x)dx + \int\limits_0^b f^{-1}(x)dx$.
For the case $b \ge f(a)$ the proof is similar.

14.10. (a)

$$\begin{aligned}&x_1(y_1-y_2)+(x_1+x_2)(y_2-y_3)+\cdots\\&\quad+(x_1+\cdots+x_{n-1})(y_{n-1}-y_n)\\&\qquad+(x_1+\cdots+x_n)y_n = x_1y_1+(x_1+x_2-x_1)y_2\\&\qquad\quad+\cdots+((x_1+\cdots+x_n)-(x_1+\cdots+x_{n-1}))y_n\\&\qquad= x_1y_1+x_2y_2+\cdots+x_ny_n.\end{aligned}$$

(b) We have

$$\begin{aligned}x_1y_{i_1}+\cdots+x_ny_{i_n} &= y_{i_1}(x_1-x_2)+(y_{i_1}+y_{i_2})(x_2-x_3)+\cdots+(y_{i_1}+\cdots+y_{i_{n-1}})(x_{n-1}-x_n)\\&\quad+(y_{i_1}+\cdots+y_{i_n})x_n \le y_1(x_1-x_2)+(y_1+y_2)(x_2-x_3)\\&\quad+\cdots+(y_1+\cdots+y_{n-1})(x_{n-1}-x_n)+(y_1+\cdots+y_n)x_n = x_1y_1+\cdots+x_ny_n.\end{aligned}$$

The second inequality can be proved in a similar way.

14.11. (a) Let $x_1 = \frac{a_1}{a_2}, x_2 = \frac{a_2}{a_3}, \ldots, x_{n-1} = \frac{a_{n-1}}{a_n}$, where $a_1 > 0, \ldots, a_n > 0$. In this case, $x_n = \frac{a_n}{a_1}$. Using Problem 14.10(b) for the numbers $a_1, a_2 \ldots, a_n$ and $\frac{1}{a_1}, \frac{1}{a_2}, \ldots, \frac{1}{a_n}$, we obtain $a_1\cdot\frac{1}{a_1}+a_2\cdot\frac{1}{a_2}+\cdots+a_n\cdot\frac{1}{a_n} \le a_1\cdot\frac{1}{a_2}+a_2\cdot\frac{1}{a_3}+\cdots+a_n\cdot\frac{1}{a_1}$, or $n \le x_1+x_2+\cdots+x_n$.

(b) Note that

$$\begin{aligned}&(1+a_1\cdots a_n)\left(\frac{1}{a_1(a_2+1)}+\cdots+\frac{1}{a_n(a_1+1)}\right)\\&\quad=\frac{1}{a_1(a_2+1)}+\cdots+\frac{1}{a_n(a_1+1)}+\frac{a_2\cdots a_n}{a_2+1}+\cdots+\frac{a_1a_2\cdots a_{n-1}}{a_1+1}=A.\end{aligned}$$

For $n \ge 3$ we obtain $A \ge \frac{1}{a_1(a_2+1)}+\cdots+\frac{1}{a_n(a_1+1)}+\frac{a_1a_2}{a_1+1}+\frac{a_2a_3}{a_2+1}+\cdots+\frac{a_1a_n}{a_n+1} = B.$

One can easily prove that $\frac{1}{a}$, b and $\frac{1}{b+1}$, $\frac{a}{a+1}$, $(a, b > 0)$ have the same order; hence

$$\frac{1}{a(b+1)} + \frac{ab}{a+1} \geq \frac{1}{a} \cdot \frac{a}{a+1} + \frac{1}{b+1} \cdot b \tag{1}$$

(see Problem 14.10(b)).
By (1), we obtain

$$B \geq \left(\frac{1}{a_1+1} + \frac{a_2}{a_2+1}\right) + \left(\frac{1}{a_2+1} + \frac{a_3}{a_3+1}\right) + \cdots + \left(\frac{1}{a_n+1} + \frac{a_1}{a_1+1}\right)$$
$$= \left(\frac{1}{a_1+1} + \frac{a_1}{a_1+1}\right) + \cdots + \left(\frac{1}{a_n+1} + \frac{a_n}{a_n+1}\right) = n.$$

Therefore, $(1 + a_1 \cdots a_n)\left(\frac{1}{a_1(a_2+1)} + \cdots + \frac{1}{a_n(a_1+1)}\right) \geq n$.
For $n = 2$, we have $a_1, a_2 \leq 1$; hence

$$(1 + a_1a_2)\left(\frac{1}{a_1(a_2+1)} + \frac{1}{a_2(a_1+1)}\right) = \frac{a_1}{a_1+1} + \frac{a_2}{a_2+1} + \frac{1}{a_1(a_2+1)}$$
$$+ \frac{1}{a_2(a_1+1)} \geq \frac{a_1}{a_1+1} + \frac{a_2}{a_2+1} + \frac{1}{a_1+1} + \frac{1}{a_2+1} = 2.$$

(c) It is sufficient to prove that $\frac{3+A+M+S+\frac{1}{A}+\frac{1}{M}+\frac{1}{S}+\frac{A}{M}+\frac{M}{S}+\frac{S}{A}}{(A+1)(M+1)(S+1)} = \frac{1}{A(M+1)} + \frac{1}{M(S+1)} + \frac{1}{S(A+1)}$; then it is sufficient to use inequality of Problem 14.11(b) for $n = 3$.

(d) Let $\frac{1}{a(1+b)} + \frac{1}{b(1+c)} + \frac{1}{c(1+a)} \leq \frac{1}{a(1+c)} + \frac{1}{b(1+a)} + \frac{1}{c(1+b)}$. Then since $\frac{1}{a(1+b)} + \frac{1}{b(1+c)} + \frac{1}{c(1+a)} \leq \frac{1}{a(1+a)} + \frac{1}{b(1+b)} + \frac{1}{c(1+c)}$ (see Problem 14.10(b)). Hence

$$\left(\frac{1}{a} + \frac{1}{b} + \frac{1}{c}\right)\left(\frac{1}{1+a} + \frac{1}{1+b} + \frac{1}{1+c}\right)$$
$$\geq 3\left(\frac{1}{a(1+b)} + \frac{1}{b(1+c)} + \frac{1}{c(1+a)}\right) \geq \frac{9}{1+abc}$$

(see Problem 14.10(b)).

(e) It is sufficient to note that $\frac{2ab}{\sqrt{a+b}}$, $\frac{2bc}{\sqrt{b+c}}$, $\frac{2ac}{\sqrt{a+c}}$ and $\frac{1}{\sqrt{a+b}}$, $\frac{1}{\sqrt{b+c}}$, $\frac{1}{\sqrt{c+a}}$ have opposite order. Thus it follows that

$$\left(\frac{a}{\sqrt{a+b}} + \frac{b}{\sqrt{b+c}} + \frac{c}{\sqrt{c+a}}\right)^2 = \frac{a^2}{a+b} + \frac{b^2}{b+c} + \frac{c^2}{c+a} + \frac{2ab}{\sqrt{a+b}} \cdot \frac{1}{\sqrt{b+c}}$$
$$+ \frac{2bc}{\sqrt{b+c}} \cdot \frac{1}{\sqrt{c+a}} + \frac{2ac}{\sqrt{a+c}} \cdot \frac{1}{\sqrt{a+b}} \geq \frac{a^2}{a+b} + \frac{b^2}{b+c} + \frac{c^2}{c+a} + \frac{2ab}{\sqrt{a+b}} \cdot \frac{1}{\sqrt{a+b}}$$
$$+ \frac{2bc}{\sqrt{b+c}} \cdot \frac{1}{\sqrt{b+c}} + \frac{2ac}{\sqrt{a+c}} \cdot \frac{1}{\sqrt{c+a}} = \frac{a+b+c}{2} + \left(\frac{a+b}{4} + \frac{ab}{a+b}\right) + \left(\frac{b+c}{4} + \frac{bc}{b+c}\right)$$
$$+ \left(\frac{c+a}{4} + \frac{ca}{c+a}\right) \geq \frac{a+b+c}{2} + \sqrt{ab} + \sqrt{bc} + \sqrt{ca} = \left(\frac{\sqrt{a}+\sqrt{b}+\sqrt{c}}{\sqrt{2}}\right)^2,$$

$$\text{or} \frac{a}{\sqrt{a+b}} + \frac{b}{\sqrt{b+c}} + \frac{c}{\sqrt{c+a}} \geq \frac{\sqrt{a}+\sqrt{b}+\sqrt{c}}{\sqrt{2}}.$$

14.12. Let us begin by proving that $\varphi(u) \geq \varphi(x)+(u-x)\varphi'(x)$. Consider the function $f(x) = \varphi(u) - \varphi(x) + (x-u)\varphi'(x)$. Then $f'(x) = (x-u)\varphi''(x)$, and hence in the case $x > u$, it follows that $f(x)$ is increasing, and for $x < u$ it follows that $f(x)$ is decreasing. Thus, it follows that $f(x) \geq f(u) = 0$, or $\varphi(u) \geq \varphi(x) + (u-x)\varphi'(x)$, and therefore $\varphi(x_1) \geq \varphi(y_1) + (x_1 - y_1)\varphi'(y_1), \ldots, \varphi(x_n) \geq \varphi(y_n) + (x_n - y_n)\varphi'(y_n)$.
We obtain $\varphi(x_1) + \cdots + \varphi(x_n) \geq \varphi(y_1) + \cdots + \varphi(y_n) + (x_1 - y_1)\varphi'(y_1) + \cdots + (x_n - y_n)\varphi'(y_n)$.
Now let us prove that $(x_1 - y_1)\varphi'(y_1) + \cdots + (x_n - y_n)\varphi'(y_n) \geq 0$.
By Problem 14.10(a), we have

$$\begin{aligned}(x_1 - y_1)\varphi'(y_1) + \cdots + (x_n - y_n)\varphi'(y_n) &= (x_1 - y_1)\big(\varphi'(y_1) - \varphi'(y_2)\big)\\ &+ (x_1 - y_1 + x_2 - y_2)\big(\varphi'(y_2) - \varphi'(y_3)\big) + \cdots + (x_1 - y_1 + x_2 - y_2 + \cdots + x_n - y_n)\varphi'(y_n) \geq 0,\end{aligned}$$

since $x_1 + x_2 + \cdots + x_k \geq y_1 + y_2 + \cdots + y_k$ and $\varphi'(y_{k-1}) - \varphi'(y_k) \geq 0$ $(k = 1, \ldots, n-1)$ ($\varphi'(x)$ is increasing).
We deduce that $\varphi(x_1) + \cdots + \varphi(x_n) \geq \varphi(y_1) + \cdots + \varphi(y_n)$.

14.13. Consider the numbers $\ln a_1, \ldots, \ln a_n$ and $\ln b_1, \ldots, \ln b_{n-1}, \ln b_n'$, where $b_n' = \frac{a_1 \cdots a_n}{b_1 \cdots b_{n-1}}$, and the function $\varphi(x) = e^x$ Then from Problem 14.12, it follows that $\varphi(\ln b_1) + \cdots + \varphi\big(\ln b_n'\big) \geq \varphi(\ln a_1) + \cdots + \varphi(\ln a_n)$, or $b_1 + \cdots + b_{n-1} + b_n' \geq a_1 + \cdots + a_n$.
On the other hand, $b_n' = \frac{a_1 \cdots a_n}{b_1 \cdots b_{n-1}} \leq b_n$, and therefore, $b_1 + \cdots + b_n \geq a_1 + \cdots + a_n$.

14.14. Consider the function $f(x) = -\sin\frac{x}{2}$ in $\left(0, \frac{\pi}{2}\right)$.
We have $f''(x) = \frac{1}{4}\sin\frac{x}{2} > 0$ if $x \in \left(0, \frac{\pi}{2}\right)$.
Let $\alpha \geq \beta \geq \gamma$. Then we have $\alpha \geq \frac{\pi}{3}$, $\alpha + \beta \geq \frac{\pi}{3} + \frac{\pi}{3}$, $\alpha + \beta + \gamma = \frac{\pi}{3} + \frac{\pi}{3} + \frac{\pi}{3}$, and therefore, by the inequality of Problem 14.12, we have $f(\alpha) + f(\beta) + f(\gamma) \geq f\left(\frac{\pi}{3}\right) + f\left(\frac{\pi}{3}\right) + f\left(\frac{\pi}{3}\right)$, or $\sin\frac{\alpha}{2} + \sin\frac{\beta}{2} + \sin\frac{\gamma}{2} \leq \frac{3}{2}$.
On the other hand, $\frac{\pi}{2} > \alpha$, $\frac{\pi}{2} + \frac{\pi}{2} > \alpha + \beta$, $\frac{\pi}{2} + \frac{\pi}{2} + 0 = \alpha + \beta + \gamma$, whence $f\left(\frac{\pi}{2}\right) + f\left(\frac{\pi}{2}\right) + f(0) > f(\alpha) + f(\beta) + f(\gamma)$, or $\sqrt{2} < \sin\frac{\alpha}{2} + \sin\frac{\beta}{2} + \sin\frac{\gamma}{2}$.

14.15. The proof of the inequality $\sqrt{\sin\alpha} + \sqrt{\sin\beta} + \sqrt{\sin\gamma} > 0$ is obvious. Let us prove the second inequality. Consider the function $f(x) = -\sqrt{\sin x}$ in $(0, \pi)$. Therefore $f''(x) = \frac{2\sin^2 x + \cos^2 x}{4\sin x\sqrt{\sin x}} > 0$ if $x \in (0, \pi)$.
Since in the case $\alpha \geq \beta \geq \gamma$ one has $\alpha > \frac{\pi}{2}$, $\alpha + \beta > \frac{\pi}{2} + \frac{\pi}{4}$, $\alpha + \beta + \gamma = \frac{\pi}{2} + \frac{\pi}{4} + \frac{\pi}{4}$, it follows by the inequality of Problem 14.12 that $f(\alpha) + f(\beta) + f(\gamma) > f\left(\frac{\pi}{2}\right) + f\left(\frac{\pi}{4}\right) + f\left(\frac{\pi}{4}\right)$, or $\sqrt{\sin\alpha} + \sqrt{\sin\beta} + \sqrt{\sin\gamma} < 1 + \sqrt[4]{8}$.
Alternative proof. Note that $x + y \leq \sqrt{2(x^2 + y^2)}$. Therefore for $\alpha > \frac{\pi}{2}$, we have

$$\begin{aligned}\sqrt{\sin\alpha} + \sqrt{\sin\beta} + \sqrt{\sin\gamma} &< 1 + \sqrt{\sin\beta} + \sqrt{\sin\gamma} \leq 1\\ &+ \sqrt{2(\sin\beta + \sin\gamma)} \leq 1 + \sqrt{4\sin\frac{\beta+\gamma}{2}} < 1 + \sqrt{2\sqrt{2}} = 1 + \sqrt[4]{8}.\end{aligned}$$

14.16. We have

$$1+\frac{1}{\sqrt{2}}+\frac{1}{\sqrt{3}}+\cdots+\frac{1}{\sqrt{n}}>\frac{2}{1+\sqrt{2}}+\frac{2}{\sqrt{2}+\sqrt{3}}+\cdots+\frac{2}{\sqrt{n}+\sqrt{n+1}}$$
$$=2\left(\sqrt{2}-1\right)+2\left(\sqrt{3}-\sqrt{2}\right)+\cdots+2\left(\sqrt{n+1}-\sqrt{n}\right)=2\sqrt{n+1}-2>2\sqrt{n}-2.$$

14.17. Consider the polynomial $P(x) = (x-a)(x-b)(x-c)(x-d)$. Since $P(x)=x^4-S_1x^3+S_2x^2-S_3x+S_4$ and $P(a)=0$, we must have

$$a^4 = S_1a^3 - S_2a^2 + S_3a - S_4. \tag{1}$$

If $a \leq 0$, then $S_1a^3 - S_2a^2 + S_3a - S_4 < 0$, and $a^4 \geq 0$, which leads to a contradiction with (1). Therefore, $a > 0$. In a similar way one can prove that $b > 0,\ c > 0,\ d > 0$.

14.18. Consider the sequence $a_n = \frac{n!}{n\left(\frac{n}{e}\right)^n}$.
Thus, it follows that

$$\frac{a_{n+1}}{a_n} = \frac{e}{\left(1+\frac{1}{n}\right)^{n+1}} < \frac{1}{\sqrt{1+\frac{1}{n}}}$$

(see Problem 9.32(d)), and therefore, $a_{n+1} < \frac{\sqrt{n}}{\sqrt{n+1}} \cdot a_n$.
We have

$$a_n < \frac{\sqrt{n-1}}{\sqrt{n}} \cdot a_{n-1} < \frac{\sqrt{n-1}}{\sqrt{n}} \cdot \frac{\sqrt{n-2}}{\sqrt{n-1}} \cdot a_{n-2}$$
$$< \cdots < \frac{\sqrt{n-1}}{\sqrt{n}} \cdot \frac{\sqrt{n-2}}{\sqrt{n-1}} \cdots \frac{\sqrt{1}}{\sqrt{2}} \cdot a_1 = \frac{a_1}{\sqrt{n}} = \frac{e}{\sqrt{n}}.$$

Hence $a_n < \frac{e}{\sqrt{n}}$, and therefore, if $n \geq 8$, then $a_n < \frac{e}{\sqrt{8}} < 1$.

14.19. By induction, one can easily prove that $2^n \geq n^2 - 1$, $n \in \mathbb{N}$.
It follows that

$$\left(1+\frac{1}{2^1}\right)\cdots\left(1+\frac{1}{2^n}\right) \leq \left(1+\frac{1}{2}\right)\cdot\left(1+\frac{1}{2^2-1}\right)\cdots\left(1+\frac{1}{n^2-1}\right)$$
$$= \frac{3}{2}\cdot\frac{2^2}{1\cdot 3}\cdots\frac{n^2}{(n-1)\cdot(n+1)} = \frac{3n}{n+1} < 3.$$

14.20. Let $a, b, c \in [0, a_1]$ and $a \geq b \geq c$. If $b \leq \frac{a+c}{2}$, then the following conditions hold: $a \geq a-b+c, a+c = (a-b+c)+b$ and $a-b+c \geq b$. Therefore, by Problem 2.8, it follows that $g(a)+g(c) \geq g(a-b+c)+g(b)$, or $g(a)-g(b)+g(c) \geq g(a-b+c)$.
If $b \geq \frac{a+c}{2}$, then $a \geq b$, $a+c = b+(a-b+c)$ and $b \geq a-b+c$. Therefore, $g(a)+g(c) \geq g(a-b+c)+g(b)$.

It follows that $g(a) - g(b) + g(c) \geq g(a - b + c)$.
Hence using Problem 7.5, we obtain the proof of the given inequality (taking $f(x) = g(x) - g(0)$).

14.21. (a), (b) Let $p(1) = a + b + c = m,\ \ p(-1) = a - b + c = n$. Then $|m| \leq 1,\ \ |n| \leq 1$, and $|c| = |p(0)| \leq 1$.
Note that $b = \frac{m-n}{2},\quad a = \frac{m+n}{2} - c$, and therefore,

$$|a| + |b| + |c| = \left|\frac{m+n}{2} - c\right| + \left|\frac{m-n}{2}\right| + |c|$$
$$\leq \left|\frac{m+n}{2}\right| + \left|\frac{m-n}{2}\right| + 2|c| \leq \max(|m|, |n|) + 2|c| \leq 3.$$

We have

$$a^2 + b^2 + c^2 = \frac{m^2 + n^2}{2} + 2c^2 - mc - nc$$
$$\leq \frac{m^2 + n^2}{2} + 2c^2 + \frac{m^2 + c^2}{2} + \frac{n^2 + c^2}{2} = m^2 + n^2 + 3c^2 \leq 5.$$

(c) Consider the polynomial $F(x) = a + bx + cx^2 + dx^3$. Then by Lagrange's formula, [1]
we have

$$F(x) = F(-2) \cdot \frac{(x+1)(x-1)(x-2)}{-12} + F(-1) \cdot \frac{(x+2)(x-1)(x-2)}{6}$$
$$+ F(1) \cdot \frac{(x+2)(x+1)(x-2)}{-6} + F(2) \cdot \frac{(x+2)(x+1)(x-1)}{12}.$$

Hence $a = F(0) = \frac{F(-2)}{-6} + \frac{2}{3}F(-1) + \frac{2}{3}F(1) - \frac{1}{6}F(2)$.
On the other hand, $F(x) = x^3 p\left(\frac{1}{x}\right),\ \ x \neq 0$, and therefore $F(2) = 8p\left(\frac{1}{2}\right),\ \ F(-2) = -8p\left(-\frac{1}{2}\right),\ \ F(1) = p(1),\ \ F(-1) = -p(-1)$.
Thus, $a = \frac{4}{3}p\left(-\frac{1}{2}\right) - \frac{2}{3}p(-1) + \frac{2}{3}p(1) - \frac{4}{3}p\left(\frac{1}{2}\right)$, and it follows that $|a| \leq \frac{4}{3}\left|p\left(-\frac{1}{2}\right)\right| + \frac{2}{3}|p(-1)| + \frac{2}{3}|p(1)| + \frac{4}{3}\left|p\left(\frac{1}{2}\right)\right| \leq \frac{4}{3} + \frac{2}{3} + \frac{2}{3} + \frac{4}{3} = 4$. Note that the polynomial $F(x) = 4x^3 - 3x$ satisfies the assumption of the problem (if we set $x = cos\alpha$, then $F(x) = cos3\alpha$).

14.22. Let the roots of the polynomial $P(x)$ be the numbers $-b_1, -b_2, \ldots, -b_n$, where $b_i > 0\ i = 1, 2, \ldots, n$.
We have $P(x) = (x + b_1) \cdots (x + b_n)$, and hence $a_0 = b_1 \cdots b_n$ and $a_1 = a_0\left(\frac{1}{b_1} + \cdots + \frac{1}{b_n}\right)$. Thus, we need to prove that

[1] If $P(x)$ is a polynomial of degree n, and $x_1, x_2, \ldots, x_n, x_{n+1}$ are distinct real numbers, then
$$P(x) = P(x_1) \cdot \frac{(x - x_2) \cdots (x - x_{n+1})}{(x_1 - x_2) \cdots (x_1 - x_{n+1})} + P(x_2) \cdot \frac{(x - x_1)(x - x_3) \cdots (x - x_{n+1})}{(x_2 - x_1)(x_2 - x_3) \cdots (x_2 - x_{n+1})}$$
$$+ \cdots + P(x_{n+1}) \cdot \frac{(x - x_1) \cdots (x - x_n)}{(x_{n+1} - x_1) \cdots (x_{n+1} - x_n)}.$$

$$\left(\frac{1}{b_1}+\cdots+\frac{1}{b_n}\right)(1+b_1)(1+b_2)\cdots(1+b_n)\geq 2n^2, n=2,3,\ldots.$$

By Cauchy's inequality, it follows that $\frac{1}{b_1}+\cdots+\frac{1}{b_n}\geq\frac{n}{\sqrt[n]{b_1b_2\cdots b_n}}$, and $1+b_i=\underbrace{\frac{1}{n-1}+\ldots+\frac{1}{n-1}}_{n-1}+b_i\geq n\cdot\sqrt[n]{\frac{b_i}{(n-1)^{n-1}}}$, $i=1,\ldots,n$. Therefore,

$$\left(\frac{1}{b_1}+\cdots+\frac{1}{b_n}\right)(1+b_1)\cdots(1+b_n)\geq n^2\left(1+\frac{1}{n-1}\right)^{n-1}$$
$$\geq n^2\left(1+(n-1)\cdot\frac{1}{n-1}\right)=2n^2.$$

In order to complete the proof, it is sufficient to use Bernoulli's inequality.

14.23. We call two segments having a common vertex a *bird*, and the segments themselves *wings*. The number of non-single-color triangles is equal to $C_n^3-t(n)$. Each non-single-color triangle contains *birds* with *wings* of distinct colors. Therefore, the number of birds with *wings* of distinct colors is equal to $2(C_n^3-t(n))$.

On the other hand, if through some vertex, k segments of the first color and $n-1-k$ segments of the second color pass, then the number of birds with *wings* of distinct colors containing that vertex (that is, both *wings* contain that vertex) is equal to $k(n-1-k)$.

Note that

$$k(n-1-k)\leq\begin{cases}\left(\frac{n-1}{2}\right)^2, & \text{if } n \text{ is even,}\\ \frac{n}{2}\left(\frac{n}{2}-1\right), & \text{if } n \text{ is odd.}\end{cases}$$

Thus, it follows that

$$2\left(C_n^3-t(n)\right)\leq\begin{cases}\frac{(n-1)^2n}{4}, & \text{if } n \text{ is odd,}\\ \frac{n^2}{2}\left(\frac{n}{2}-1\right), & \text{if } n \text{ is even.}\end{cases}$$

Hence, we obtain

if $n=2k$, then $t(n)\geq\frac{n(n-1)(n-2)}{6}-\frac{n^2(n-2)}{8}=\frac{k(k-1)(k-2)}{3}$,

if $n=4k+1$, then $t(n)\geq\frac{n(n-1)(n-2)}{6}-\frac{n(n-1)^2}{8}=\frac{2}{3}k(k-1)(4k+1)$,

if $n=4k+3$, then $t(n)\geq\frac{n(n-1)(n-2)}{6}-\frac{n(n-1)^2}{8}=\frac{2}{3}k(k+1)(4k-1)$.

14.24. We proceed by induction.

If $n=3,4$, then the statement of the problem is true.

Assume that the statement holds for $n\leq k-1$ points. Let us prove that it holds for $n=k$ points ($k\geq 5$).

Consider the following two cases.

(a) Every edge is part of a triangle.
In this case, the number of triangles formed is greater than $\frac{k^2}{12}$, and therefore not less than $\left[\frac{k}{2}\right]$, $k = 5, 6, \ldots$.

(b) There exists an edge AB that does not belong to any triangle. Let us denote all other vertices of the graph by $A_1, A_2, \ldots, A_{k-2}$. Hence, each vertex A_i is not connected to both A and B $(i = 1, 2, \ldots, k-2)$. Therefore, the number of edges with vertices $A_1, A_2, \ldots, A_{k-2}$ is greater than $\frac{k^2}{4} - (k - 1) = \frac{(k-2)^2}{4}$, and hence the number of triangles formed by these vertices is at least $\left[\frac{k-2}{2}\right] = \left[\frac{k}{2}\right] - 1$.

Let one of the triangles be $A_1A_2A_3$. If none of the vertices A_i is connected to A or B, then we will not consider the segment A_1A_2. Then the points $A_1, A_2, \ldots, A_{k-2}$ are connected to $\frac{(k-2)^2}{4}$ segments. Therefore, there exist at least $\left[\frac{k}{2}\right] - 1$ triangles, and with $A_1A_2A_3$, there exist $\left[\frac{k}{2}\right]$ triangles. It is left to consider the case in which A is connected to the vertices $A_1, A_2, \ldots, A_m$, and B is connected to the rest of them. If the segment A_iA_j is an edge for some (i, j), $i < j \le m$ or $m < i < j$, then this ends the proof. Otherwise, the number of edges with vertices $A_1, A_2, \ldots, A_{k-2}$ is not greater than $m(k - 2 - m)$, which leads to a contradiction, since $m(k - 2 - m) \le \frac{(k-2)^2}{4}$.

14.25. Note that the numbers $\{n\alpha\}$ are distinct numbers. We argue by contradiction. Assume that $\{i\alpha\} = \{j\alpha\}$, $i \ne j$. Then $i\alpha = j\alpha + k$, $k \in \mathbb{Z}$, and hence $\alpha = \frac{k}{i-j} \in \mathbb{Q}$, which leads to a contradiction.
Let $m \in \mathbb{N}$ and $\frac{1}{m} < \min(a, b - a, 1 - b)$. There exist positive integers $i > j$ such that $|\{i\alpha\} - \{j\alpha\}| < \frac{1}{m}$ (it is sufficient to divide the segment $[0, 1]$ into $m + 1$ equal parts and use Dirichlet's principle). Therefore, $\left|(i-j)\alpha - [i\alpha] + [j\alpha]\right| < \frac{1}{m}$, and hence $\{(i-j)\alpha\} < \frac{1}{m}$, or $\{(i-j)\alpha\} > 1 - \frac{1}{m}$. Let $\{(i-j)\alpha\} < \frac{1}{m}$, $(i-j)\alpha = x + k$, $k \in \mathbb{Z}$, $x = \{(i-j)\alpha\}$. Consider the numbers $x, 2x, 3x, \ldots$. One of these numbers belongs to (a, b), so suppose $a < lx < b$, $l \in \mathbb{N}$. We have that $l(i-j)\alpha = lx + kl$; hence $\{l(i-j)\alpha\} = lx \in (a, b)$.
Let $\{(i-j)\alpha\} > 1 - \frac{1}{m}$, $(i-j)\alpha = x + k$, $k \in \mathbb{Z}$, $x = \{(i-j)\alpha\}$. We have $1 - x < \frac{1}{m}$.
Consider the numbers $1 - x$, $2(1 - x), \ldots$. One of these numbers belongs to $(1 - b, 1 - a)$, so let $1 - b < l(1 - x) < 1 - a$, $l(i-j)\alpha = lx + lk = -l(1-x) + l + lk$. Therefore, $l(k+1) + a - 1 < l(i-j)\alpha < l(k+1) + b - 1$, or $a < \{l(i-j)\} < b$.

14.26. (a) We have $\sum\limits_{i=1}^{n}(x_i - m)(M - x_i) \ge 0$, and therefore $\sum\limits_{i=1}^{n} x_i^2 \le -nmM$.

(b) Note that $M \ge 0 \ge m$, whence $\sum\limits_{i=1}^{n}(M(x_i - m)(M^3 - x_i^3) - m(x_i^3 - m^3)(M - x_i)) \ge 0$.

Thus, we deduce that $(M-m)\sum_{i=1}^{n} x_i^4 \le -n\big(M^4 m - m^4 M\big)$. Therefore, if $M \ne m$, then we have $\sum_{i=1}^{n} x_i^4 \le -mMn\big(m^2+M^2+mM\big)$. If $m=M=0$, then it follows that $\sum_{i=1}^{n} x_i^4 = 0 = -mMn\big(m^2+M^2+mM\big)$.

14.27. We have

$$\left(x^2+y^2\right)^m - x^{2m} - y^{2m} = \frac{C_m^1\left(x^{2m-2}y^2 + x^2y^{2m-2}\right) + \ldots + C_m^{m-1}\left(x^2y^{2m-2} + y^2x^{2m-2}\right)}{2}$$
$$\ge C_m^1 x^m y^m + C_m^2 x^m y^m + \cdots + C_m^{m-1} x^m y^m = (2^m-2)x^m y^m.$$

Here we have used that $\left(x^2\right)^k \cdot \left(y^2\right)^{m-k} + \left(x^2\right)^{m-k} \cdot \left(y^2\right)^k \ge 2\sqrt{x^{2m}\cdot y^{2m}} \ge 2x^m \cdot y^m$, $k=1,\cdots,m-1$.
Therefore, $\left(x^2+y^2\right)^m - x^{2m} - y^{2m} \ge (2^m-2)x^m y^m$, or $\left(x^2+y^2\right)^m \ge (x^m-y^m)^2 + 2^m x^m y^m$.
This ends the proof.

14.28. If $\alpha \le \frac{\pi}{2}$, $\beta \le \frac{\pi}{2}$, $\gamma \le \frac{\pi}{2}$, then we have $\sin\alpha\,\cos(\beta-\gamma) = \frac{1}{2}(\sin 2\gamma + \sin 2\beta) \ge \sqrt{\sin 2\gamma\,\sin 2\beta}$, and thus $\cos(\beta-\gamma) \ge \frac{\sqrt{\sin 2\gamma\,\sin 2\beta}}{\sin\alpha}$.
Therefore,

$$\cos(\alpha-\beta)\cos(\beta-\gamma)\cos(\gamma-\alpha) \ge \frac{\sqrt{\sin 2\alpha\,\sin 2\beta}}{\sin\gamma}\cdot\frac{\sqrt{\sin 2\gamma\,\sin 2\beta}}{\sin\alpha}\cdot\frac{\sqrt{\sin 2\alpha\,\sin 2\gamma}}{\sin\beta}$$
$$= 8\cos\alpha\,\cos\beta\,\cos\gamma.$$

Let $\alpha \le \beta < \frac{\pi}{2} < \gamma$.
If $\gamma \le \frac{\pi}{2} + \alpha$ or $\gamma \ge \frac{\pi}{2} + \beta$, then $\cos(\alpha-\beta)\cos(\beta-\gamma)\cos(\gamma-\alpha) \ge 0 > 8\cos\alpha\,\cos\beta\,\cos\gamma$.
If $\frac{\pi}{2}+\alpha < \gamma < \frac{\pi}{2}+\beta$, then $0 < -\cos(\gamma-\alpha) < -\cos\gamma$ and $0 < \cos(\alpha-\beta)\cos(\gamma-\beta) < \cos(\beta-\alpha) < 4\cos(\beta-\alpha) + 4\cos(\alpha+\beta) = 8\cos\alpha\,\cos\beta$; hence $\cos(\alpha-\beta)\cos(\beta-\gamma)\cos(\gamma-\alpha) > 8\cos\alpha\,\cos\beta\,\cos\gamma$.

14.29. Consider the following two cases.

(a) if $\sin\frac{x}{2} = 0$, we have $x = 2\pi n$, $n \in \mathbb{Z}$ and $\left|\sin x + \frac{\sin 2x}{2} + \cdots + \frac{\sin nx}{n}\right| = 0 < 3$.
(b) if $\sin\frac{x}{2} \ne 0$, then there exists a positive integer m such that $\frac{1}{m+1} < \left|\sin\frac{x}{2}\right| \le \frac{1}{m}$.

For $n \le m$ we have

$$\left|\sin x + \frac{\sin 2x}{2} + \cdots + \frac{\sin nx}{n}\right| \le |\sin x| + \frac{|\sin 2x|}{2} + \cdots + \frac{|\sin nx|}{n}$$
$$\le \underbrace{|\sin x| + |\sin x| + \cdots + |\sin x|}_{n} \le m|\sin x| \le 2m\left|\sin\frac{x}{2}\right| \le 2 \qquad (1)$$

(see Problems 1.13 and 7.4(a)).

Now let us estimate the following expression: $\left|\frac{\sin(m+1)x}{m+1} + \cdots + \frac{\sin nx}{n}\right|$ for $n > m$. By Problem 14.10, we have that

$$\frac{\sin(m+1)x}{m+1} + \cdots + \frac{\sin nx}{n} = \sin(m+1)x\left(\frac{1}{m+1} - \frac{1}{m+2}\right)$$
$$+ +(\sin(m+1)x + \sin(m+2)x)\left(\frac{1}{m+2} - \frac{1}{m+3}\right)$$
$$+ \cdots + (\sin(m+1)x + \cdots + \sin(n-1)x)\left(\frac{1}{n-1} - \frac{1}{n}\right)$$
$$+ (\sin(m+1)x + \cdots + \sin nx)\frac{1}{n}$$

Thus, it follows that

$$\left|\frac{\sin(m+1)x}{m+1} + \cdots + \frac{\sin nx}{n}\right| \leq |\sin(m+1)x|\left(\frac{1}{m+1} - \frac{1}{m+2}\right)$$
$$+ |\sin(m+1)x + \sin(m+2)x|\left(\frac{1}{m+2} - \frac{1}{m+3}\right)$$
$$+ \cdots + |\sin(m+1)x + \cdots + \sin(n-1)x|\left(\frac{1}{n-1} - \frac{1}{n}\right) + |\sin(m+1)x + \cdots + \sin nx|\frac{1}{n}$$
$$\leq \frac{1}{\left|\sin\frac{x}{2}\right|}\left(\frac{1}{m+1} - \frac{1}{m+2} + \cdots + \frac{1}{n-1} - \frac{1}{n} + \frac{1}{n}\right) = \frac{1}{(m+1)\left|\sin\frac{x}{2}\right|} < 1.$$

Hence

$$\left|\frac{\sin(m+1)x}{m+1} + \cdots + \frac{\sin nx}{n}\right| < 1. \tag{2}$$

Here we have used the following inequality:

$$|\sin kx + \sin(k+1)x + \cdots + \sin px| \leq \frac{1}{\left|\sin\frac{x}{2}\right|}.$$

Let us prove this inequality. We have

$$|\sin kx + \sin(k+1)x + \cdots + \sin px|$$
$$= \frac{\left|2\sin\frac{x}{2}\sin kx + \cdots + 2\sin\frac{x}{2}\sin px\right|}{2\left|\sin\frac{x}{2}\right|}$$
$$= \frac{\left|\cos\left(k-\frac{1}{2}\right)x - \cos\left(k+\frac{1}{2}\right)x + \cos\left(k+\frac{1}{2}\right)x - \cos\left(k+1+\frac{1}{2}\right)x + \cdots + \cos\left(p-\frac{1}{2}\right)x - \cos\left(p+\frac{1}{2}\right)x\right|}{2\left|\sin\frac{x}{2}\right|}$$
$$= \frac{\left|\cos\left(k-\frac{1}{2}\right)x - \cos\left(p+\frac{1}{2}\right)x\right|}{2\left|\sin\frac{x}{2}\right|} \leq \frac{2}{2\left|\sin\frac{x}{2}\right|} = \frac{1}{\left|\sin\frac{x}{2}\right|}.$$

From inequalities (1) and (2), it follows that

$$\left|\sin x + \frac{\sin 2x}{2} + \cdots + \frac{\sin nx}{n}\right| < 3.$$

14.30. We need to prove that

$$A = \frac{(a_1 + a_2 + a_3)(a_2 + a_3 + a_4)\cdots(a_n + a_1 + a_2)}{(a_1 + a_2)(a_2 + a_3)\cdots(a_n + a_1)} \geq \left(\frac{3}{2}\right)^n. \tag{1}$$

Let us prove (1) by mathematical induction.

(a) For $n = 3$, we have
$a_1 + a_2 + a_3 = \frac{a_1+a_2}{2} + \frac{a_2+a_3}{2} + \frac{a_3+a_1}{2} \geq 3\sqrt[3]{\frac{a_1+a_2}{2} \cdot \frac{a_2+a_3}{2} \cdot \frac{a_3+a_1}{2}}$. Therefore,
$\frac{(a_1+a_2+a_3)(a_2+a_3+a_1)(a_3+a_1+a_2)}{(a_1+a_2)(a_2+a_3)(a_3+a_1)} \geq \frac{3^3 \cdot \frac{(a_1+a_2)}{2}\frac{(a_2+a_3)}{2}\frac{(a_3+a_1)}{2}}{(a_1+a_2)(a_2+a_3)(a_3+a_1)} = \left(\frac{3}{2}\right)^3.$

(b) Let $n \geq 4$ and suppose that inequality (1) holds for $n - 1$ numbers. Let us first prove inequality (1) if $a_1 = a_2$ or $a_{n-1} = a_n$.

If $a_1 = a_2$, we have

$$\begin{aligned} A &= \frac{(2a_2 + a_3)(a_2 + a_3 + a_4)\cdots(a_{n-1} + a_n + a_2)(a_n + 2a_2)}{2a_2(a_2 + a_3)\cdots(a_{n-1} + a_n)(a_n + a_2)} \\ &= \frac{(2a_2 + a_3)(a_n + 2a_2)}{2a_2(a_n + a_2 + a_3)} \cdot \frac{(a_2 + a_3 + a_4)\cdots(a_{n-1} + a_n + a_2)(a_n + a_2 + a_3)}{(a_2 + a_3)\cdots(a_{n-1} + a_n)(a_n + a_2)} \\ &\geq \frac{(2a_2 + a_3)(a_n + 2a_2)}{2a_2(a_n + a_2 + a_3)} \cdot \left(\frac{3}{2}\right)^{n-1} \geq \frac{3}{2} \cdot \left(\frac{3}{2}\right)^{n-1} \geq \left(\frac{3}{2}\right)^n, \end{aligned}$$

since the inequality $(2a_2 + a_3)(a_n + 2a_2) \geq 3a_2(a_n + a_2 + a_3)$ is equivalent to the inequality $(a_2 - a_3)(a_2 - a_n) \geq 0$.
If $a_{n-1} = a_n$, the proof is similar.
Now let us prove that $a_1 < a_2$ and $a_{n-1} < a_n$. If we substitute the sequence of numbers $a_1, a_2, \ldots, a_{n-1}, a_n$ by the numbers $a_1' = a_1 + x, a_2, \ldots, a_{n-1}, a_n - x = a_n'$, where $0 \leq x \leq \min(a_2 - a_1, a_n - a_{n-1})$, then the value of A does not increase, and therefore,

$$A \geq \frac{(a_1 + a_2 + a_3 + x)(a_2 + a_3 + a_4)\cdots(a_{n-2} + a_{n-1} + a_n - x)(a_{n-1} + a_n + a_1)(a_n + a_1 + a_2)}{(a_1 + x + a_2)(a_2 + a_3)\cdots(a_{n-1} + a_n - x)(a_n + a_1)},$$

or

$$\frac{(a_1 + a_2 + a_3)(a_{n-2} + a_{n-1} + a_n)}{(a_1 + a_2)(a_{n-1} + a_n)} \geq \frac{(a_1 + a_2 + a_3 + x)(a_{n-2} + a_{n-1} + a_n - x)}{(a_1 + x + a_2)(a_{n-1} + a_n - x)}. \tag{2}$$

Note that the last inequality can be rewritten in the following way: $ax^2+bx+c \leq 0$, where $a = \frac{(a_1+a_2+a_3)(a_{n-2}+a_{n-1}+a_n)}{(a_1+a_2)(a_{n-1}+a_n)} - 1 > 0$, $c = 0$, and therefore, in order to prove (2) it is sufficient to prove it for the values $x = 0$ and $x = a_4 - a_1$

$(n=4)$, $x=a_5+a_4-a_2-a_1$ $(n=5)$, and $x=a_2-a_1$ $(n\geq 6)$, since $\min(a_2-a_1, a_n-a_{n-1})\leq a_2-a_1\leq a_4-a_1\leq a_4-a_1+a_5-a_2$.
If $x=0$, the proof of (2) is obvious.
If $n=4$, then $x=a_4-a_1$, and we need to prove that $\frac{(a_1+a_2+a_3)(a_2+a_3+a_4)}{(a_1+a_2)(a_3+a_4)}\geq\frac{(a_2+a_3+a_4)(a_2+a_3+a_1)}{(a_2+a_4)(a_1+a_3)}$, or $(a_4-a_1)(a_3-a_2)\geq 0$.
If $n=5$, then $x=a_5+a_4-a_2-a_1$, and the proof of (2) is obvious.
For $n\geq 6$, then $x=a_2-a_1$, and we need to prove that $ax+b\leq 0$. Therefore,

$$\left(\frac{(a_1+a_2+a_3)(a_{n-2}+a_{n-1}+a_n)}{(a_1+a_2)(a_{n-1}+a_n)}-1\right)(a_2-a_1)$$
$$+\left(a_n+a_{n-1}+a_{n-2}-a_1-a_2-a_3-\frac{(a_1+a_2+a_3)(a_{n-2}+a_{n-1}+a_n)}{a_1+a_2}\right.$$
$$\left.+\frac{(a_1+a_2+a_3)(a_{n-2}+a_{n-1}+a_n)}{a_{n-1}+a_n}\right)\leq 0,$$

or

$$\frac{(a_3a_{n-2}+a_3(a_{n-1}+a_n)+a_{n-2}(a_1+a_2))(a_2-a_1)}{(a_1+a_2)(a_{n-1}+a_n)}$$
$$+\left(a_{n-2}-a_3-\frac{(a_{n-1}+a_n-a_1-a_2)(a_3a_{n-2}+a_3(a_{n-1}+a_n)+a_{n-2}(a_1+a_2))}{(a_1+a_2)(a_{n-1}+a_n)}\right)\leq 0,$$

$(a_3a_{n-2}+a_3(a_{n-1}+a_n)+a_{n-2}(a_1+a_2))(2a_2-a_n-a_{n-1})+(a_{n-2}-a_3)(a_1+a_2)(a_{n-1}+a_n)\leq 0$,
$2a_2a_{n-2}(a_1+a_2+a_3)+(a_{n-1}+a_n)(a_2a_3-a_3a_{n-2}-a_3a_{n-1}-a_3a_n-a_3a_1)\leq 0$,
$2a_2a_{n-2}(a_1+a_2+a_3)\leq(a_3a_{n-1}+a_3a_n)(a_{n-2}+a_{n-1}+a_n+a_1-a_2)$, where the last inequality holds because $2a_2a_{n-2}\leq a_3a_{n-2}+a_3a_{n-2}\leq a_3a_{n-1}+a_3a_n$ and $a_1+a_2+a_3\leq a_{n-2}+a_{n-1}+a_n+a_1-a_2$.
Taking $x=\min(a_2-a_1, a_n-a_{n-1})$, we obtain for the numbers $a_1', a_2, \ldots, a_{n-1}, a_n'$ the case $a_1'=a_2$ or $a_{n-1}=a_n'$, such that $a_1'\leq a_2\leq\cdots\leq a_{n-1}\leq a_n'$ Thus, it follows that $A\geq\frac{(a_1'+a_2+a_3)(a_2+a_3+a_4)\cdots(a_n'+a_1+a_2)}{(a_1'+a_2)(a_2+a_3)\cdots(a_n'+a_1)}\geq\left(\frac{3}{2}\right)^n$.
This ends the proof of the given inequality.

14.31. (a) Note that $yz<1$, $xz<1$, $xy<1$; therefore $\frac{x}{1-yz}+\frac{y}{1-zx}+\frac{z}{1-xy}\geq x+y+z\geq x^2+y^2+z^2=1$, and hence $\frac{x}{1-yz}+\frac{y}{1-zx}+\frac{z}{1-xy}\geq 1$.
Note that $yz\leq\frac{y^2+z^2}{2}$, $xz\leq\frac{z^2+x^2}{2}$, $xy\leq\frac{x^2+y^2}{2}$; therefore,

$$\frac{x}{1-yz}+\frac{y}{1-zx}+\frac{z}{1-xy}\leq\frac{x}{1-\frac{y^2+z^2}{2}}+\frac{y}{1-\frac{z^2+x^2}{2}}+\frac{z}{1-\frac{x^2+y^2}{2}}$$
$$=\frac{2x}{1+x^2}+\frac{2y}{1+y^2}+\frac{2z}{1+z^2}.$$

On the other hand, note that

$$\left(\frac{2}{1+x^2}-\frac{2}{1+y^2}\right)\left(\frac{x}{1+x^2}-\frac{y}{1+y^2}\right)=-\frac{2(x-y)^2(x+y)(1-xy)}{(1+x^2)^2(1+y^2)^2}\le 0,$$

and thus from inequality (8.5.1) (see Chapter 8) we obtain

$$\frac{2x}{1+x^2}+\frac{2y}{1+y^2}+\frac{2z}{1+z^2}\le\frac{6(x+y+z)}{3+x^2+y^2+z^2}\le\frac{3}{2}\sqrt{3(x^2+y^2+z^2)}=\frac{3\sqrt{3}}{2}.$$

It follows that $\frac{x}{1-yz}+\frac{y}{1-zx}+\frac{z}{1-xy}\le\frac{3\sqrt{3}}{2}$ (see also Problem 5.17).

(b) We have $\frac{x}{1+yz}+\frac{y}{1+zx}+\frac{z}{1+xy}=\frac{x^2}{x+xyz}+\frac{y^2}{y+xyz}+\frac{z^2}{z+xyz}$, $xyz\neq 0$.
By inequality (8.4), we have

$$\frac{x}{1+yz}+\frac{y}{1+zx}+\frac{z}{1+xy}\ge\frac{(x+y+z)^2}{x+y+z+3xyz}=\frac{(1+xy+yz+zx)+xy+yz+zx}{x+y+z+3xyz}$$
$$=\frac{x+y+z+(1-x)(1-y)(1-z)+xyz+xy+yz+zx}{x+y+z+3xyz}\ge\frac{x+y+z+xyz+xy+yz+zx}{x+y+z+3xyz}>1;$$

hence $\frac{x}{1+yz}+\frac{y}{1+zx}+\frac{z}{1+xy}>1$.
If $xyz=0$, for example $x=0$, so that $y^2+z^2=1$ and $\frac{x}{1+yz}+\frac{y}{1+zx}+\frac{z}{1+xy}=y+z\ge y^2+z^2=1$, let us prove that if $x,y,z\ge 0$ and $x^2+y^2+z^2=1$, then $\frac{x}{1+yz}+\frac{y}{1+zx}+\frac{z}{1+xy}\le\sqrt{2}$.
Note that if $a\ge 0$, then $\frac{1}{1+a}\le 1-a+a^2$, and therefore,

$$\frac{x}{1+yz}+\frac{y}{1+zx}+\frac{z}{1+xy}\le x(1-yz+y^2z^2)+y(1-xz+x^2z^2)+x(1-xy+x^2y^2)$$
$$=x+y+z-3xyz+xyz(yz+xz+xy).$$

Moreover, we have $xy+yz+zx\le x^2+y^2+z^2$, whence $\frac{x}{1+yz}+\frac{y}{1+zx}+\frac{z}{1+xy}\le x+y+z-2xyz$.
Let us prove that $x+y+z-2xyz\le\sqrt{2}$, where $x,y,z\ge 0$ and $x^2+y^2+z^2=1$.
Indeed, let $\max(x,y,z)=z$. Then $3z^2\ge x^2+y^2+z^2=1$; therefore, $2z^2+2\sqrt{2}z-1\ge 2z^2+2\sqrt{2}z^2-1>3z^2-1\ge 0$, and thus from the inequality $4z^4-8z^2+4\sqrt{2}z-1=(\sqrt{2}z-1)^2(2z^2+2\sqrt{2}z-1)\ge 0$, it follows that $\left(\sqrt{z}(x+y)-\frac{1}{2\sqrt{z}}\right)^2+\frac{4z^4-8z^2+4\sqrt{2}z-1}{4z}\ge 0$, or $z(x+y)^2-(x+y)+z^3-2z+\sqrt{2}\ge 0$, $x+y+z-\left((x+y)^2+z^2-1\right)z\le\sqrt{2}$, $x+y+z-2xyz\le\sqrt{2}$.

14.32. Let us prove that if inequality (14.1) holds for all positive numbers a, b, c, then $\lambda\ge 8$ or $\lambda=0$.
Indeed, if $\lambda\neq 0$, then for $a=1,\ b=c=\frac{1}{n}$, from (14.1) we obtain

$$\frac{1}{\sqrt{1+\lambda/n^2}}+\frac{1}{\sqrt{1+n\lambda}}+\frac{1}{\sqrt{1+n\lambda}}\ge\frac{3}{\sqrt{1+\lambda}}.$$

For $n \to +\infty$ we have that $1 \geq \frac{3}{\sqrt{1+\lambda}}$, and therefore, $\lambda \geq 8$ (if $\lambda < 0$, then the expression $\sqrt{1+n\lambda}$ is not defined for $n > -\frac{1}{\lambda}$).
Let us prove that for $\lambda \geq 8$ and $a > 0,\ b > 0,\ c > 0$ inequality (14.1) holds.

Lemma 1 *If $x > 0,\ y > 0$, and $xy \leq 9$, then $\frac{1}{\sqrt{1+x}} + \frac{1}{\sqrt{1+y}} \geq 1$.*
Indeed, we have $\left(\sqrt{1+x}+\sqrt{1+y}\right)^2 = 2+x+y+2\sqrt{(1+x)(1+y)} \geq 2+x+y+2+2\sqrt{xy} \geq (1+x)(1+y)$, and therefore, $\sqrt{1+x}+\sqrt{1+y} \geq \sqrt{1+x}\sqrt{1+y}$.

Lemma 2 *If $x > 0,\ y > 0$, and $xy > 9$, then $\frac{1}{\sqrt{1+x}} + \frac{1}{\sqrt{1+y}} \geq \frac{2}{\sqrt{1+\sqrt{xy}}}$.*
We need to prove that $\left(\sqrt{1+x}+\sqrt{1+y}\right)^2(1+\sqrt{xy}) \geq 4(1+x)(1+y)$.
We have $\left(\sqrt{1+x}+\sqrt{1+y}\right)^2(1+\sqrt{xy}) \geq \left(4+x+y+2\sqrt{xy}\right)\left(1+\sqrt{xy}\right)$ (see the proof of Lemma 1).
Now, let us prove the following inequality: $\left(4+x+y+2\sqrt{xy}\right)\left(1+\sqrt{xy}\right) \geq 4(1+x)(1+y)$. Note that it is equivalent to the following obvious inequality: $\left(\sqrt{xy}-3\right)\left(\sqrt{x}-\sqrt{y}\right)^2 \geq 0$.
Define $\left(\frac{2}{1+x^2} - \frac{2}{1+y^2}\right)\left(\frac{x}{1+x^2} - \frac{y}{1+y^2}\right) = -\frac{2(x-y)^2(x+y)(1-xy)}{(1+x^2)^2(1+y^2)^2} \leq 0$. Then $x > 0,\ y > 0,\ z > 0,\ xyz = \lambda^3$. Let us prove that for $\lambda \geq 8$ the following inequality holds:

$$\frac{1}{\sqrt{1+x}} + \frac{1}{\sqrt{1+y}} + \frac{1}{\sqrt{1+z}} \geq \frac{3}{\sqrt{1+\lambda}}.$$

Let $\max(x, y, z) = z$. Then $z \geq \lambda$. If $xy \leq 9$, then by Lemma 1 we have

$$\frac{1}{\sqrt{1+x}} + \frac{1}{\sqrt{1+y}} + \frac{1}{\sqrt{1+z}} > \frac{1}{\sqrt{1+x}} + \frac{1}{\sqrt{1+y}} \geq 1 \geq \frac{3}{\sqrt{1+\lambda}}.$$

If $xy > 9$, then by Lemma 2, we have $\frac{1}{\sqrt{1+x}} + \frac{1}{\sqrt{1+y}} \geq \frac{2}{\sqrt{1+\sqrt{xy}}}$, $\frac{1}{\sqrt{1+z}} + \frac{1}{\sqrt{1+\lambda}} \geq \frac{2}{\sqrt{1+\sqrt{z\lambda}}}$ (since $z\lambda \geq \lambda^2 \geq 64 > 9$).
Summing these inequalities, we obtain

$$\frac{1}{\sqrt{1+x}} + \frac{1}{\sqrt{1+y}} + \frac{1}{\sqrt{1+z}} + \frac{1}{\sqrt{1+\lambda}} \geq \frac{2}{\sqrt{1+\sqrt{xy}}} + \frac{2}{\sqrt{1+\sqrt{z\lambda}}}.$$

Since $\sqrt{xy} \geq 3$ and $\sqrt{z\lambda} \geq \lambda$, we have $\sqrt{xy} \cdot \sqrt{z\lambda} \geq 3\lambda \geq 24 > 9$, and therefore, by Lemma 2, we have $\frac{2}{\sqrt{1+\sqrt{xy}}} + \frac{2}{\sqrt{1+\sqrt{z\lambda}}} > \frac{4}{\sqrt{1+\sqrt[4]{xyz\lambda}}} = \frac{4}{\sqrt{1+\lambda}}$.
Hence $\frac{1}{\sqrt{1+x}} + \frac{1}{\sqrt{1+y}} + \frac{1}{\sqrt{1+z}} + \frac{1}{\sqrt{1+\lambda}} > \frac{4}{\sqrt{1+\lambda}}$, which ends the proof of inequality (14.1).

Remark In a similar way, one can prove that $n \geq 2, \quad a_1 > 0, \ldots, a_n > 0$. Then

$$\sqrt{\frac{a_1^{n-1}}{a_1^{n-1} + (n^2-1)a_2 \cdots a_n}} + \cdots + \sqrt{\frac{a_n^{n-1}}{a_n^{n-1} + (n^2-1)a_1 \cdots a_{n-1}}} \geq 1.$$

14.33. Let us first prove the following lemma.

Lemma *If x, y, z are the lengths of some triangle, then*

$$\sqrt{x^2+xy+y^2}\sqrt{x^2+xz+z^2} \leq x^2 + x\frac{y+z}{2} + \left(\frac{y+z}{2}\right)^2.$$

Indeed, we need to prove that
$(x^2+xy+y^2)(x^2+xz+z^2) \leq \left(x^2 + x\frac{y+z}{2} + \left(\frac{y+z}{2}\right)^2\right)^2$. *This inequality holds because it is equivalent to the following obvious inequality:* $\left(\frac{y-z}{2}\right)^2\left(\left(\frac{y+z}{2}\right)^2 + yz + x(y+z-x)\right) \geq 0$.
This ends the proof of the lemma.
From the lemma, we have

$$\begin{aligned}
&\left(\sqrt{a^2+ab+b^2} + \sqrt{b^2+bc+c^2} + \sqrt{c^2+ac+a^2}\right)^2 \\
&\quad = 2a^2+2b^2+2c^2+ab+bc+ca+2\sqrt{a^2+ab+b^2} + \sqrt{b^2+bc+c^2} \\
&\qquad + 2\sqrt{a^2+ab+b^2}\sqrt{c^2+ac+a^2} + 2\sqrt{b^2+bc+c^2}\sqrt{c^2+ac+a^2} \\
&\quad \leq 2a^2+2b^2+2c^2+ab+bc+ca+2b^2+b(a+c)+\frac{(a+c)^2}{2} \\
&\qquad + 2a^2+a(b+c)+\frac{(b+c)^2}{2}+2c^2+c(a+b)+\frac{(a+b)^2}{2} \\
&\quad = 5a^2+5b^2+5c^2+4ab+4bc+4ca.
\end{aligned}$$

It follows that

$$\begin{aligned}
&\sqrt{a^2+ab+b^2} + \sqrt{b^2+bc+c^2} + \sqrt{c^2+ac+a^2} \\
&\quad \leq \sqrt{5a^2+5b^2+5c^2+4ab+4bc+4ca}.
\end{aligned}$$

Remark 1. One can prove that if $a, b, c \geq 0$, then

$$\begin{aligned}
&\sqrt{a^2+ab+b^2} + \sqrt{b^2+bc+c^2} + \sqrt{c^2+ca+a^2} \\
&\quad \leq \sqrt{5a^2+5b^2+5c^2+4ab+4bc+4ca}.
\end{aligned}$$

2. If $a = -1, b = c = 1$, then $1 + \sqrt{3} + 1 > \sqrt{11}$.

14.34. (a) Let us begin by proving that if the given inequality holds for $n = 3$, then it holds for all $n = 4, 5, \ldots$.
Indeed, we have

$$\frac{x_1^3 + \cdots + x_n^3}{x_1 \cdots x_n} = \frac{1}{3}\left(\frac{x_1^3 + x_2^3 + x_3^3}{x_1 x_2 x_3} \cdot \frac{1}{x_4 \cdots x_n} + \cdots + \frac{x_n^3 + x_1^3 + x_2^3}{x_n x_1 x_2} \cdot \frac{1}{x_3 \cdots x_{n-1}}\right)$$
$$\geq \frac{1}{3}\left(\frac{(1-x_1)^3 + (1-x_2)^3 + (1-x_3)^3}{(1-x_1)(1-x_2)(1-x_3)} \cdot \frac{1}{x_4 \cdots x_n} + \cdots + \frac{(1-x_n)^3 + (1-x_1)^3 + (1-x_2)^3}{(1-x_n)(1-x_1)(1-x_2)} \cdot \frac{1}{x_3 \cdots x_{n-1}}\right)$$
$$\geq \frac{1}{3}\left(\frac{(1-x_1)^3 + (1-x_2)^3 + (1-x_3)^3}{(1-x_1)\cdots(1-x_n)} + \cdots + \frac{(1-x_n)^3 + (1-x_1)^3 + (1-x_2)^3}{(1-x_1)\cdots(1-x_n)}\right)$$
$$= \frac{(1-x_1)^3 + \cdots + (1-x_n)^3}{(1-x_1)\cdots(1-x_n)},$$

and therefore, $\frac{x_1^3+\cdots+x_n^3}{x_1\cdots x_n} \geq \frac{(1-x_1)^3+\cdots+(1-x_n)^3}{(1-x_1)\cdots(1-x_n)}$.
Let us prove that the given inequality holds for $n = 3$.

$$\frac{x_1^3 + \cdots + x_n^3}{x_1 \cdots x_n} = \frac{1}{3}\left(\frac{x_1^3 + x_2^3 + x_3^3}{x_1 x_2 x_3} \cdot \frac{1}{x_4 \cdots x_n} + \cdots + \frac{x_n^3 + x_1^3 + x_2^3}{x_n x_1 x_2} \cdot \frac{1}{x_3 \cdots x_{n-1}}\right)$$
$$\geq 3 + \frac{1}{2}\left(\frac{1}{x_1' x_2'} + \frac{1}{x_2' x_3'} + \frac{1}{x_3' x_1'}\right)\left((x_1' - x_2')^2 + (x_2' - x_3')^2 + (x_3' - x_1')^2\right)$$
$$= \frac{(1-x_1)^3 + (1-x_2)^3 + (1-x_3)^3}{(1-x_1)(1-x_2)(1-x_3)},$$

where $x_i' = 1 - x_i$, $i = 1, 2, 3$. Therefore, $\frac{x_1^3+x_2^3+x_3^3}{x_1x_2x_3} \geq \frac{(1-x_1)^3+(1-x_2)^3+(1-x_3)^3}{(1-x_1)(1-x_2)(1-x_3)}$.

(b) It is sufficient to prove the given inequality for $n = 4$ (see the proof of Problem 14.34(a)).
Let $x_1 \geq x_2 \geq x_3 \geq x_4$; then $x_1x_1' \geq x_2x_2' \geq x_3x_3' \geq x_4x_4'$ and $\frac{x_1}{x_1'} \geq \frac{x_2}{x_2'} \geq \frac{x_3}{x_3'} \geq \frac{x_4}{x_4'}$, where $x_i' = 1 - x_i$, $i = 1, 2, 3, 4$. It follows that

$$\frac{x_1^4 + x_2^4 + x_3^4 + x_4^4}{x_1x_2x_3x_4} = \frac{(x_1 - x_2)^4}{x_1x_2x_3x_4} + \frac{4(x_1 - x_2)^2}{x_3x_4} + 2\left(\frac{x_1x_2}{x_3x_4} + \frac{x_3x_4}{x_1x_2}\right) + \frac{(x_3 - x_4)^4}{x_1x_2x_3x_4} + \frac{4(x_3 - x_4)^2}{x_1x_2}$$
$$\geq \frac{(x_1' - x_2')^4}{x_1'x_2'x_3'x_4'} + \frac{4(x_1' - x_2')^2}{x_3'x_4'} + 2\left(\frac{x_1'x_2'}{x_3'x_4'} + \frac{x_3'x_4'}{x_1'x_2'}\right) + \frac{(x_3' - x_4')^4}{x_1'x_2'x_3'x_4'} + \frac{4(x_3' - x_4')^2}{x_1'x_2'}$$
$$= \frac{(1-x_1)^4 + (1-x_2)^4 + (1-x_3)^4 + (1-x_4)^4}{(1-x_1)(1-x_2)(1-x_3)(1-x_4)},$$

since

$$\frac{x_1x_2}{x_3x_4} + \frac{x_3x_4}{x_1x_2} - \frac{x_1'x_2'}{x_3'x_4'} - \frac{x_3'x_4'}{x_1'x_2'}$$
$$= (x_1x_1' \cdot x_2x_2' - x_3x_3' \cdot x_4x_4')\left(\frac{1}{x_1'x_2'x_3x_4} - \frac{1}{x_1x_2x_3'x_4'}\right) \geq 0.$$

Another proof of this inequality can be obtained using the following identity:

$$\frac{x_1^4+x_2^4+x_3^4+x_4^4}{x_1x_2x_3x_4}=\frac{1}{2}\left(\frac{(x_1-x_2)^4+(x_2-x_3)^4+(x_3-x_4)^4+(x_4-x_1)^4}{x_1x_2x_3x_4}\right)$$
$$+\frac{2(x_1-x_2)^2}{x_3x_4}+\frac{2(x_2-x_3)^2}{x_1x_4}+\frac{2(x_3-x_4)^2}{x_1x_2}+\frac{2(x_4-x_1)^2}{x_2x_3}+\left(\frac{(x_1-x_3)^2}{x_1x_3}+2\right)\left(\frac{(x_2-x_4)^2}{x_2x_4}+2\right).$$

14.35. We need to prove that $B=a_1^{2015}+\cdots+a_n^{2015}-A(a_1^{2000}+\cdots+a_n^{2000})\geq 0$, where $A=\frac{1^{2015}+\cdots+n^{2015}}{1^{2000}+\cdots+n^{2000}}$.
If $\{a_1,\ldots,a_n\}=\{1,\ldots,n\}$, then $B=0$.
If $\{a_1,\ldots,a_n\}\neq\{1,\ldots,n\}$, then there exist numbers i and j $(i,j\in\{1,\ldots,n\})$ such that $a_i\geq n+1$ and $a_k\neq j$, for $k=1,\ldots,n$.
Then $a_i^{2015}-Aa_i^{2000}>j^{2015}-Aj^{2000}$, as $a_i^{2000}>j^{2000}$ and $a_i^{15}\geq(n+1)^{15}>A$.
Therefore, on substituting a_i by j, we see that the value of B is decreasing. After several such steps we obtain that $B\geq 0$.

14.36. We need to prove that $\sqrt{1+x^2}+\sqrt{1+y^2}+\sqrt{(1-x)^2+(1-y)^2}-(1-xy)\geq\sqrt{5}(1-xy)$.
Note that

$$\sqrt{1+x^2}+\sqrt{1+y^2}+\sqrt{(1-x)^2+(1-y)^2}-(1-xy)=xy-\left(\sqrt{1+x^2}-1\right)\left(\sqrt{1+y^2}-1\right)$$
$$+\sqrt{(1-x)^2+(1-y)^2}+\sqrt{1+x^2}\cdot\sqrt{1+y^2}\geq\sqrt{(1-x)^2+(1-y)^2}+\sqrt{1+x^2}\cdot\sqrt{1+y^2},$$

as $a\geq\sqrt{1+a^2}-1\geq 0$, for $a\geq 0$.
Let us prove that $\sqrt{(1-x)^2+(1-y)^2}+\sqrt{1+x^2}\cdot\sqrt{1+y^2}\geq\sqrt{5}(1-xy)$.
If $xy\geq\frac{1}{2}$, then

$$\sqrt{(1-x)^2+(1-y)^2}+\sqrt{1+x^2}\cdot\sqrt{1+y^2}\geq\sqrt{1+x^2}\cdot\sqrt{1+y^2}$$
$$\geq 1+xy\geq\frac{3}{2}>\frac{\sqrt{5}}{2}\geq\sqrt{5}(1-xy).$$

If $xy<\frac{1}{2}$, then by inequality (14.1), we have

$$\sqrt{(1-x)^2+(1-y)^2}+\sqrt{1+x^2}\cdot\sqrt{1+y^2}=\sqrt{\left(\sqrt{1-2xy}\right)^2+(1-x-y)^2}+\sqrt{(1-xy)^2+(x+y)^2}$$
$$\geq\sqrt{\left(\sqrt{1-2xy}+1-xy\right)^2+1}=\sqrt{3-4xy+x^2y^2+2(1-xy)\sqrt{1-2xy}}$$
$$\geq\sqrt{3-4xy+x^2y^2+2(1-xy)(1-2xy)}=\sqrt{5}(1-xy),$$

and therefore, $\sqrt{(1-x)^2+(1-y)^2}+\sqrt{1+x^2}\cdot\sqrt{1+y^2}\geq\sqrt{5}(1-xy)$.

14.37. (a) For $n=2$ we have $\frac{1}{\sqrt{1+x_1}}+\frac{1}{\sqrt{1+x_2}}=\frac{1+\sqrt{x_1}}{\sqrt{1+x_1}}\leq\sqrt{2}$.
For $n=3$ let us prove that there exists an acute triangle with angles $\alpha,\ \beta,\ \gamma$ such that $x_1=\frac{\tan\alpha}{\tan\beta}$, $x_2=\frac{\tan\beta}{\tan\gamma}$, and then $x_3=\frac{\tan\gamma}{\tan\alpha}$.

Indeed, it is sufficient to take $\tan\alpha = \sqrt{(1+x_2+x_1x_2)x_1}$, $\tan\beta = \sqrt{\frac{1+x_2+x_1x_2}{x_1}}$, $\tan\gamma = \sqrt{\frac{1+x_2+x_1x_2}{x_1}} \cdot \frac{1}{x_2}$.

We need to prove that $\sqrt{\frac{\sin\beta\cos\alpha}{\sin\gamma}} + \sqrt{\frac{\sin\gamma\cos\beta}{\sin\alpha}} + \sqrt{\frac{\sin\alpha\cos\gamma}{\sin\beta}} \le \frac{3}{\sqrt{2}}$, or equivalently,

$$\sin\beta\sqrt{\sin(2\alpha)} + \sin\gamma\sqrt{\sin(2\beta)} + \sin\alpha\sqrt{\sin(2\gamma)} \le 3\sqrt{\sin\alpha\,\sin\beta\,\sin\gamma}.$$

According to the Cauchy–Bunyakovsky–Schwarz inequality, we have

$$\begin{aligned}&\sin\beta\sqrt{\sin(2\alpha)} + \sin\gamma\sqrt{\sin(2\beta)} + \sin\alpha\sqrt{\sin(2\gamma)}\\ &\quad\le \sqrt{\left(\sin^2\beta + \sin^2\gamma + \sin^2\alpha\right)(\sin(2\alpha) + \sin(2\beta) + \sin(2\gamma))}\\ &\quad= 2\sqrt{\left(\sin^2\beta + \sin^2\gamma + \sin^2\alpha\right)\sin\alpha\,\sin\beta\,\sin\gamma} \le 3\sqrt{\sin\alpha\,\sin\beta\,\sin\gamma},\end{aligned}$$

and therefore, $\sin\beta\sqrt{\sin(2\alpha)} + \sin\gamma\sqrt{\sin(2\beta)} + \sin\alpha\sqrt{\sin(2\gamma)} \le 3\sqrt{\sin\alpha\,\sin\beta\,\sin\gamma}$

(b) The proof is by induction.

For $n = 4$, from the inequality $\frac{1}{\sqrt{1+a}} + \frac{1}{\sqrt{1+b}} < 1 + \frac{1}{1+ab}$, where $a > 0,\ b > 0$, it follows that

$$\frac{1}{\sqrt{1+x_1}} + \frac{1}{\sqrt{1+x_2}} + \frac{1}{\sqrt{1+x_3}} + \frac{1}{\sqrt{1+x_4}} < 1 + \frac{1}{1+x_1x_2} + 1 + \frac{1}{1+x_3x_4} = 3.$$

Assume that $n \ge 5$ and that the inequality holds for $n-1$ numbers. Let us prove that it holds for n numbers.

We have

$$\begin{aligned}&\frac{1}{\sqrt{1+x_1}} + \cdots + \frac{1}{\sqrt{1+x_{n-1}}} + \frac{1}{\sqrt{1+x_n}} < \frac{1}{\sqrt{1+x_1}} + \cdots + \frac{1}{\sqrt{1+x_{n-2}}} + 1 + \frac{1}{1+x_{n-1}x_n}\\ &\quad< \frac{1}{\sqrt{1+x_1}} + \cdots + \frac{1}{\sqrt{1+x_{n-2}}} + \frac{1}{\sqrt{1+x_{n-1}x_n}} + 1 < n-2+1 = n-1,\end{aligned}$$

whence

$$\frac{1}{\sqrt{1+x_1}} + \cdots + \frac{1}{\sqrt{1+x_n}} < n-1.$$

Now let us prove that $\frac{1}{\sqrt{1+a}} + \frac{1}{\sqrt{1+b}} < 1 + \frac{1}{1+ab}$, where $a > 0,\ b > 0$.

Define $\sqrt{1+a} = 1+x,\ \sqrt{1+b} = 1+y$, and note that $x > 0,\ y > 0$. We need to prove that $(2+x+y)(1+x^2y^2+2x^2y+2xy^2+4xy) < (1+x+y+xy)(2+x^2y^2+2x^2y+2xy^2+4xy)$,

or equivalently,

$$2xy + 2x^2y + 2xy^2 < x + y + x^3y^3 + 2x^2y^3 + 2x^3y^2 + 3x^2y^2. \qquad (1)$$

Indeed, we have

$$\frac{x}{2}+2x^3y^2 \geq 2x^2y, \quad \frac{y}{2}+2x^2y^3 \geq 2xy^2,$$
$$\frac{x}{2}+\frac{y}{2}+3x^2y^2 \geq 3\sqrt[3]{\frac{3}{4}}\cdot xy > 2xy, \quad x^3y^3 > 0.$$

Summing these inequalities, we obtain inequality (1).

14.38. Consider the following two cases.

(a) $1 \leq \alpha \leq 2+\frac{1}{n-1}$. Then note that

$$\frac{x_i^\alpha - x_i}{x_1+\cdots+x_n+x_i^\alpha - x_i} = \frac{x_i - x_i^{2-\alpha}}{x_i+\frac{x_1+\cdots+x_n}{x_i^{\alpha-1}} - x_i^{2-\alpha}} \geq \frac{x_i - x_i^{2-\alpha}}{x_1+\cdots+x_n}, \quad i=1,\ldots,n.$$

Summing these inequalities, we obtain

$$\sum_{i=1}^n \frac{x_i^\alpha - x_i}{x_1+\cdots+x_n+x_i^\alpha - x_i} \geq \sum_{i=1}^n \frac{x_i - x_i^{2-\alpha}}{x_1+\cdots+x_n}. \tag{1}$$

If $1 \leq \alpha \leq 2-\frac{1}{n-1}$, then by Problem 3.42(a), it follows that

$$\sum_{i=1}^n \frac{x_i - x_i^{2-\alpha}}{x_1+\cdots+x_n} \geq 0.$$

If $2-\frac{1}{n-1} \leq \alpha \leq 2+\frac{1}{n-1}$, then by Problem 3.42(b), again it follows that

$$\sum_{i=1}^n \frac{x_i - x_i^{2-\alpha}}{x_1+\cdots+x_n} \geq 0. \tag{2}$$

Hence, from (1) and (2) we have $\sum_{i=1}^n \frac{x_i^\alpha - x_i}{x_1+\cdots+x_n+x_i^\alpha - x_i} \geq 0$.

(b) $\alpha \geq 2+\frac{1}{n-1}$.

This solution was proposed by N.Nikolov.

If we prove that there exists a number $\gamma \geq 1$ such that

$$\frac{x_i^\alpha - x_i}{x_1+\cdots+x_n+x_i^\alpha - x_i} \geq \frac{nx_i^\gamma - \sum_{i=1}^n x_i^\gamma}{(n-1)\sum_{i=1}^n x_i^\gamma}, \quad i=1,\ldots,n, \tag{3}$$

then summing these inequalities, we will obtain $\sum_{i=1}^n \frac{x_i^\alpha - x_i}{x_1+\cdots+x_n+x_i^\alpha - x_i} \geq 0$.

Proving inequality (3) is equivalent to proving the following inequality:

$$\frac{nx_i(x_i^{\alpha-1}-1)}{x_1+\cdots+x_n} \geq \frac{(n-1)x_i^{\gamma}}{\left(\sum\limits_{i=1}^{n} x_i^{\gamma}\right)-x_i^{\gamma}} - 1. \tag{4}$$

It is sufficient to prove that

$$\frac{nx_i}{x_1+\cdots+x_n}\left(\frac{x_i^{\alpha-1}}{x_1^{\frac{\alpha-1}{n}}\cdots x_n^{\frac{\alpha-1}{n}}}-1\right) \geq \frac{(n-1)x_i^{\gamma}}{\left(\sum\limits_{i=1}^{n} x_i^{\gamma}\right)-x_i^{\gamma}} - 1, \tag{5}$$

since $x_1^{\frac{\alpha-1}{n}}\cdots x_n^{\frac{\alpha-1}{n}} \geq 1$.
Without loss of generality, one can assume that $x_i = 1$.
Now let us prove that there exists a number $\gamma \geq 1$ such that the following inequality holds

$$\frac{n}{x_1+\cdots+x_n}\left(\frac{1}{G}-1\right) \geq \frac{1}{A}-1, \tag{6}$$

where $A = \frac{\sum\limits_{i=1}^{n} x_i^{\gamma}-1}{n-1}$, $\quad G = x_1^{\frac{\alpha-1}{n}}\cdots x_n^{\frac{\alpha-1}{n}}$.
Taking $\gamma = \frac{(n-1)(\alpha-1)}{n}$, we obtain that $A \geq G$.
If $G \geq 1$, then $0 \geq \frac{1}{G}-1 \geq \frac{1}{A}-1$ and $\frac{n}{x_1+\cdots+x_n} \leq 1$, which ends the proof of inequality (6).
If $G \leq 1$ and $A \leq 1$, then $\frac{1}{G}-1 \geq \frac{1}{A}-1 \geq 0$ and $\frac{n}{x_1+\cdots+x_n} \geq 1$, since by Jensen's inequality, we have $1 \geq A \geq \left(\frac{x_1+\cdots+x_n-1}{n-1}\right)^{\gamma}$, which completes the proof of inequality (6).
If $G \leq 1$ and $A > 1$, then $\frac{n}{x_1+\cdots+x_n}\left(\frac{1}{G}-1\right) \geq 0 > \frac{1}{A}-1$.

Problems for Independent Study

Prove the following inequalities (1–16).

1. $\sqrt{3x^2+2x+1}+\sqrt{3x^2-4x+2} \geq \frac{\sqrt{51}}{3}$.
2. $\left(\sqrt{2+\sqrt{2}}-1\right)^2 \leq x^2+y^2 \leq 1$ and $(2-\sqrt{2+\sqrt{2}})^2 \leq (x-\sqrt{2}/2)^2+(y-\sqrt{2}/2)^2 \leq 2-\sqrt{2}$, where $(x+\sqrt{2}/2)^2+(y+\sqrt{2}/2)^2 = 2+\sqrt{2}$, $x \in [0,1]$.
3. $\sqrt{a_1^2+(1-a_2)^2}+\sqrt{a_2^2+(1-a_3)^2}+\cdots+\sqrt{a_n^2+(1-a_1)^2} \geq n\sqrt{2}/2$, where $n \geq 2$.
4. $\sqrt{2}+\sqrt{4-2\sqrt{2}}+\cdots+\sqrt{2n-2\sqrt{(n-1)n}} \geq \sqrt{n(n+1)}$, where $n \in \mathbb{N}$.
5. (a) $\frac{(a-b)^2}{8a} \leq \frac{a+b}{2}-\sqrt{ab} \leq \frac{(a-b)^2}{8b}$, where $a \geq b > 0$.

(b) $\sqrt{(a+c)^2+(b+d)^2} \le \sqrt{a^2+b^2}+\sqrt{c^2+d^2} \le \sqrt{(a+c)^2+(b+d)^2}+\frac{|ad-bc|}{\sqrt{(a+c)^2+(b+d)^2}}$, where $a>0,\ b>0,\ c>0,\ d>0$.

6. $0<\sqrt{4n+2}-\sqrt{n}-\sqrt{n+1}<1/16\sqrt{n^3}$, where $n\in\mathbb{N}$.
7. (a) $\sqrt{x^2+xy+y^2}+\sqrt{y^2+yz+z^2}\ge\sqrt{x^2+xz+z^2}$;
   (b) $c\sqrt{a^2-ab+b^2}+a\sqrt{b^2-bc+c^2}\ge b\sqrt{a^2+ac+c^2}$, where $a>0,\ b>0,\ c>0$.
8. $\sqrt{1-\cos(x_3-x_2)}+\sqrt{1-\cos(x_2-x_1)}\ge\sqrt{1-\cos(x_3-x_1)}$.
9. $9<\int\limits_0^3\sqrt[4]{x^4+1}dx+\int\limits_0^3\sqrt[4]{x^4-1}dx<9.0001$.
10. $\frac{x_1^3}{x_1^2+x_1x_2+x_2^2}+\frac{x_2^3}{x_2^2+x_2x_3+x_3^2}+\dots+\frac{x_n^3}{x_n^2+x_nx_1+x_1^2}\ge\frac{1}{3}(x_1+\ldots+x_n)$, where $n\ge3,\ x_1>0,\ldots,x_n>0$.
11. $\frac{x_1^7}{x_1^4+2x_1^3x_2+2x_1x_2^3+x_2^4}+\frac{x_2^7}{x_2^4+2x_2^3x_3+2x_2x_3^3+x_3^4}+\dots+\frac{x_n^7}{x_n^4+2x_n^3x_1+2x_nx_1^3+x_1^4}\ge\frac{1}{6}(x_1^3+\dots+x_n^3)$, where $n\ge3,\ x_1>0,\ldots,x_n>0$.
12. $3^n+4^n+\dots+(n+2)^n<(n+3)^n$, where $n\ge6$ and $n\in\mathbb{N}$.
13. $\frac{1}{2015}<\ln\frac{2015}{2014}<\frac{1}{2014}$.
14. $\tan x\ge x^3$, where $x\in\left(0,\frac{\pi}{2}\right)$.
15. $\sum\limits_{i=1}^{n}\tan\alpha_i\ge(n-1)\sum\limits_{i=1}^{n}\cot\alpha_i$, where $n\ge3,\quad 0<\alpha_i<\frac{\pi}{2},\quad i=1,\ldots,n$ and $\sum\limits_{i=1}^{n}\cos^2\alpha_i=1$.
16. $a^3+b^3+c^3+6abc>\frac{1}{4}(a+b+c)^3$, where $a>0,\ b>0,\ c>0$.
17. Compare the numbers $\left(\frac{1}{3}\right)^{100}+\left(\frac{2}{3}\right)^{100}$ and $\left(\frac{1}{\sqrt{2}}\right)^{100}+\left(1-\frac{1}{\sqrt{2}}\right)^{100}$.
18. Prove that if $a+b+c+d=0$ and $a^{2015}+b^{2015}+c^{2015}+d^{2015}=0$, then $a^3+b^3+c^3+d^3=0$.
19. Prove that if $\alpha\sqrt{\frac{a^2+b^2}{2}}+(1-\alpha)\sqrt{ab}\le\frac{a+b}{2}$ holds for every pair of positive numbers $a,\ b$, then $\alpha\le\frac{1}{2}$.
20. Prove that if $|a_1|+\dots+|a_n|<\varepsilon<\frac{1}{2}$, then $|1-(1+a_1)\cdots(1+a_n)|<2\varepsilon$.
21. The function f is defined on the set Q and $f(0)\cdot f(1)<0$. Prove that there exist rational numbers r_1 and r_2, such that $f(r_1)-f(r_2)>(r_1-r_2)^2$.
22. Given that $a\ge0,\ b\ge0,\ c\ge0$ and $a+b+c\le1$, find the greatest value of the expression $|a-b|\cdot|b-c|\cdot|c-a|$.
23. Let $\alpha,\ \beta,\ \gamma$ be the angles of some triangle, and $a,\ b,\ c$ be its sides. Prove that

   (a) $0<\sin\alpha+\sin\beta+\sin\gamma\le\frac{3\sqrt{3}}{2}$,
   (b) $0<\sqrt{\sin\alpha}+\sqrt{\sin\beta}+\sqrt{\sin\gamma}\le3\cdot\left(\frac{3}{4}\right)^{\frac{1}{4}}$,
   (c) $0<\sin\alpha\ \sin\beta\ \sin\gamma\le\frac{3\sqrt{3}}{8}$,
   (d) $2<\cos\frac{\alpha}{2}+\cos\frac{\beta}{2}+\cos\frac{\gamma}{2}\le\frac{3\sqrt{3}}{2}$,
   (e) $\cos\alpha\ \cos\beta\ \cos\gamma\le\frac{1}{8}$,
   (f) $0<\cos\frac{\alpha}{2}\cos\frac{\beta}{2}\cos\frac{\gamma}{2}\le\frac{3\sqrt{3}}{8}$,
   (g) $\frac{1}{3}\le\frac{a^2+b^2+c^2}{(a+b+c)^2}<\frac{1}{2}$,

(h) $\frac{1}{4} \le \frac{ab+bc+ca}{(a+b+c)^2} \le \frac{1}{3}$,
(i) $\frac{1}{4} \le \frac{(a+b)(b+c)(c+a)}{(a+b+c)^2} \le \frac{8}{27}$.

24. Let α, β, γ be the angles of an acute triangle. Prove that

(a) $2 < \sin\alpha + \sin\beta + \sin\gamma \le \frac{3\sqrt{3}}{2}$,
(b) $2 < \sqrt{\sin\alpha} + \sqrt{\sin\beta} + \sqrt{\sin\gamma} \le 3 \cdot \left(\frac{3}{4}\right)^{\frac{1}{4}}$,
(c) $\frac{3}{4} \le \sin^2\frac{\alpha}{2} + \sin^2\frac{\beta}{2} + \sin^2\frac{\gamma}{2} < 1$,
(d) $1 < \cos\alpha + \cos\beta + \cos\gamma \le \frac{3}{2}$,
(e) $2 < \cos^2\frac{\alpha}{2} + \cos^2\frac{\beta}{2} + \cos^2\frac{\gamma}{2} \le \frac{9}{4}$,
(f) $\frac{1}{2} < \cos\frac{\alpha}{2}\cos\frac{\beta}{2}\cos\frac{\gamma}{2} \le \frac{3\sqrt{3}}{8}$,
(g) $\tan^m\alpha + \tan^m\beta + \tan^m\gamma \ge 3^{\frac{m+2}{2}}$, where $m \ge 1$.

25. Prove that for an obtuse triangle with angles α, β, γ and sides a, b, c the following inequalities hold:

(a) $0 < \sin\alpha + \sin\beta + \sin\gamma \le 1 + \sqrt{2}$,
(b) $0 < \sin\alpha\, \sin\beta\, \sin\gamma \le \frac{1}{2}$,
(c) $0 < \cos\frac{\alpha}{2}\cos\frac{\beta}{2}\cos\frac{\gamma}{2} \le \frac{1+\sqrt{2}}{4}$,
(d) $\frac{1+\cos\alpha\, \cos\beta\, \cos\gamma}{\sin\alpha\, \sin\beta\, \sin\gamma} \ge 2$,
(e) $\frac{1}{3} \le \frac{a^2+b^2+c^2}{(a+b+c)^2} \le \frac{3}{8}$,
(f) $\frac{5}{16} < \frac{ab+bc+ca}{(a+b+c)^2} \le \frac{1}{3}$,
(g) $\frac{9}{32} < \frac{(a+b)(b+c)(c+a)}{(a+b+c)^2} \le \frac{8}{27}$.

26. Prove that if $0 < a_1 \le \ldots \le a_n$ and $0 < b_1 \le \ldots \le b_n$, then $(a_1 + b_1)\cdots(a_n + b_n) \le (a_1 + b_{i_1})\cdots(a_n + b_{i_n}) \le (a_1 + b_n)\cdots(a_n + b_1)$, where the numbers $i_1, \ldots, i_n$ are a permutation of the numbers $1, \ldots, n$.

27. Given that the assumptions of Problem 26 hold for the numbers $a_i, b_i,\ i = 1, \ldots, n$, and g is a nondecreasing convex function (see Chapter §11). Prove that

$$g\left(\frac{b_1}{a_1}\right) + \cdots + g\left(\frac{b_n}{a_n}\right) \le g\left(\frac{b_{i_1}}{a_1}\right) + \cdots + g\left(\frac{b_{i_n}}{a_1}\right) \le g\left(\frac{b_n}{a_1}\right) + \cdots + g\left(\frac{b_1}{a_n}\right).$$

28. Let g be a convex function on $[0, a_1]$ and $g(0) \le 0$. Prove that $\sum_{i=1}^{n}(-1)^{i-1}g(a_i) \ge g\left(\sum_{i=1}^{n}(-1)^{i-1}a_i\right)$, where $0 < a_n \le \ldots \le a_1$.

29. Prove that, if $a_1 > 0, \ldots, a_n > 0$, then $\left(\frac{a_1}{a_2}\right)^n + \cdots + \left(\frac{a_{n-1}}{a_n}\right)^n + \left(\frac{a_n}{a_1}\right)^n \ge \frac{a_1}{a_2} + \cdots + \frac{a_{n-1}}{a_n} + \frac{a_n}{a_1}$, where $n \ge 2$.

30. Suppose that for the polynomial $p(x) = ax^2 + bx + c$ the following assumptions hold:

$0 \le p(-1) \le 1,\ \ 0 \le p(0) \le 1,\ \ 0 \le p(1) \le 1$. Prove that for all $x \in [0, 1]$ the following inequality holds: $p(x) \le \frac{9}{8}$.

31. Given that for a trinomial $p(x) = ax^2 + bx + c$ on $[0, 1]$ the inequality $|p(x)| \le 1$ holds, prove that $|b| \le 8$.
32. Given that for a trinomial $p(x) = ax^2+bx+c$ on $[-1, 1]$ the inequality $|p(x)| \le 1$ holds, prove that $\left|p'(x)\right| \le 4$ on $[-1, 1]$.
Remark. For every polynomial $p(x)$ the following inequality holds: $\left|p'(x)\right| \le \frac{2n^2}{b-a} \max\limits_{[a;b]} |p(x)|$, where $x \in [a, b]$ and n is the degree of $p(x)$.
33. Given that the degree of the polynomial $p(x)$ is not greater than $2n$ and that for every integer $k \in [-n, n]$ the inequality $|p(k)| \le 1$ holds, prove that for all $x \in [-n, n]$ the following inequality holds: $|p(x)| \le 2^{2n}$.
34. Let $x_0 < x_1 < \cdots < x_n$ be integers. Prove that at least one of the numbers $|p(x_0)|, |p(x_1)|, \ldots, |p(x_n)|$ is not smaller than $\frac{n!}{2^n}$, where $p(x) = x^n + a_1x^{n-1} + \cdots + a_n$.
35. Let $x_1 < \ldots < x_k,\ \ k \ge 3$, be given numbers belonging to $[-1, 1]$. Prove that

$$\frac{1}{|(x_1 - x_2)\cdots(x_1 - x_k)|} + \cdots + \frac{1}{|(x_k - x_1)\cdots(x_k - x_{k-1})|} \ge 2^{k-2}.$$

36. Prove that if $a > 0,\ b > 0,\ c > 0$, then

$$\frac{a}{a + \sqrt{(a+b)(a+c)}} + \frac{b}{b + \sqrt{(b+a)(b+c)}} + \frac{c}{c + \sqrt{(c+a)(c+b)}} \le 1.$$

37. Let $p(x) = ax^3 + bx^2 + cx + d$ and suppose that for all x belonging to $[-1, 1]$ the following inequality holds: $|p(x)| \le 1$. Prove that $|a| + |b| + |c| + |d| \le 7$.
38. Prove that
   (a) $x^\lambda(x-y)(x-z) + y^\lambda(y-x)(y-z) + z^\lambda(z-x)(z-y) \ge 0$, where $x > 0,\ y > 0,\ z > 0$;
   (b) $g(x)(f(x)-f(y))(f(x)-f(z)) + g(y)(f(y)-f(x))(f(y) - f(z)) + g(z)(f(z)-f(x))(f(z)-f(y)) \ge 0$,
   where f and g are monotonic functions and $E(g) \subseteq [0, +\infty)$.
39. Let each of the edges of the complete graph on n vertices be colored in one of three colors. Prove that there exists a single-color connected subgraph having at least $\frac{n}{2}$ vertices.
40. Prove that $\sqrt{a_1} + \cdots + \sqrt{a_n} \le \sqrt{a_1 + C(n)} \cdots \sqrt{a_n + C(n)}$, where $n \ge 2$,

$$a_1 \ge 0, \ldots, a_n \ge 0, \quad C(n) = \frac{n-1}{n^{\frac{n-2}{n-1}}}.$$

41. Prove that $0 \le \sin x + \ldots + \frac{\sin(nx)}{n} < 2\sqrt{2}$, where $0 \le x \le \frac{\pi}{2}$.
Hint. If $x \ne 0$, then there exists a positive integer m such that $\frac{\sqrt{2}}{2(m+1)} < \sin\frac{x}{2} \le \frac{\sqrt{2}}{2m}$.

42. Let $f: R \to R$. Prove that f is constant if
    (a) for all real numbers x and y the following inequality holds: $|f(x) - f(y)| \leq C|x - y|^{\alpha}$, where $C > 0, \quad \alpha > 1$,
    (b) for all real numbers a, b, c, d forming an arithmetic progression, the following inequality holds: $|f(a) - f(d)| \geq A|f(c) - f(b)|$, where $A > 3$.

43. Prove that $\frac{1}{a+3} + \frac{1}{b+3} + \frac{1}{c+3} + \frac{1}{d+3} \leq \frac{23}{27} + \frac{4}{27} \cdot \frac{1}{abcd}$, where $a > 0, b > 0, c > 0, d > 0$ and $a + b + c + d = 4$.
    Hint. We have that

$$abcd\left(\frac{1}{a+3} + \frac{1}{b+3} + \frac{1}{c+3} + \frac{1}{d+3}\right) \leq \frac{3a+1}{16} \cdot bcd + \frac{3b+1}{16} \cdot acd$$
$$+ \frac{3c+1}{16} \cdot abd + \frac{3d+1}{16} \cdot abc \leq \frac{3}{4}abcd + \frac{4}{27} + \frac{11}{108}abcd = \frac{4}{27} + \frac{23}{27}abcd$$

(see Problem 2.5).

Selected Inequalities

Problems

Prove the following inequalities.

1. $(ab + bc + ca)\left|1 - \frac{1}{abc}\right| \leq \left|ab - \frac{1}{c}\right| + \left|bc - \frac{1}{a}\right| + \left|ca - \frac{1}{b}\right|$, where $abc \neq 0$.
2. $\sin^2\beta \sin\frac{\alpha}{2} \sin\left(\alpha + \frac{\beta}{2}\right) > \sin^2\alpha \sin\frac{\beta}{2} \sin\left(\beta + \frac{\alpha}{2}\right)$, where $0 < \alpha < \beta$ and $\alpha + \beta < \pi$.
3. $\left|\frac{a+b}{a-b}\right|^{ab} \geq 1$, where $a \neq b$.
4. $\left(\frac{a-b}{a+b}\right)^{11} + \left(\frac{b-c}{b+c}\right)^{11} + \left(\frac{c-a}{c+a}\right)^{11} < 1$, where $a > 0, b > 0, c > 0$.
5. $x_0 + \frac{1}{x_0 - x_1} + \frac{1}{x_1 - x_2} + \cdots + \frac{1}{x_{n-1} - x_n} \geq x_n + 2n$, where $n \in \mathbb{N}$ and $x_0 > x_1 > \ldots > x_n$.
6. $S_k + S_m - S_{km} \leq 1$, where $k, m \in \mathbb{N}$ and $S_n = \sum\limits_{i=1}^{n} \frac{1}{i}$.
7. $\frac{1}{n+1}\left(1 + \frac{1}{3} + \cdots + \frac{1}{2n-1}\right) > \frac{1}{n}\left(\frac{1}{2} + \frac{1}{4} + \cdots + \frac{1}{2n}\right)$, where $n \geq 2, \quad n \in \mathbb{N}$.
8. $2^{\frac{1}{2}} \cdot 2^{\frac{2}{2^2}} \cdots 2^{\frac{n}{2^n}} < 4$, where $n \in \mathbb{N}$.
9.
$$(a_1 - a_2)(a_1 - a_3)(a_1 - a_4)(a_1 - a_5) + \cdots$$
$$+ (a_5 - a_1)(a_5 - a_2)(a_5 - a_3)(a_5 - a_4) \geq 0.$$

10. $a^m + a^n \geq m^m + n^n$, where $m, n \in \mathbb{N}$ and $a = \frac{m^{m+1} + n^{n+1}}{m^m + n^n}$.
11. $2^n \cdot n! \leq \frac{(m+n)!}{(m-n)!} \leq (m^2 + m)^n$, where $m, n \in \mathbb{N}$ and $m \geq n$.
12. $\left|\sqrt{2} - \frac{p}{q}\right| > \frac{1}{3q^2}$, where $p \in \mathbb{Z}, \quad q \in \mathbb{N}$.
13. $\sqrt{7} - \frac{m}{n} > \frac{1}{mn}$, where $m, n \in \mathbb{N}$ and $\sqrt{7} - \frac{m}{n} > 0$.
14. $\sqrt{(x+y+z)^2 - 3(xy+yz+zx)} \leq \max(x,y,z) - \min(x,y,z) \leq \frac{\sqrt{6}}{2}\sqrt{(x+y+z)^2 - 3(xy+yz+zx)}$.
15. $2\left(\frac{a}{b} + \frac{b}{c} + \frac{c}{a}\right) \geq a + b + c + \frac{1}{a} + \frac{1}{b} + \frac{1}{c}$, where $a > 0, b > 0, c > 0$ and $abc = 1$.
16. $(n - 0,5)(1 + a^{2n}) \geq a^{2n-1} + a^{2n-2} + \cdots + a^2 + a$, where $n \in \mathbb{N}$.

17. $x_1^4 + \cdots + x_n^4 \geq x_1^3 + \cdots + x_n^3$, where $x_1 + \cdots + x_n \geq n$.
18. $(-a_1 + a_2 + a_3 + a_4)(a_1 - a_2 + a_3 + a_4)(a_1 + a_2 - a_3 + a_4)(a_1 + a_2 + a_3 - a_4) \leq 8(a_1^2 a_4^2 + a_2^2 a_3^2)$.
19. $\sin^2\alpha_1 + \cdots + \sin^2\alpha_n \leq \frac{9}{4}$, where $\alpha_1 > 0, \ldots, \alpha_n > 0$ and $\alpha_1 + \cdots + \alpha_n = \pi$.
20. $\sqrt{a+b-c} + \sqrt{b+c-a} + \sqrt{c+a-b} \leq \sqrt{a} + \sqrt{b} + \sqrt{c}$, where a, b, c are the sides of some triangle.
21. $a^3+b^3+c^3 \geq a^2(2c-b)+b^2(2a-c)+c^2(2b-a)$, where $a > 0,\ b > 0,\ c > 0$.
22. $\frac{1}{a^3+b^3+abc} + \frac{1}{b^3+c^3+abc} + \frac{1}{c^3+a^3+abc} \leq \frac{1}{abc}$, where $a > 0,\ b > 0,\ c > 0$.
23. $\sqrt{12uvw + u^2 + v^2 + w^2} \leq 1$, where $u \geq 0,\ v \geq 0,\ w \geq 0$ and $u+v+w = 1$.
24. $\frac{n(n-1)}{2} \cdot \sum\limits_{1 \leq i < j \leq n} \frac{1}{a_i \cdot a_j} \geq 4 \cdot \left(\sum\limits_{1 \leq i < j \leq n} \frac{1}{a_i + a_j} \right)^2$, where $n \geq 2$, $a_1 > 0, \ldots, a_n > 0$.
25. $7(pq + qr + rp) \leq 2 + 9pqr$, where $p > 0,\ q > 0,\ r > 0$ and $p + q + r = 1$.
26. $\sqrt{a_1^2 + a_2^2 + a_3^2} + \sqrt{a_2^2 + a_3^2 + a_4^2} + \cdots + \sqrt{a_n^2 + a_1^2 + a_2^2} \geq \sqrt{3}(a_1 + \cdots + a_n)$.
27. $\sqrt{a_1^2 + 2a_2^2} + 3\sqrt{a_2^2 + 2a_3^2} + 7\sqrt{2a_1^2 + a_2^2} \geq (5a_1 + 4a_2 + 2a_3)\sqrt{3}$.
28. $\sqrt{a^2 + ab + b^2} + \sqrt{b^2 + bc + c^2} + \sqrt{c^2 + ca + a^2} \geq \sqrt{3} \cdot |a + b + c|$.
29. $\left| \frac{a^3-b^3}{a+b} + \frac{b^3-c^3}{b+c} + \frac{c^3-a^3}{c+a} \right| \leq \frac{(a-b)^2+(b-c)^2+(c-a)^2}{6}$, where $a > 0,\ b > 0,\ c > 0$.
30. $\frac{1}{x^k} + \frac{1}{y^k} + \frac{1}{z^k} \geq x^k + y^k + z^k$, where $k \geq 2$, $k \in \mathbb{N}$, $x > 0,\ y > 0,\ z > 0$ and $\frac{1}{x} + \frac{1}{y} + \frac{1}{z} \geq x + y + z$, $xyz = 1$.
31. $x^2 + y^2 \leq 2$, where $x > 0,\ y > 0$ and $x^2 + y^3 \geq x^3 + y^4$.
32. $\log_a(\log_a b) + \log_b(\log_b c) + \log_c(\log_c a) > 0$, where $c > b > a > 1$.
33. $\sin(n\alpha) \leq 0$, where $n \in \mathbb{N}$, $0 < \alpha < \pi$ and $\sin\alpha + \ldots + \sin(n\alpha) \leq 0$.
34. $\sin((n+1)\alpha) \geq 0$, where $n \in \mathbb{N}$, $0 < \alpha < \pi$ and $\sin\alpha + \ldots + \sin(n\alpha) \leq 0$.
35. $\frac{x_1^3}{x_2^5+x_3^5+x_4^5+x_5^5+x_6^5+5} + \cdots + \frac{x_6^3}{x_1^5+x_2^5+x_3^5+x_4^5+x_5^5+5} \leq \frac{3}{5}$, where $0 \leq x_i \leq 1$, $i = 1, \ldots, 6$.
36. $(ab + ac + ad + bc + bd + cd)^2 + 12 \geq 6(abc + abd + acd + bcd)$, where $a + b + c + d = 0$.
37. $(y^3 + x)(z^3 + y)(x^3 + z) \geq 125xyz$, where $x \geq 2,\ y \geq 2,\ z \geq 2$.
38. $(a + 3b)(b + 4c)(c + 2a) \geq 60abc$, where $c \geq b \geq a \geq 0$.
39. $(a + b + 3c)(a + 3b + c)(3a + b + c) + 8(ab + bc + ac + abc) \geq \frac{349}{27}\sqrt{abc}$, where $a > 0,\ b > 0,\ c > 0$ and $\sqrt{a} + \sqrt{b} + \sqrt{c} = 1$.
40. $\left(\frac{p}{k}\right)^k \cdot \left(\frac{q}{l}\right)^l \leq 1$, where $p,\ k,\ q,\ l \in \mathbb{N}$ and $p + q \leq k + l$.
41. $\sqrt[n]{x_1 \cdot \ldots \cdot x_n} \geq (n-1)a$, where $n \geq 2$, $a > 0$, $x_1 > 0, \ldots, x_n > 0$ and $\frac{1}{a+x_1} + \cdots + \frac{1}{a+x_n} \leq \frac{1}{a}$.
42. $\tan\alpha_1 \cdots \tan\alpha_{n+1} \geq n^{n+1}$, where $\alpha_1, \ldots, \alpha_{n+1} \in \left(0, \frac{\pi}{2}\right)$ and $\tan\left(\alpha_1 - \frac{\pi}{4}\right) + \cdots + \tan\left(\alpha_{n+1} - \frac{\pi}{4}\right) \geq n - 1$.
43. $\frac{1}{n-1+x_1} + \cdots + \frac{1}{n-1+x_n} \leq 1$, where $x_1 > 0, \ldots, x_n > 0$ and $x_1 \cdots x_n = 1$.
44. $a^4+b^4+c^4-2(a^2b^2+b^2c^2+c^2a^2)+3(abc)^{\frac{4}{3}} \geq 0$, where $a \geq 0,\ b \geq 0,\ c \geq 0$.
45. $a^4 + b^4 + c^4 - 2(a^2b^2 + b^2c^2 + c^2a^2) + abc(a + b + c) \geq 0$, where $a \geq 0,\ b \geq 0,\ c \geq 0$.
46. $\frac{a^2+b}{b+c} + \frac{b^2+c}{c+a} + \frac{c^2+a}{a+b} \geq 2$, where $a > 0,\ b > 0,\ c > 0$ and $a + b + c = 1$.

47. $\sqrt{\frac{a^2+bc}{a+1}}+\sqrt{\frac{b^2+cd}{b+1}}+\sqrt{\frac{c^2+da}{c+1}}+\sqrt{\frac{d^2+ab}{d+1}}\geq 4$, where $a>0,\ b>0,\ c>0,\ d>0$ and $abcd=1$.
48. $\sqrt{3(ab+bc+ca)}\big(9(a+b+c)^2+ab+bc+ca\big)\leq 9\big((a+b+c)^3+abc\big)$, where $a>0,\ b>0,\ c>0$.
49. $4(x^3+y^3+z^3+xyz)^2\geq\left(x^2+y^2+z^2+\left(\frac{x+y+z}{3}\right)^2\right)^3$, where $x>0,\ y>0,\ z>0$.
50. $2,25(x^2-x+1)(y^2-y+1)(z^2-z+1)\geq(xyz)^2-xyz+1$.
51. $\left|ab(a^2-b^2)+bc(b^2-c^2)+ca(c^2-a^2)\right|\leq\frac{1}{4}(a^2+b^2+c^2)^2$, where $a>0,\ b>0,\ c>0$.
52. $\sqrt{a-1}+\sqrt{b-1}+\sqrt{c-1}\leq\sqrt{abc+c}$, where $a\geq 1,\ b\geq 1,\ c\geq 1$.
53. $(p(xy))^2\leq p(x^2)p(y^2)$, where $p(t)=a_0t^n+a_1t^{n-1}+\ldots+a_n$ and $a_0\geq 0,\ a_1\geq 0,\ldots,a_n\geq 0$.
54. $x_1^2+\cdots+x_n^2\geq\frac{2n+1}{3}.(x_1+\cdots+x_n)$, where $x_1<\cdots<x_n$ and $x_1,\ldots,x_n\in\mathbb{N}$.
55. $\frac{1}{(1+a)^2}+\frac{1}{(1+b)^2}+\frac{1}{(1+c)^2}+\frac{1}{(1+d)^2}\geq 1$, where $a>0,\ b>0,\ c>0,\ d>0$ and $abcd=1$.
56. $1<\sum\limits_{i=1}^{n}\frac{x_i}{\sqrt{1+x_0+\cdots+x_{i-1}}\cdot\sqrt{x_i+\cdots+x_n}}<\frac{\pi}{2}$, where $n\geq 2,\quad x_0=1,x_1>0,\ldots,x_n>0$ and $x_1+\cdots+x_n=1$.
57. $0,785n^2-n<\sqrt{n^2-1^2}+\cdots+\sqrt{n^2-(n-1)^2}<0,79n^2$, where $n\geq 2,\ n\in\mathbb{N}$.
58. $\frac{a^2}{(a+b)(a+c)}+\frac{b^2}{(b+c)(b+a)}+\frac{c^2}{(c+a)(c+b)}\geq\frac{3}{4}$, where $a>0,\ b>0,\ c>0$.
59. $\frac{1}{1+ab}+\frac{1}{1+bc}+\frac{1}{1+ca}\geq\frac{3}{2}$, where $a>0,\ b>0,\ c>0$ and $a^2+b^2+c^2=3$.
60. $\frac{a}{b+2c}+\frac{b}{c+2a}+\frac{c}{a+2b}\geq 1$, where $a>0,\ b>0,\ c>0$.
61. $2\sum\limits_{i<j}x_ix_j\leq\frac{n-2}{n-1}+\sum\limits_{i=1}^{n}\frac{a_ix_i^2}{1-a_i}$, where $n\geq 2,\quad a_1>0,\ldots,a_n>0,x_1>0,\ldots,x_n>0$ and $a_1+\cdots+a_n=x_1+\cdots+x_n=1$.
62. $a^3+b^3+c^3\geq a+b+c$, where $a>0,\ b>0,\ c>0$ and $ab+bc+ca\leq 3abc$.
63. $a^2x+b^2y+c^2z>d^2$, where $x>0,\ y>0,\ z>0,\frac{1}{x}+\frac{1}{y}+\frac{1}{z}\leq 1$ and $a,\ b,\ c,\ d$ are the sides of some quadrilateral.
64. $\sqrt{x+y+z}\geq\sqrt{x-1}+\sqrt{y-1}+\sqrt{z-1}$, where $x\geq 1,\ y\geq 1,\ z\geq 1$ and $\frac{1}{x}+\frac{1}{y}+\frac{1}{z}=2$.
65. $x_1^{k-1}+\cdots+x_n^{k-1}\geq(n-1)\Big(\frac{1}{x_1}+\ldots+\frac{1}{x_n}\Big)$, where $\frac{1}{1+x_1^k}+\cdots+\frac{1}{1+x_n^k}=1,\quad k\geq 2,\quad k\in\mathbb{N}$ and $x_1\geq 1,\ldots,x_n\geq 1$.
66. $\sqrt{a^2+ab+b^2}+\sqrt{b^2+bc+c^2}+\sqrt{c^2+ca+a^2}\leq\sqrt{5a^2+5b^2+5c^2+4ab+4bc+4ac}$, where $a>0,\ b>0,\ c>0$.
67. $\sqrt{2\sqrt[3]{3\sqrt[4]{4\ldots\sqrt[n]{n}}}}<2$, where $n\geq 2,\ \ n\in\mathbb{N}$.
68. $\sqrt{1+\sqrt{2+\sqrt{3+\sqrt{\cdots+\sqrt{n}}}}}<2$, where $n\in\mathbb{N}$.
69. $n!\leq\left(\frac{7n+9}{16}\right)^n$, where $n\in\mathbb{N}$.
70. $\frac{\sqrt{2}}{2}\cdot\frac{1}{\sqrt{2n}}<\frac{1}{2}\cdot\frac{3}{4}\cdots\frac{2n-1}{2n}<\frac{\sqrt{3}}{2}\cdot\frac{1}{\sqrt{2n}}$, where $n\geq 2,\ \ n\in\mathbb{N}$.
71. *Maclaurin's inequality*: $\sqrt[k]{b_k}\geq\sqrt[k+1]{b_{k+1}},\quad k=2,\ldots,n-1$, where

$$b_k = \frac{1}{C_n^k}(a_1 \cdots a_{k-1}a_k + a_1 \cdots a_{k-1}a_{k+1} + \cdots + a_1 \cdots a_{k-1}a_n + \cdots + a_{n-k+1} \cdots a_{n-1}a_n),$$

where $n \geq 2$, $a_1, a_2, \ldots, a_n > 0$.

72. $a_1 + \frac{a_2}{2} + \cdots + \frac{a_n}{n} \geq a_n$, where $n \geq 2$ and $a_{i+j} \leq a_i + a_j$, $i, j \in \mathbb{N}$, $i + j \leq n$.
73. $a_1a_2^4 + a_2a_3^4 + \cdots + a_na_1^4 \geq a_2a_1^4 + a_3a_2^4 + \cdots + a_1a_n^4$, where $n \geq 3$ and $a_1 < a_2 < \cdots < a_n$.
74. $u_{n+2} + u_n \geq 2 + \frac{u_{n+1}^2}{u_n}$, where $n \in \mathbb{N}$ and $u_1 = 1$, $u_2 = 2$, $u_{n+2} = 3u_{n+1} - u_n$, $n = 1, 2, \ldots$.
75. $a_i \geq 0$, $i = 1, 2, \ldots, n$, where $n \geq 3$ and $a_{i-1} + a_{i+1} \leq 2a_i$, $i = 2, \ldots, n-1$.
76. $\max(a_1, a_2, \ldots, a_n) \geq 2$, where $n \geq 4$ and $a_1 + \cdots + a_n \geq n$, $a_1^2 + \cdots + a_n^2 \geq n^2$.
77. $\frac{a_0 + \cdots + a_n}{n+1} \cdot \frac{a_1 + \cdots + a_{n-1}}{n-1} \geq \frac{a_0 + \cdots + a_{n-1}}{n} \cdot \frac{a_1 + \cdots + a_n}{n}$, where $n \geq 2$ and $a_i > 0$, $i = 0, 1, \ldots, n$, $a_{i-1}a_{i+1} \leq a_i^2$, $i = 1, \ldots, n-1$.
78. $\frac{1 - x^{n+1}}{n+1} > \frac{1 - x^n}{n} \cdot \sqrt{x}$, where $n \in \mathbb{N}$, $0 < x < 1$.
79. $\sin(x_1 + x_2) + \ldots + \sin(x_{n-1} + x_n) + \sin(x_n + x_1) > 2$, where $n = 3, 4, \ldots$ and $x_1 + \cdots + x_n = \frac{\pi}{2}$.
80. $x^y + y^x > 1$, where $x > 0$, $y > 0$.
81. $a_1^2 + \cdots + a_n^2 - n \geq \frac{2n}{n - \sqrt{n-1}} \cdot (a_1 + \cdots + a_n - n)$, where $n \geq 2$ and $a_1 \cdots a_n = 1$.
82. $x_1 \arctan x_1 + \ldots + x_n \arctan x_n \geq n \ln 2$, where $n \geq 2$ and $x_1 > 0, \ldots, x_n > 0$, $x_1 \cdots x_n = 1$.
83. $\sum\limits_{i=1}^{n} \frac{a_i^p}{b_i} \geq n^{2-p} \cdot \frac{\left(\sum\limits_{i=1}^{n} a_i\right)^p}{\sum\limits_{i=1}^{n} b_i}$, where $a_i > 0$, $b_i > 0$, $i = 1, \ldots, n$, $p \geq 2$.
84. $x^{2^x} + \cdots + x^{n^x} \geq 2^{x^2} + \cdots + n^{x^n}$, where $x \geq n \geq 2$, $n \in \mathbb{N}$.
85. $\frac{1}{2-a} + \frac{1}{2-b} + \frac{1}{2-c} \geq 3$, where $a \geq 0$, $b \geq 0$, $c \geq 0$ and $a^2 + b^2 + c^2 = 3$.
86. $\frac{a}{4+a^2} + \frac{b}{4+b^2} + \frac{c}{4+c^2} + \frac{d}{4+d^2} + \frac{e}{4+e^2} \leq 1$, where $a \geq 0$, $b \geq 0$, $c \geq 0$, $d \geq 0$, $e \geq 0$ and $\frac{1}{4+a} + \frac{1}{4+b} + \frac{1}{4+c} + \frac{1}{4+d} + \frac{1}{4+e} = 1$.
87. *Hardy's inequality*: $\sum\limits_{k=1}^{n} \left(\frac{a_1 + \ldots + a_k}{k}\right)^p \leq \left(\frac{p}{p-1}\right)^p \sum\limits_{k=1}^{n} a_k^p$, where $p > 1$ and $a_i > 0$, $i = 1, \ldots, n$.
88. *Abel's inequality*: $\left|\sum\limits_{i=1}^{n} a_k b_k\right| \leq B(|a_1| + 2|a_n|)$, where $|b_1 + \ldots + b_k| \leq B$, $k = 1, \ldots, n$, $a_1 \leq \ldots \leq a_n$ or $a_1 \geq \ldots \geq a_n$.
89. *Abel's inequality*: $\left|\sum\limits_{i=1}^{n} a_k b_k\right| \leq Ba_1$, where $|b_1 + \cdots + b_k| \leq B$, $k = 1, \ldots, n$, $a_1 \geq \cdots \geq a_n \geq 0$.
90. $\sum\limits_{i=1}^{n} a_i^2 > \frac{1}{4}\left(1 + \frac{1}{2} + \cdots + \frac{1}{n}\right)$, where $a_1 > 0, \ldots, a_n > 0$ and $\sum\limits_{i=1}^{k} a_i > \sqrt{k}$, $k = 1, \ldots, n$.
91. $\frac{1}{x_1} + \cdots + \frac{1}{x_n} \leq \frac{1}{y_1} + \cdots + \frac{1}{y_n}$, where $n \geq 2$, $x_i > 0$, $y_i > 0$, $x_1 + \cdots + x_i \geq y_1 + \cdots + y_i$, $i = 1, \ldots, n$ and $x_1y_1 < x_2y_2 < \cdots < x_ny_n$.
92. $\left|\sum\limits_{k=1}^{n-1} \frac{\cos(n-k)x}{k} - \sum\limits_{k=1}^{n-1} \frac{\cos(n+k)x}{k}\right| < 6$, where $n \geq 2$, $n \in \mathbb{N}$.

93. $\left|\sum_{k=1}^{n-1}\frac{\cos(n-k)x}{k}-\sum_{k=1}^{n-1}\frac{\cos(n+k)x}{k}\right|<4\sqrt{2}$, where $n\geq 2,\ \ n\in\mathbb{N}$ and $0\leq x\leq\frac{\pi}{2}$.

94. $a_1^{\frac{1}{k}}+\cdots+a_n^{\frac{1}{k}}\leq b_1^{\frac{1}{k}}+\cdots+b_n^{\frac{1}{k}}$, where $k\geq 2,\ \ k\in\mathbb{N},\ \ a_i>0,\ \ b_i>0,\ \ b_1+\cdots+b_i\geq a_1+\cdots+a_i,\ i=1,\ldots,n$ and $b_1\leq\cdots\leq b_n$.

95. $\frac{1}{x_1}+\cdots+\frac{1}{x_n}\leq\frac{1}{u_1}+\cdots+\frac{1}{u_n}$, where $u_1=2,\ \ u_2=3,\ \ u_{k+1}=u_1\cdots u_k+1,\ \ k=2,3,\ldots$, and $x_1,\ldots,x_n\in\mathbb{N},\ \ \frac{1}{x_1}+\cdots+\frac{1}{x_n}<1$.

96. $A(b-c)^2+B(c-a)^2+C(a-b)^2\geq 0$, if at least one of the following conditions holds:

(a) $a\geq b\geq c$ and $B\geq 0,\ \ B+A\geq 0,\ \ B+C\geq 0$,
(b) $a\geq b\geq c$ and $A\geq 0,\ \ C\geq 0,\ \ A+2B\geq 0,\ \ C+2B\geq 0$,
(c) $a\geq b\geq c>0$ and $B\geq 0,\ \ C\geq 0,\ \ a^2B+b^2A\geq 0$,
(d) $A+B+C\geq 0,\ \ AB+BC+CA\geq 0$,
(e) $a\geq b\geq c,\ \ A\geq 0,\ \ B\geq 0,\ \ b^2B+c^2C\geq 0$, and a,b,c are the sides of some triangle.

97. $(xy+yz+zx)\left(\frac{1}{(x+y)^2}+\frac{1}{(y+z)^2}+\frac{1}{(z+x)^2}\right)\geq\frac{9}{4}$, where $x>0,\ \ y>0,\ \ z>0$.

98. $\frac{1-h}{2}<\sum_{i=1}^{n}x_{2i}(x_{2i+1}-x_{2i-1})<\frac{1+h}{2}$, where $0=x_1<x_2<\cdots<x_{2n+1}=1$ and $x_{i+1}-x_i\leq h,\ \ i=1,2,\ldots,2n$.

99. $\sum_{1\leq i<j\leq n}\left(|a_i-a_j|+|b_i-b_j|\right)\leq\sum_{1\leq i,j\leq n}|a_i-b_j|$.

100. *Carlson's inequality*: $\left(\sum_{i=1}^{n}a_i\right)^4\leq\pi^2\left(\sum_{i=1}^{n}a_i^2\right)\left(\sum_{i=1}^{n}i^2a_i^2\right)$.

Hints

1. $(ab+bc+ca)\left|1-\frac{1}{abc}\right|\leq\left|ab+bc+ca-\frac{1}{c}-\frac{1}{a}-\frac{1}{b}\right|$.
2. $\frac{\sin^2\beta\sin\frac{\alpha}{2}\sin(\alpha+\frac{\beta}{2})-\sin^2\alpha\sin\frac{\beta}{2}\sin(\beta+\frac{\alpha}{2})}{\sin\frac{\alpha}{2}\sin\frac{\beta}{2}}=2\sin\frac{\beta-\alpha}{2}\cos\frac{\alpha+\beta}{2}\sin(\alpha+\beta)$.
3. $ab\cdot\left(\left|\frac{a+b}{a-b}\right|-1\right)\geq 0$.
4. Let $\max(a,b,c)=a$, and then $\frac{a-b}{a+b}<1,\ \ \frac{b-c}{b+c}\leq\frac{a-c}{c+a}$.
5. $x_0-x_n=(x_0-x_1)+\cdots+(x_{n-1}-x_n),\ \ n\geq 2$.
6. $S_{km}=S_k+\left(\frac{1}{k+1}+\cdots+\frac{1}{2k}\right)+\cdots+\left(\frac{1}{(m-1)k+1}+\cdots+\frac{1}{mk}\right)\geq S_k+k\cdot\frac{1}{2k}+\cdots+k\cdot\frac{1}{mk}$.
7. $\frac{1}{n+1}\left(1+\frac{1}{3}+\cdots+\frac{1}{2n-1}\right)-\frac{1}{n}\left(\frac{1}{2}+\frac{1}{4}+\cdots+\frac{1}{2n}\right)=\frac{1}{2n(n+1)}\left(2n\left(\frac{1}{n+1}+\cdots+\frac{1}{2n}\right)-\left(1+\cdots+\frac{1}{n}\right)\right)$.
8. $\frac{1}{2}+\frac{2}{2^2}+\cdots+\frac{n}{2^n}<\left(\frac{1}{2}+\frac{1}{2^2}+\cdots\right)+\left(\frac{1}{2^2}+\frac{1}{2^3}+\cdots\right)+\cdots=1+\frac{1}{2}+\frac{1}{2^2}+\cdots=2$.
9. Let $a_1\leq a_2\leq\ldots\leq a_5$, then $(a_3-a_1)(a_3-a_2)(a_3-a_4)(a_3-a_5)\geq 0$, $(a_1-a_2)(a_1-a_3)(a_1-a_4)(a_1-a_5)\geq(a_1-a_2)(a_2-a_3)(a_2-a_4)(a_2-a_5)$, and $(a_5-a_1)(a_5-a_2)(a_5-a_3)(a_5-a_4)\geq(a_4-a_1)(a_4-a_2)(a_4-a_3)(a_5-a_4)$.

10. Let $m \geq n$, then $m \geq a \geq n$ and $a - n = \frac{m^m}{n^n} \cdot (m-a) \geq \frac{m^{m-1}+\dots+a^{m-1}}{n^{n-1}+\dots+a^{n-1}} \cdot (m-a)$.
11. $\frac{(m+n)!}{(m-n)!} = (m-n+1)\cdots(m+n) \geq (2n)!$ and $k(2m+1-k) \leq m^2+m, \quad k = m-n+1, \dots, m$.
12. If $p \leq 0$, then $\left|\sqrt{2} - \frac{p}{q}\right| \geq \sqrt{2} > \frac{1}{3} \geq \frac{1}{3q^2}$, and if $0 < p < (3-\sqrt{2})q$, then $\left|\sqrt{2} - \frac{p}{q}\right| = \frac{|2q^2-p^2|}{q(\sqrt{2}q+p)} \geq \frac{1}{q(\sqrt{2}q+p)} > \frac{1}{3q^2}$. It is left to consider the case $p \geq (3-\sqrt{2})q$.
13. $m^2 \leq 7n^2 - 3$.
14. $(x+y+z)^2 - 3(xy+yz+zx) = \frac{1}{2}\left((x-y)^2+(y-z)^2+(z-x)^2\right)$.
15. $\frac{x}{y} + \frac{y}{z} \geq x + \frac{1}{z}$, where $x > 0, \ y > 0, \ z > 0$ and $xyz = 1$.
16. $1 + a^{2n} \geq |a|^k + |a|^{2n-k}$, where $k \in \{1, 2, \dots, n\}$.
17. $x^4 - x^3 \geq |x^3|(|x|-1) \geq |x| - 1$.
18. $8(a_1^2a_4^2 + a_2^2a_3^2) - (-a_1+a_2+a_3+a_4)(a_1-a_2+a_3+a_4)(a_1+a_2-a_3+a_4)(a_1+a_2+a_3-a_4) = (a_1^2+a_4^2-a_2^2-a_3^2)^2 + 4(a_1a_4-a_2a_3)^2$.
19. $\sin^2\alpha + \sin^2\beta \leq \sin^2(\alpha+\beta)$, where $\alpha > 0, \ \beta > 0$ and $\alpha + \beta \leq \frac{\pi}{2}$.
20. $\left(\frac{\sqrt{a+b-c}+\sqrt{b+c-a}}{2}\right)^2 \leq \frac{(\sqrt{a+b-c})^2+(\sqrt{b+c-a})^2}{2} = b$.
21. $a^3 + c^2a \geq 2\sqrt{a^3 \cdot c^2a} = 2a^2c$.
22. $\frac{abc}{a^3+b^3+abc} = \frac{abc}{(a+b)(a^2-ab+b^2)+abc} \leq \frac{abc}{(a+b)ab+abc} = \frac{c}{a+b+c}$.
23. $(uv+vw+wu)^2 \geq 3(uv \cdot vw + vw \cdot wu + wu \cdot uv) = 3uvw$.
24. $\frac{1}{a_i \cdot a_j} \geq 4 \cdot \left(\frac{1}{a_i+a_j}\right)^2$.
25. Let $\min(p,q,r) = r$, then

$$7(pq+qr+rp) - 9pqr = (7-9r)pq + 7r(p+q) \leq$$
$$\leq (7-9r)\frac{(p+q)^2}{4} + 7r(p+q) = 2 - \frac{1}{4}(3r-1)^2(r+1).$$

26. $\sqrt{x^2+y^2+z^2} \geq \frac{\sqrt{3}}{3}(x+y+z)$.
27. $\sqrt{x^2+2y^2} \geq \frac{\sqrt{3}}{3}(x+2y)$.
28. $\sqrt{x^2+xy+y^2} = \sqrt{\left(x+\frac{y}{2}\right)^2 + \left(\frac{\sqrt{3}}{2}y\right)^2}$.
29. $\left|\frac{a^3-b^3}{a+b} + \frac{b^3-c^3}{b+c} + \frac{c^3-a^3}{c+a}\right| = \frac{|a-b|\cdot|b-c|\cdot|c-a|}{(a+b)(b+c)(c+a)}(ab+bc+ca)$.
Let $a \leq b \leq c$, then $\frac{|c-a|}{c+a} < 1$, $\frac{ab+bc+ca}{(a+b)(b+c)} < 1$ and

$$6|a-b|\cdot|b-c| = 2|a-b|\cdot|b-c| + 4|a-b|\cdot|b-c| \leq \left((a-b)^2+(b-c)^2\right)$$
$$+ ((b-a)+(c-b))^2.$$

30. $(1-x)(1-y)(1-z) \geq 0$, and therefore $(1-x^k)(1-y^k)(1-z^k) \geq 0$.
31. $x^2 - x^3 \geq y^4 - y^3 \geq y^3 - y^2$ and $(x+y)(x^3+y^3) \geq (x^2+y^2)^2 \geq (x^3+y^3)^2$.
32. $\log_a(\log_a b) > \log_b(\log_a b)$.
33. $\sin\alpha + \dots + \sin(n\alpha) = \sin^2\frac{n\alpha}{2} \cdot \cot\frac{\alpha}{2} + \frac{1}{2}\sin(n\alpha)$.

34. $\sin\alpha + \cdots + \sin(n\alpha) = \sin^2 \frac{(n+1)\alpha}{2} \cdot \cot\frac{\alpha}{2} - \frac{1}{2}\sin((n+1)\alpha)$.
35. $\frac{x_i^3}{x_1^5+x_2^5+x_3^5+x_4^5+x_5^5+x_6^5-x_i^5+5} \leq \frac{x_i^3}{x_1^5+x_2^5+x_3^5+x_4^5+x_5^5+x_6^5+4}$ and $3x_i^5 - 5x_i^3 + 2 = (x_i-1)^2(3x_i^3 + 6x_i^2 + 4x_i + 2)$.
36. Set $a = x+\alpha, b = y+\alpha, c = z+\alpha, d = t+\alpha, X = xy+xz+xt+yz+yt+zt,\ \ Y = xyz + xyt + xzt + yzt$. In this case, the given inequality can be rewritten as $\left(X - 6\alpha^2\right)^2 + 12 \geq 6\left(Y + 2\alpha X - 8\alpha^3\right)$, where $x + y + z + t = -4\alpha$.
Let us choose the number α such that $36\alpha^4 + 12 = -48\alpha^3$, for example $\alpha = -1$.
We need to prove that $X^2 \geq 6Y$, where $x + y + z + t = 4$.
If $Y \leq 0$, then $X^2 \geq 0 \geq 6Y$. And if $Y > 0$, then $X^2 = 2Y(x + y + z + t) + (xy - zt)^2 + (xz)^2 + (xt)^2 + (yz)^2 + (yt)^2 \geq 8Y > 6Y$. Therefore $X^2 > 6Y$.
37. $y^3 + x \geq 4y + x \geq 5\sqrt[5]{xy^4}$.
38. $(a+3b)(b+4c)(c+2a) \geq 4\sqrt[4]{ab^3}\cdot 5\sqrt[5]{bc^4}\cdot 3\sqrt[3]{ca^2} \geq 60\sqrt[4]{ab^3}\cdot\sqrt[5]{bc^4}\cdot\sqrt[3]{c^{3/5}b^{2/5}a^2}$.
39. $(a+b+3c)(a+3b+c)(3a+b+c)+8(ab+bc+ac+abc) \geq (\frac{1}{3}+2c)(\frac{1}{3}+2b)(\frac{1}{3}+2a) + 8(ab + bc + ac) + 8abc \geq \frac{1}{9} + \frac{28}{3}(ab + bc + ac) + 16abc \geq \frac{1}{9} + \frac{28}{3}(\sqrt{ab}\cdot\sqrt{bc} + \sqrt{bc}\cdot\sqrt{ca} + \sqrt{ca}\cdot\sqrt{ab}) + 16abc = \frac{1}{9} + \frac{28}{3}\sqrt{abc} + 16abc \geq \frac{349}{27}\sqrt{abc}$.
40. $k + l \geq p + q = \frac{p}{k} + \cdots + \frac{p}{k} + \frac{q}{l} + \cdots + \frac{q}{l} \geq (k + l)\cdot\left(\frac{p}{k}\cdots\frac{p}{k}\cdot\frac{q}{l}\cdots\frac{q}{l}\right)^{\frac{1}{k+l}}$.
41. $x_i \geq a\left(\frac{a+x_i}{a+x_1} + \cdots + \frac{a+x_i}{a+x_n} - 1\right) \geq (n-1)a\left(\frac{a+x_i}{a+x_1}\cdots\frac{a+x_i}{a+x_n}\right)^{\frac{1}{n-1}}$.
42. Let $\tan\alpha_i = x_i$, then $\frac{1}{1+x_1} + \cdots + \frac{1}{1+x_n} \leq 1$.
43. $\frac{x_i}{n-1+x_i} = \frac{x_i}{(n-1)x_1^\alpha\cdots x_n^\alpha+x_i} = \frac{x_i^{1-\alpha}}{(n-1)\frac{x_1^\alpha\cdots x_n^\alpha}{x_i^\alpha}+x_i^{1-\alpha}} \geq \frac{x_i^{1-\alpha}}{x_1^{1-\alpha}+\cdots+x_n^{1-\alpha}}$, where $(n-1)\alpha = 1-\alpha$.
44. $\left(a^{\frac{4}{3}}\right)^3 + \left(b^{\frac{4}{3}}\right)^3 + \left(c^{\frac{4}{3}}\right)^3 + 3a^{\frac{4}{3}}\cdot b^{\frac{4}{3}}\cdot c^{\frac{4}{3}} \geq a^{\frac{8}{3}}b^{\frac{4}{3}} + a^{\frac{4}{3}}b^{\frac{8}{3}} + b^{\frac{8}{3}}c^{\frac{4}{3}} + b^{\frac{4}{3}}c^{\frac{8}{3}} + c^{\frac{8}{3}}a^{\frac{4}{3}} + c^{\frac{4}{3}}a^{\frac{8}{3}} \geq 2\sqrt{a^{\frac{8}{3}}b^{\frac{4}{3}}\cdot a^{\frac{4}{3}}b^{\frac{8}{3}}} + 2\sqrt{b^{\frac{8}{3}}c^{\frac{4}{3}}\cdot b^{\frac{4}{3}}c^{\frac{8}{3}}} + 2\sqrt{c^{\frac{8}{3}}a^{\frac{4}{3}}\cdot c^{\frac{4}{3}}a^{\frac{8}{3}}}$.
45. $abc(a + b + c) \geq 3(abc)^{\frac{4}{3}}$.
46. $\frac{a^2+b}{b+c} = \frac{a+b}{b+c} - a$.
47. $a^2 + bc \geq \frac{(a+1)^3}{2a(d+1)}$.
48. Let $a + b + c = p,\ \ ab + bc + ca = \frac{p^2-q^2}{3}$, where $q \geq 0$. Then if $q < p \leq 2q$, then $\min(abc) = 0$, and if $p > 2q$, then $\min(abc) = \left(\frac{p+q}{3}\right)^2 \cdot \frac{p-2q}{3}$.
49. Let $x + y + z = 3u,\ \ xy + yz + zx = 3v^2,\ \ xyz = w^3$. Then one needs to prove that $4(27u^3 - 27uv^2 + 4w^3)^2 \geq (9u^2 - 6v^2 + u^2)^3$.
50. $1.5(x^2 - x + 1)(y^2 - y + 1) \geq ((xy)^2 - xy + 1)$.
51. Let $a \geq b \geq c$. If $a > c$, then without loss of generality one can assume that $(a-c)^2 + (b-c)^2 = 1$. Therefore $a = c + \cos\alpha,\ \ b = c + \sin\alpha,\ \ \alpha \in \left[0, \frac{\pi}{4}\right]$.
52. $\sqrt{x\cdot 1} + \sqrt{1\cdot y} \leq \sqrt{(x+1)(1+y)}$, where $x \geq 0,\ y \geq 0$.
53. $p(x^2) = \left(\sqrt{a_0}\cdot x^n\right)^2 + \left(\sqrt{a_1}\cdot x^{n-1}\right)^2 + \cdots + \left(\sqrt{a_n}\right)^2$.
54. Let $y_i = x_i - i$, then $0 \leq y_1 \leq \cdots \leq y_n$.

55. Let $\frac{1}{(1+a)^2}+\frac{1}{(1+b)^2}+\frac{1}{(1+c)^2}+\frac{1}{(1+d)^2}=R^2$, then

$$\frac{1}{1+a}=R\cos\alpha\,\cos\beta,\ \frac{1}{1+b}=R\cos\alpha\,\sin\beta,$$
$$\frac{1}{1+c}=R\sin\alpha\,\cos\gamma,\quad \frac{1}{1+d}=R\sin\alpha\,\sin\gamma,\quad \alpha,\beta,\gamma\in\left(0,\frac{\pi}{2}\right).$$

56. $$\sum_{i=1}^{n}\frac{x_i}{\sqrt{1+x_0+\cdots+x_{i-1}}\cdot\sqrt{x_i+\cdots+x_n}}=x_1+\frac{x_2}{\sqrt{1-x_1^2}}$$
$$+\cdots+\frac{x_n}{\sqrt{1-(x_1+\cdots+x_{n-1})^2}}>x_1+x_2+\cdots+x_n=1.$$

Let $x_1+\cdots+x_i=\cos\alpha_i$, $\alpha_i\in\left[0,\frac{\pi}{2}\right)$, $i=1,2,\ldots,n$. Then $\frac{x_i}{\sqrt{1-(x_1+\cdots+x_{i-1})^2}}<\sin(\alpha_{i-1}-\alpha_i)$, $i=2,3,\ldots,n$.

57. $\frac{\pi n^2}{4}-n<\sqrt{n^2-1^2}+\cdots+\sqrt{n^2-(n-1)^2}<\frac{\pi n^2}{4}$.

58. $\frac{a^2}{(a+b)(a+c)}+\frac{b^2}{(b+c)(b+a)}+\frac{c^2}{(c+a)(c+b)}\geq\frac{(a+b+c)^2}{a^2+b^2+c^2+3ab+3bc+3ca}$.

59. $\frac{1}{1+ab}+\frac{1}{1+bc}+\frac{1}{1+ca}\geq\frac{3^2}{3+ab+bc+ca}$.

60. $\frac{a}{b+2c}=\frac{a^2}{ab+2ac}$.

61. $\sum\limits_{i=1}^{n}\frac{a_ix_i^2}{1-a_i}=\sum\limits_{i=1}^{n}\frac{x_i^2}{1-a_i}-\sum\limits_{i=1}^{n}x_i^2$.

62. $a^3+b^3+c^3=\frac{a^2}{1/a}+\frac{b^2}{1/b}+\frac{c^2}{1/c}\geq\frac{(a+b+c)^2}{1/a+1/b+1/c}$.

63. $a^2x+b^2y+c^2z=\frac{a^2}{1/x}+\frac{b^2}{1/y}+\frac{c^2}{1/z}$.

64. $1=\frac{(\sqrt{x-1})^2}{x}+\frac{(\sqrt{y-1})^2}{y}+\frac{(\sqrt{z-1})^2}{z}$.

65. $n-1=\frac{x_1^{k-1}\cdot 1}{\frac{1}{x_1}+x_1^{k-1}}+\cdots+\frac{x_n^{k-1}\cdot 1}{\frac{1}{x_n}+x_n^{k-1}}\leq\frac{(x_1^{k-1}+\cdots+x_n^{k-1})\cdot n}{\frac{1}{x_1}+x_1^{k-1}+\cdots+\frac{1}{x_n}+x_n^{k-1}}$, as $\left(\frac{x_i^{k-1}}{\frac{1}{x_i}+x_i^{k-1}}-\frac{x_j^{k-1}}{\frac{1}{x_j}+x_j^{k-1}}\right)\cdot\left(\frac{1}{\frac{1}{x_i}+x_i^{k-1}}-\frac{1}{\frac{1}{x_j}+x_j^{k-1}}\right)\leq 0$.

66. $\frac{a^2+ab+b^2}{a+b}+\frac{b^2+bc+c^2}{b+c}+\frac{c^2+ca+a^2}{c+a}\leq\frac{5a^2+5b^2+5c^2+4ab+4bc+4ac}{2(a+b+c)}$.

67. $\sqrt[m]{m\sqrt[m+1]{(m+1)\ldots\sqrt[n]{n}}}<2$, where $2\leq m\leq n$.

68. $\sqrt{m+\sqrt{m+1+\sqrt{\cdots+\sqrt{n}}}}<m+1$, where $1\leq m\leq n$.

69. If $n\geq 5$, then

$$\left(\frac{7n+16}{16}\right)^{n+1}:\left(\frac{7n+9}{16}\right)^{n}=\frac{7n+16}{16}\cdot\left(1+\frac{7}{7n+9}\right)^{n}$$
$$\geq\frac{7n+16}{16}\cdot\left(1+\frac{7n}{7n+9}+\frac{n(n-1)}{2}\cdot\frac{49}{(7n+9)^2}\right)$$
$$>\frac{1}{16}\cdot\left(7n+16+7n+\frac{49n(n-1)}{14n+18}\right)>n+1.$$

70. $\frac{2n+1}{2n+2}>\frac{\sqrt{n}}{\sqrt{n+1}}$ and $\frac{1}{2}\cdot\frac{3}{4}\cdots\frac{2n-1}{2n}\leq\frac{1}{\sqrt{3n+1}}$.

71. If $2 \le k \le n-2$, then $\frac{\sqrt[k+1]{b_{k+1}}}{\sqrt[k+2]{b_{k+2}}} = \frac{\sqrt[k+1]{b_{k+1}}\cdot\sqrt[k+2]{b_k}}{\sqrt[k+2]{b_{k+2}\cdot b_k}} \ge \frac{\sqrt[k+1]{b_{k+1}}\cdot\sqrt[k+2]{b_k}}{\sqrt[k+2]{b_{k+1}^2}} = \left(\frac{\sqrt[k]{b_k}}{\sqrt[k+1]{b_{k+1}}}\right)^{\frac{k}{k+2}}$.

72.
$$(k+1)\left(a_1+\frac{a_2}{2}+\cdots+\frac{a_{k+1}}{k+1}\right) = (a_1+a_k)+\left(\left(a_1+\frac{a_2}{2}\right)+a_{k-1}\right)+\cdots$$
$$+\left(\left(a_1+\frac{a_2}{2}+\cdots+\frac{a_k}{k}\right)+a_1\right)+a_{k+1}$$
$$\ge (a_1+a_k)+(a_2+a_{k-1})+\cdots+(a_k+a_1)+a_{k+1} \ge (k+1)a_{k+1}.$$

73. $(a_2+a_1)(f(a_2)-f(a_1))+\cdots+(a_n+a_{n-1})(f(a_n)-f(a_{n-1})) \ge (a_n+a_1)(f(a_n)-f(a_1))$, where
$f(x)=x^4$ is convex in $\mathbb{R}$.

74. $2u_n \le u_{n+1} < 3u_n, \quad n=1,2,\ldots$.

75. Let $A=\{i|a_i=\min(a_1,\ldots,a_n)\}$. Then if $1\notin A$ and $n\notin A$, then from $k\in A$ it follows that $k-1\in A$.

76. Let $A=\{i|a_i\ge 0\}$ and $|A|=m$. If $\max(a_1,a_2,\ldots,a_n)<2$, then
$4m+(2m-n)^2 > a_1^2+\cdots+a_n^2 \ge n^2$, and therefore $m=n$.

77. Let $\frac{a_i}{a_{i+1}} = b_i, \quad i=0,1,\ldots,n-1, \quad S=a_1+\cdots+a_{n-1}$. Then $b_0\le b_1\le \ldots \le b_{n-1}$, and therefore,

$$\frac{a_0}{a_i} = b_0\cdots b_{i-1} \le b_{n-i}\cdots b_{n-1} = \frac{a_{n-i}}{a_n}, \quad 1\le i\le n, \quad S\ge (n-1)^2 a_0 a_n.$$

78. $\frac{1-x^{n+1}}{n+1}\cdot\frac{1-x^{n-1}}{n-1} > \left(\frac{1-x^n}{n}\right)^2, \quad n=2,3,\ldots$

79. $\sin x > \frac{2}{\pi}\cdot x$, where $0<x<\frac{\pi}{2}$.

80. Let $0<x<1, \ 0<y<1$. Then $x^y = \frac{1}{\left(1+\frac{1}{x}-1\right)^y} > \frac{1}{1+y\left(\frac{1}{x}-1\right)} > \frac{x}{x+y}$.

81. $c_2 > c_3 > \cdots > c_n > \cdots$ and $c_n \ge \frac{2n}{n-\sqrt{n-1}}, \quad n=2,3,\ldots$, where $c_n = \min\limits_{(0,1]} \frac{(t+1)^2((t^{n-2})^2+2(t^{n-3})^2+\cdots+n-1)}{t(t^{n-2}+2t^{n-3}+\cdots+n-1)}$.

82. Let $f(x)=x\arctan x$. Then $f''(x)=\frac{2}{(1+x^2)^2}$, and therefore, $x_1\arctan x_1+\cdots+x_n\arctan x_n \ge n\cdot\frac{x_1+\cdots+x_n}{n}\arctan\frac{x_1+\cdots+x_n}{n}$.

83.
$$\left(\sum_{i=1}^n \frac{a_i^p}{b_i}\right)\cdot\left(\sum_{i=1}^n b_i\right) \ge \left(\sum_{i=1}^n a_i^{\frac{p}{2}}\right)^2$$
$$\ge \left(\sum_{i=1}^n n\left(\frac{\sum\limits_{i=1}^n a_i}{n}\right)^{\frac{p}{2}}\right)^2 = n^{2-p}\left(\sum_{i=1}^n a_i\right)^p.$$

84. Let $f(x)=x\ln k+\ln(\ln x)-k\ln x-\ln(\ln k)$. If $x\ge k\ge 3$, then $f'(x)=\ln k+\frac{1}{\ln x}\cdot\frac{1}{x}-\frac{k}{x} > \ln k-1>0$, and if $x\ge k=2$, then

$$f'(x) = \ln 2 + \frac{1}{\ln x} \cdot \frac{1}{x} - \frac{2}{x} = \frac{1}{x}\left(x \ln 2 + \frac{1}{\ln x} - 2\right)$$
$$\geq \frac{1}{x}\left(2\sqrt{x \ln 2 \cdot \frac{1}{\ln x}} - 2\right) > \frac{1}{x}\left(2\sqrt{2 \ln 2} - 2\right) > 0.$$

Hence if $x \geq k \geq 2$, $k \in \mathbb{N}$, then $f(x) \geq f(k)$, and therefore $x^{k^x} \geq k^{x^k}$.

85. Let $a^2 = x$ and $f(x) = \frac{1}{2-\sqrt{x}}$. If $0 \leq x \leq 3$, then $f(x) \geq f(1) + f'(1)(x-1)$.
86. Let $\frac{1}{4+a} = x$ and $f(x) = \frac{x-4x^2}{17x^2-8x+1}$. If $0 < x \leq \frac{1}{4}$, then $f(x) \leq f\left(\frac{1}{5}\right) + f'\left(\frac{1}{5}\right) \cdot \left(x - \frac{1}{5}\right)$.
87. See Problem 11.25(c).
88.

$$\left|\sum_{i=1}^{n} a_k b_k\right| = |(a_1 - a_2)b_1 + (a_2 - a_3)(b_1 + b_2)$$
$$+ \cdots + (a_{n-1} - a_n)(b_1 + \cdots + b_{n-1}) + a_n(b_1 + \cdots + b_n)|$$
$$\leq |a_1 - a_2|B + |a_2 - a_3|B + \cdots + |a_{n-1} - a_n|B + |a_n|B$$
$$= |a_1 - a_n|B + |a_n|B \leq B(|a_1| + 2|a_n|).$$

89.

$$\left|\sum_{i=1}^{n} a_k b_k\right| = |(a_1 - a_2)b_1 + (a_2 - a_3)(b_1 + b_2)$$
$$+ \ldots + (a_{n-1} - a_n)(b_1 + \cdots + b_{n-1}) + a_n(b_1 + \cdots + b_n)|$$
$$\leq |a_1 - a_2|B + |a_2 - a_3|B + \cdots + |a_{n-1} - a_n|B + |a_n|B = Ba_1.$$

90. Without loss of generality one can assume that $a_1 \geq a_2 \geq \cdots \geq a_n$.

$$\sum_{i=1}^{n} a_i^2 = (a_1 - a_2)a_1 + (a_2 - a_3)(a_1 + a_2)$$
$$+ \cdots + (a_{n-1} - a_n)(a_1 + \cdots + a_{n-1}) + a_n(a_1 + \cdots + a_n)$$
$$> (a_1 - a_2)\sqrt{1} + (a_2 - a_3)\sqrt{2} + \cdots + (a_{n-1} - a_n)\sqrt{n-1} + a_n\sqrt{n}$$
$$> \frac{1}{2}\left(a_1 \cdot \frac{1}{\sqrt{1}} + \cdots + a_n \cdot \frac{1}{\sqrt{n}}\right) > \frac{1}{4} \cdot \left(\left(\frac{1}{\sqrt{1}}\right)^2 + \cdots + \left(\frac{1}{\sqrt{n}}\right)^2\right),$$

91. since $2\sqrt{k} > 1 + \cdots + \frac{1}{\sqrt{k}}$, $k = 1, \ldots, n$.

$$\frac{1}{x_1 y_1} \cdot (x_1 - y_1) + \cdots + \frac{1}{x_n y_n} \cdot (x_n - y_n)$$
$$= \left(\frac{1}{x_1 y_1} - \frac{1}{x_2 y_2}\right) \cdot (x_1 - y_1) + \cdots + \left(\frac{1}{x_{n-1} y_{n-1}} - \frac{1}{x_n y_n}\right)$$
$$\cdot (x_1 - y_1 + \cdots + x_{n-1} - y_{n-1}) + \frac{1}{x_n y_n} \cdot (x_1 - y_1 + \cdots + x_n - y_n) \geq 0.$$

92. $$\left|\sum_{k=1}^{n-1}\frac{\cos(n-k)x}{k}-\sum_{k=1}^{n-1}\frac{\cos(n+k)x}{k}\right| = 2|\sin(nx)|\left|\sum_{k=1}^{n-1}\frac{\sin(kx)}{k}\right|$$
$$\leq 2\left|\sum_{k=1}^{n-1}\frac{\sin(kx)}{k}\right| < 6.$$

93. If $x = 0$, then $\left|\sum_{k=1}^{n-1}\frac{\cos(n-k)x}{k}-\sum_{k=1}^{n-1}\frac{\cos(n+k)x}{k}\right| = 0 < 4\sqrt{2}$. If $0 < x \leq \frac{\pi}{2}$, then there exists a positive integer m such that $\frac{1}{\sqrt{2}(m+1)} < \left|\sin\frac{x}{2}\right| \leq \frac{1}{\sqrt{2}m}$.
For $n \leq m+1$ we have $\left|\sum_{k=1}^{n-1}\frac{\cos(n-k)x}{k}-\sum_{k=1}^{n-1}\frac{\cos(n+k)x}{k}\right| \leq 2\left|\sum_{k=1}^{n-1}\frac{\sin(kx)}{k}\right| \leq 2m|\sin x| < 2\sqrt{2}$.
Prove that if $n \geq m+2$, then $\left|\frac{\sin(m+1)x}{m+1}+\ldots+\frac{\sin((n-1)x)}{n}\right| < \frac{1}{(m+1)\left|\sin\frac{x}{2}\right|} < \sqrt{2}$.

94. Without loss of generality one can prove that $a_1 \leq a_2 \leq \cdots \leq a_n$. Let $A_i = \left(b_i^{\frac{1}{k}}\right)^{k-1} + \left(b_i^{\frac{1}{k}}\right)^{k-2}\cdot a_i^{\frac{1}{k}} + \cdots + \left(a_i^{\frac{1}{k}}\right)^{k-1}$. Then $A_i \leq A_{i+1}$, $i = 1, \ldots, n-1$, and

$$b_1^{\frac{1}{k}} + \cdots + b_n^{\frac{1}{k}} - \left(a_1^{\frac{1}{k}} + \ldots + a_n^{\frac{1}{k}}\right)$$
$$= \frac{b_1 - a_1}{A_1} + \cdots + \frac{b_n - a_n}{A_n} = (b_1 - a_1)\left(\frac{1}{A_1} - \frac{1}{A_2}\right)$$
$$+ \cdots + (b_1 - a_1 + \cdots + b_{n-1} - a_{n-1})\left(\frac{1}{A_{n-1}} - \frac{1}{A_n}\right)$$
$$+ (b_1 - a_1 + \cdots + b_n - a_n)\frac{1}{A_n} \geq 0.$$

95. Let $x_1 \leq \cdots \leq x_n$. Then one can prove that $\frac{1}{x_1} \leq \frac{1}{u_1}, \ldots, \frac{1}{x_1} + \cdots + \frac{1}{x_{n-1}} \leq \frac{1}{u_1} + \cdots + \frac{1}{u_{n-1}}$.
If $\frac{1}{x_1} + \ldots + \frac{1}{x_n} > \frac{1}{u_1} + \ldots + \frac{1}{u_n}$, then

$$\frac{1}{x_1 \cdot \ldots \cdot x_n} \leq 1 - \left(\frac{1}{x_1} + \ldots + \frac{1}{x_n}\right) < 1 - \left(\frac{1}{u_1} + \ldots + \frac{1}{u_n}\right)$$
$$= \frac{1}{u_{n+1} - 1} = \frac{1}{u_1 \cdot \ldots \cdot u_n}.$$

96. (a) $A(b-c)^2 + B(c-a)^2 + C(a-b)^2 \geq A(b-c)^2 + B(a-b)^2 + B(b-c)^2 + C(a-b)^2 \geq (B+A)(b-c)^2 + (B+C)(a-b)^2 \geq 0$.
(b) $A(b-c)^2 + B(c-a)^2 + C(a-b)^2 \geq \min(A, C)\big((b-c)^2 + (a-b)^2\big) + B(c-a)^2 \geq \min(A, C)\cdot\frac{(a-c)^2}{2} + B(c-a)^2 = \frac{1}{2}(\min(A, C) + 2B)(a-c)^2 \geq 0$.
(c) $A(b-c)^2 + B(c-a)^2 + C(a-b)^2 \geq A(b-c)^2 + B(c-a)^2 = (b-c)^2\left(A + \frac{(c-a)^2}{(b-c)^2}B\right) \geq (b-c)^2\left(A + \frac{a^2}{b^2}B\right) \geq 0$.

(d) Let $A \geq B \geq C$. Then $A + B \geq \frac{2}{3}(A + B + C) \geq 0$. If $A + B > 0$, then

$$\begin{aligned} &A(b-c)^2 + B(c-a)^2 + C(a-b)^2 \\ &\quad = (A+B)c^2 - 2c(bA + aB) + (A+C)b^2 + (B+C)a^2 - 2Cab \geq 0, \end{aligned}$$

since $D = -(AB + BC + CA)(a - b)^2 \leq 0$.

If $A + B = 0$, then $AB + BC + CA = AB \geq 0$, and therefore, $A = B = C = 0$.

(e) $A(b - c)^2 + B(c - a)^2 + C(a - b)^2 \geq B(c - a)^2 + C(a - b)^2 = (a - b)^2\left(B\frac{(c-a)^2}{(a-b)^2} + C\right) \geq (a - b)^2\left(B\frac{b^2}{c^2} + C\right) \geq 0$.

97. Let $x + y = a,\ \ y + z = b,\ \ z + x = c$ and $a \geq b \geq c$. Then one needs to prove that $A(b - c)^2 + B(c - a)^2 + C(a - b)^2 \geq 0$, where a, b, c are the sides of some triangle and $A = \frac{2}{bc} - \frac{1}{a^2},\ \ B = \frac{2}{ac} - \frac{1}{b^2},\ \ C = \frac{2}{ab} - \frac{1}{c^2}$. We have $A = \frac{2}{bc} - \frac{1}{a^2} \geq \frac{2}{a^2} - \frac{1}{a^2} > 0,\ \ B = \frac{2}{ac} - \frac{1}{b^2} \geq \frac{2}{ac} - \frac{1}{bc} > 0$,

$$\begin{aligned} Bb^2 + Cc^2 &= \frac{2(b^3 + c^3) - 2abc}{abc} \\ &\geq \frac{2(b + c)bc - 2abc}{abc} > 0. \end{aligned}$$

98.
$$\begin{aligned} &\left|\sum_{i=1}^{n} x_{2i}(x_{2i+1} - x_{2i-1}) - \frac{1}{2}\right| \\ &= \left|\sum_{i=1}^{n} x_{2i}(x_{2i+1} - x_{2i-1}) - \sum_{i=1}^{n} \frac{x_{2i+1} + x_{2i-1}}{2}(x_{2i+1} - x_{2i-1})\right| \\ &= \left|\sum_{i=1}^{n} \frac{x_{2i+1} + x_{2i-1} - 2x_{2i}}{2}(x_{2i+1} - x_{2i-1})\right| \\ &\leq \sum_{i=1}^{n} \left|\frac{x_{2i+1} + x_{2i-1} - 2x_{2i}}{2}\right|(x_{2i+1} - x_{2i-1}) < \sum_{i=1}^{n} \frac{h}{2}(x_{2i+1} - x_{2i-1}) = \frac{h}{2}. \end{aligned}$$

99. Without loss of generality one can assume that $a_i, b_i \in [0, 1],\quad i = 1, \ldots, n$. Let $f_i(x) = 1$ if $x \in [0, a_i]$ and $f_i(x) = 0$ if $x \in (a_i, 1]$, $i = 1, \ldots, n$, $g_i(x) = 1$ if $x \in [0, b_i]$ and $g_i(x) = 0$ if $x \in (b_i, 1]$, $i = 1, \ldots, n$. Then

$$\begin{aligned} &\sum_{1 \leq i,j \leq n} \min(a_i, a_j) + \sum_{1 \leq i,j \leq n} \min(b_i, b_j) \\ &= \int_0^1 (f_1(x) + \cdots + f_n(x))^2 dx + \int_0^1 (g_1(x) + \cdots + g_n(x))^2 dx \\ &\geq \int_0^1 2(f_1(x) + \cdots + f_n(x))(g_1(x) + \cdots + g_n(x)) dx = 2 \sum_{1 \leq i,j \leq n} \min(a_i, b_j). \end{aligned}$$

100. Let $c_i^2 = t + \frac{n^2}{t}$, $i = 1, \ldots, n$, $t > 0$. Then

$$\left(\sum_{i=1}^n a_i\right)^2 \le \left(\sum_{i=1}^n (a_i c_i)^2\right)\left(\frac{1}{c_i^2}\right)$$

$$= \left(\left(\sum_{i=1}^n a_i^2\right)t + \frac{\left(\sum_{i=1}^n i^2 a_i^2\right)}{t}\right) \cdot \left(\sum_{i=1}^n \frac{t}{t^2 + i^2}\right)$$

$$\le \left(\left(\sum_{i=1}^n a_i^2\right)t + \frac{\left(\sum_{i=1}^n i^2 a_i^2\right)}{t}\right) \cdot \left(\sum_{i=1}^n t \int_{i-1}^{i} \frac{1}{t^2 + x^2} dx\right)$$

$$\le \left(\left(\sum_{i=1}^n a_i^2\right)t + \frac{\left(\sum_{i=1}^n i^2 a_i^2\right)}{t}\right) \cdot \left(t \int_{0}^{n} \frac{1}{t^2 + x^2} dx\right)$$

$$= \left(\left(\sum_{i=1}^n a_i^2\right)t + \frac{\left(\sum_{i=1}^n i^2 a_i^2\right)}{t}\right) \cdot \arctan \frac{n}{t}$$

$$\le \frac{\pi}{2} \cdot \left(\left(\sum_{i=1}^n a_i^2\right)t + \frac{\left(\sum_{i=1}^n i^2 a_i^2\right)}{t}\right) = \pi \cdot \sqrt{\left(\sum_{i=1}^n a_i^2\right)\left(\sum_{i=1}^n i^2 a_i^2\right)},$$

if $t = \sqrt{\frac{\left(\sum_{i=1}^n i^2 a_i^2\right)}{\left(\sum_{i=1}^n a_i^2\right)}}$.

Appendix—Power Sums Triangle

Let n and k be positive integers. Sums of the forms $1^k + 2^k + \cdots + n^k$ often arise in mathematics, particularly in *algebra*. These sums have been studied for hundreds of years. They have wide applications in mathematics and can also be used for proving *algebraic Inequalities*.

In 1631, the Prussian mathematician *Johann Faulhaber* (1580–1635) published (*Academiae Algebrae*, 1631) the general formula for the sums of powers of the first n positive integers. In 1713, Nicolaus Bernoulli (also spelled Niklaus), the nephew of the Swiss mathematician *Jacob Bernoulli*, 7 years after Jacob Bernoulli's death in 1705 published (the manuscript of Jacob Bernoulli called *Summae Potestatum* in the book *Ars Conjectandi* [18]) the general formula for the sums of kth powers of the first n integers as a $(k + 1)$th-degree polynomial in the variable n with coefficients involving numbers B_j, which are now known as *Bernoulli numbers*. Besides Bernoulli's work there is a wide literature in which the interested reader can find this general formula (and its proof), which has the following form:

$$\sum_{i=1}^{n} i^k = \frac{1}{k+1}\sum_{j=0}^{k}\binom{k+1}{j}B_j n^{k+1-j},$$

where $B_j = 0$ if j is odd $(j > 1)$ and

$$B_0 = 0, B_1 = \frac{1}{2}, B_2 = \frac{1}{6}, B_4 = -\frac{1}{30}, B_6 = \frac{1}{42}, B_8 = -\frac{1}{30}, B_{10} = \frac{5}{66}, \ldots.$$

A more detailed list of Bernoulli numbers can be found in the literature. A full consideration and proof of this formula is outside the scope of this book.

In studying the works of Jacob Bernoulli, Johann Faulhaber, and other authors, Hayk Sedrakyan noticed that for a high-school student to remember and apply this formula can be really challenging. So his idea was to use this result to create a **simple** and **self-constructive** *Pascal-type triangle* for sums of powers.

H. Sedrakyan and N. Sedrakyan, *Algebraic Inequalities*, Problem Books in Mathematics, https://doi.org/10.1007/978-3-319-77836-5

The "*triangle*." Let us consider the following "*triangle*," where k stands for the power of the considered sum and $n, n^2, n^3, n^4, n^5, \cdots, n^{k+1}$ stand for the terms of the $(k+1)$th-degree polynomial in the variable n that the considered sum is equal to, and the number written at the intersection of each row and each column stands for the coefficient of the corresponding term of this polynomial.

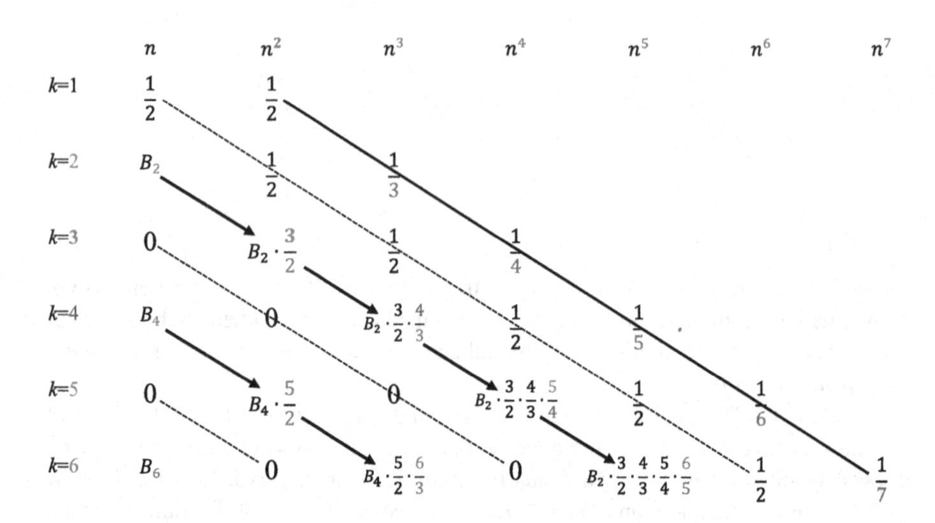

Self-construction principle. Considering this "*triangle*" diagonally, we observe that the coefficients on the dashed diagonals remain unchanged, the denominator of each coefficient on the solid line increases by one for each next row (in other words, this coefficient is equal to $\frac{1}{k+1}$), and the arrows indicate how each row can be recurrently constructed using the previous row (each time one needs to multiply the coefficient of the previous row by a fraction whose numerator is equal to the value of k written in that row and denominator is equal to the power of n written in that column). Note that $B_k = 0$ if k is odd ($k > 1$), so every odd row (starting from the third) starts with 0, and every even row $k = 2n$, $n \in \mathbb{N}$, starts with B_{2n}.

Example (application). Calculate the sum $1^5 + 2^5 + \cdots + n^5$.

Using this triangle we obtain

$$1^5 + 2^5 + \cdots + n^5 = B_4 \cdot \frac{5}{2} \cdot n^2 + B_2 \cdot \frac{3}{2} \cdot \frac{4}{3} \cdot \frac{5}{4} \cdot n^4 + \frac{1}{2} \cdot n^5 + \frac{1}{6} \cdot n^6.$$

Thus, it follows that

$$1^5 + 2^5 + \cdots + n^5 = -\frac{1}{30} \cdot \frac{5n^2}{2} + \frac{1}{6} \cdot \frac{5n^4}{2} + \frac{n^5}{2} + \frac{n^6}{6}.$$

Therefore, we deduce that

$$1^5 + 2^5 + \cdots + n^5 = -\frac{n^2}{12} + \frac{5n^4}{12} + \frac{n^5}{2} + \frac{n^6}{6}.$$

Remark Note that neither this formula nor the general formula is easy to memorize for a high-school student. Using the "*triangle,*" however, one easily derives the formula even if it has been forgotten.

Bibliography

1. Andreescu, T., Enescu, B.: Mathematical Olympiad Treasures, Birkhäuser, Boston Mathematical Notations, Dover Publications, New York (1993)
2. Andreescu, T. et al.: Old and New Inequalities, GIL Publications, Bucharest (2004)
3. Cajori, F.: A History of Mathematical Notations, Dover Publications, New York (1993)
4. Engel, A.: Problem-Solving Strategies, Springer, New York (1998)
5. Galperin, G., Tolpygo, A.: Moscow Mathematical Olympiads, Moscow Education, Moscow (1986)
6. Hardy, G. et al.: Inequalities, Cambridge University Press, Cambridge (1948)
7. Kurschak, J. et al.: Hungarian Problem Book, Budapest, Random House/L.W. Singer Company (1965)
8. Marshall, A., Olkin, I.: Inequalities: Theory of Majorization and Its Applications. Springer Publishing, New York (1979)
9. Popov, S. et al.: Mathematical Exercises, Moscow center for continuous mathematical education, Moscow (2013)
10. Rajesh, T.: 50 Greatest Mathematicians of the World, Ocean Books, New Delhi (2013)
11. Sedrakyan, H., Sedrakyan, N.: The Stair-Step Approach in Mathematics, Springer, New York (2018). 10.1007/978-3-319-70632-0
12. Sedrakyan, H., Sedrakyan, N.: Geometric Inequalities: Methods of Proving, Springer, New York (2017). 10.1007/978-3-319-55080-0
13. Sedrakyan, N., Sedrakyan, H.: Inequalities: Methods of Proving 1. Kyowoo Publishing, Seoul (2015)
14. Sedrakyan, N., Sedrakyan, H.: Inequalities: Methods of Proving 2. Kyowoo Publishing, Seoul (2015)
15. Sedrakyan, N., Avoyan, A.: Inequalities: Methods of Proving. Fizmatlit Publishing, Moscow (2002)
16. Sedrakyan, N.: Inequalities: Methods of Proving. Edit Print Publishing, Yerevan (1998)
17. Sedrakyan, N.: Kvant, J., **97**(2), 42–44, Moscow (1997). http://kvant.mccme.ru/au/sedrakyan_n.htm

H. Sedrakyan and N. Sedrakyan, *Algebraic Inequalities*, Problem Books in Mathematics, https://doi.org/10.1007/978-3-319-77836-5

Printed in the United States
By Bookmasters